First Outline of a System of the Philosophy of Nature

SUNY series in Contemporary Continental Philosophy
Dennis J. Schmidt, editor

First Outline of a System of the Philosophy of Nature

F. W. J. Schelling

Translated and with an Introduction and Notes by
Keith R. Peterson

State University of New York Press

Published by
State University of New York Press, Albany

© 2004 State University of New York

All rights reserved

Printed in the United States of America

No part of this book may be used or reproduced in any manner whatsoever without written permission. No part of this book may be stored in a retrieval system or transmitted in any form or by any means including electronic, electrostatic, magnetic tape, mechanical, photocopying, recording, or otherwise without the prior permission in writing of the publisher.

For information, address State University of New York Press,
194 Washington Avenue, Suite 305, Albany, NY 12210-2384

Production by Judith Block
Marketing by Jennifer Giovani and Susan Petrie

Library of Congress Cataloging-in-Publication Data

Schelling, Friedrich Wilhelm Joseph von, 1775–1854.
 [Erster Entwurf eines Systems der Naturphilosophie. English]
 First outline of a system of the philosophy of nature / Friedrich Wilhelm Joseph von Schelling ; translated and with an introduction and notes by Keith R. Peterson.
   p. cm. — (SUNY series in contemporary continental philosophy)
 Includes bibliographical references (p.) and index.
 ISBN-13: 978-0-7914-6003-0 (alk.paper) — 978-0-7914-6004-7 (pbk :alk.paper)
 ISBN-10: 0-7914-6003-7 (alk. paper) — 0-7914-6004-5 (pbk : alk. paper)
 1. Philosophy of nature. I. Peterson, Keith R. II. Title. III. Series

BD581.S2613 2004

2003059028

10 9 8 7 6 5 4 3 2 1

CONTENTS

Acknowledgments	vii
Abbreviations	ix
Translator's Introduction	xi
The Primacy of the Postulate	xiii
From Postulate to Deduction	xviii
Transcendental Deductions and The Idea of Nature	xxii
Logogenesis, Construction, and Potency in the Philosophy of Nature	xxvii
Conclusion	xxxii
Works Cited	xxxiii
Translator's Note	xxxvii
Title Page of Schelling 1799 Edition	1
Foreword to Schelling 1799 Edition	3
Outline of the Whole	5
First Division	13
I. The Unconditioned in Nature	13
II. The Original Qualities and Actants in Nature	19
III. Actants and Their Combinations	28
IV. Inhibition and Stages of Development	35
V. Deduction of the Dynamic Series of Stages	53
Second Division	71
First System	73
Second System	74
Third Possible System	78
Conclusions	94

Third Division • 105
    I. On the Concept of Excitability • 106
    II. Deduction of Organic Functions from the Concept of Excitability • 113
    III. The Graduated Series of Stages in Nature • 141
    Appendix to Chapter III • 158
    IV. General Theory of the Chemical Process • 172
    V. The Theater of the Dynamic Organization of the Universe • 187

Introduction to the Outline of a System of the Philosophy of Nature, or, On the Concept of Speculative Physics and the Internal Organization of a System of this Science (1799) • 193
    §1. What we call Philosophy of Nature is a Necessary Science in the System of Knowledge • 193
    §2. Scientific Character of the Philosophy of Nature • 194
    §3. Philosophy of Nature is Speculative Physics • 195
    §4. On the Possibility of Speculative Physics • 196
    §5. On a System of Speculative Physics in General • 199
    §6. Internal Organization of the System of Speculative Physics • 201

Appendix: Scientific Authors • 233

Notes • 239

English-German Glossary • 249

German-English Glossary • 253

Page Concordance • 257

Index • 261

ACKNOWLEDGMENTS

What becomes a consuming and compelling passion for us often starts off with an air of contingency—so the genesis of this translation project. A seminar on *German Idealist Philosophy of Nature*, directed by David Farrell Krell at DePaul University, supplied the first incitements to this work in an atmosphere conductive to the growth and flourishing of rare ideas and inspirations. For this inspiration, along with much encouragement and translation advice, I am indebted to David Krell. Peter Steeves and Will McNeill read a version of the Translator's Introduction, and I have benefitted from their comments. Daniel J. Selcer was kind enough to read large parts of the manuscript and provided many valuable observations. Iklim and Gordon Viol provided priceless translation advice along the way. I am, of course, solely responsible for all defects that may remain. Finally, I am deeply grateful to Dicle Türkoğlu for her tireless and determined support, and to Ela Hayal who, while her Dad was involved in the last stages of the work, slept just as much as she should have.

ABBREVIATIONS

AA Friedrich Wilhelm Joseph Schelling. *Werke: Historisch-kritische Ausgabe*. Edited by Hans Michael Baumgartner, Wilhelm G. Jacobs, and Hermann Krings. Stuttgart: Fromann-Holzboog, 1976ff. Cited by series, volume, and page number (e.g., AA I,7 301).

SW F. W. J. Schelling, *Sämmtliche Werke*. Edited by K. F. A. Schelling. 14 vols. Stuttgart: J. G. Cotta'scher Verlag, 1856ff. Cited by volume and page number (e.g., SW III 301).

TRANSLATOR'S INTRODUCTION

Since its critical and programmatic inception in the eighteenth century and acme in the nineteenth, the philosophy of nature (*Naturphilosophie*) has been defined in part negatively as the critique of mechanistic, materialistic, and deterministic science. Such critique was needed primarily because, on the one hand, mechanistic accounts threatened the Enlightenment ideal of free subjects, and on the other hand, because the reductivism of these accounts betrayed the complexity of nature. Positively, philosophy of nature focused on dynamical, organic, synthetic, and holistic accounts of the natural world, integrating human beings into that world rather than severing them from it. It becomes clear from an attention to recent trends in evolutionary and developmental biology, cosmology, ecology, critical theory, and science studies that many of the critical methods and programmatic aspirations of classical nature philosophy are once more on the rise today. A reevaluation of the philosophy of nature of F. W. J. Schelling (1775–1854) in light of these developments seems no less than obligatory.

In addition, the critical apparatus that the philosophy of nature brought to bear on the modern scientific project demanded not merely a theoretical or epistemological shift, but a reformulation of the relation between human beings and nature, often entailing novel political or ethical commitments. Early philosophy of nature met opposition in part because its ethical and political interests—not just its allegedly wild and "unverifiable" analogizing—were thought to have invalidated its "scientific" claims. Today, however, the traditional alliance between value-neutrality and objectivity, or the relationship between power and knowledge, the practical and theoretical domains, has been definitively shown to be not only theoretically untenable but politically dangerous.[1] If all knowledge, including scientific knowledge, is always "local" and always conditioned by competing interests, then there can no longer be the same prejudices against the kinds of accounts a philosophy of nature can provide, as a reading of many recent developments in science itself would indicate.[2]

A revitalized philosophy of nature would perhaps aim to tie together the wide range of disparate pursuits mentioned above, showing their implicitly and explicitly shared assumptions, and in a Schellingian spirit would raise the questions that are so often not even acknowledged by scientific theorists. Schelling's methodological and ontological insights are indispensable as points of reference for reconsidering a unified view of nature and the genuine philosophical problems involved in rethinking the natural world and humanity's deep connections with it. It is hoped that this translation of what is possibly Schelling's most important text in nature philosophy will contribute to a revitalization of the project of a philosophy of nature, and—by providing a wide audience the opportunity to read Schelling himself—at least the prospect of a reevaluation of this important phase both in Schelling's thought and in the history of philosophy and science.

While it may be tempting to explore the litany of excoriating criticisms of the philosophy of nature from Schelling's own time to the present,[3] I will not do so here. I would like to note that the received opinion regarding nature philosophy, well articulated by Bernard Cohen, seems no longer to ring true: "It has become a tradition among those who talk glibly about science that the romantic *Naturphilosophie* of Schelling and his followers represents the lowest degradation of science and that only by completely freeing themselves from that nightmare were modern biology and medical science able to resume their scientific progress. The incident has been used by empiricists as a moral to warn us against speculative philosophy in the natural sciences." More recently, historians and researchers have asked how and to what extent the school of Naturphilosophie had an impact on the history of science, and the same could be asked about its influence on later naturalistically minded, transcendental philosophers.[4] Both of these are important questions to be taken up in their own right if the full significance of the philosophy of nature is ever to be acknowledged.

Schelling always stood out among his contemporaries—he was accused of being too rational for the Romantics (e.g., Novalis, the Schlegels) and too romantic for the Rationalists (the Kantians, fideists, skeptics, and commonsensists). Schelling's proteiform work has not received the same attention as that of his two brood mates at the Tübingen seminary—G. W. F. Hegel and Friedrich Hölderlin—and it is evident that both historically and, I would suggest, in principle, Schelling has no true disciples. However, despite—or perhaps because of—its enormous variability and range, many influential thinkers have reached into the churning depths of the Schellingian corpus and extracted some choice sustenance for themselves (among them Paul Tillich, Jürgen Habermas, and Martin Heidegger, to name a few contemporary voices), and the same occurred in and around Schelling during the years when he was producing his powerful tracts in the philosophy of nature. Something resembling a school of Naturphilosophie emerged, and his ostensible influence on such reknowned figures as Henrik Steffens (1773–1845), Johann Ritter (1776–1810), Lorenz Oken (1779–1851), Hans Christian Ørsted (1777–1851), Samuel Taylor Coleridge (1772–1834), and others was always refracted through each personality such that a particular shade or emphasis of the teaching was predominant, and his precise ideas (such as they were) were seldom adopted verbatim. Even during the years of his occupation with topics in the philosophy of nature, Schelling himself undergoes a rather distinct shift in emphasis with the publication of his "Presentation of My System of Philosophy" (1801). Here it becomes apparent that while from 1796 to 1800 he understands his Fichte-inspired transcendental philosophy and self-styled nature philosophy to be two independent but mutually necessary philosophical sciences, after 1800 he increasingly understands them as two facets of a single system of "absolute identity." The work translated here was composed and published in 1799, and so belongs to the first phase; therefore, in this introduction I will only discuss the themes and emphases of this early period.[5]

Upon leaving the Tübingen Stift in 1796, the precocious, then twenty-one-year-old Schelling relocated to Leipzig, where he immersed himself in the most current scientific literature, including in his syllabus of study chemistry, physics, mathematics, natural history, and physiology. The first published product of this immersion was the text *Ideas for a Philosophy of Nature* (1797), Schelling's first comprehensive attempt to redirect the course of scientific theorizing, and somewhat of a departure from the Kantian and Fichtean themes with which he had been preoccupied until this time. More reflection on empirical researches led to the more grandly synthetic text *On the World-Soul* in 1798, which won Schelling great acclaim, including the admiration of Johann Wolfgang von Goethe, and as a result an invitation to a professorship at Jena in the same year. Exclusively as an accompaniment to his lectures on the philosophy of nature in Jena in 1799, Schelling published his *First Outline for a System of the Philosophy of Nature*, as well as a separately issued *Introduction* to the *Outline*, in which he presents his most systematic treatment of the topics explored in the earlier treatises, plus some wholly new theories treating organic life and its relation to the inorganic, and to nature as a whole. In Germany the text of the *Ideas* was issued, with amendments, a second time (1797, 1803), and *On the World-Soul* went through three editions in Schelling's lifetime (1798, 1806, 1809), but the *Outline*, being essentially a reference work for lectures, was never reissued. Nevertheless, the *First Outline of a System of the Philosophy of Nature* provides perhaps the most inclusive exhibition of Schelling's early thought and method in the philosophy of nature, displaying both its extraordinary strengths and, it must be admitted, some profound weaknesses. In the next section I will explore the genesis and development of Schelling's project of nature philosophy, with special reference to Immanuel Kant's transcendental philosophy, as well as its persistent epistemological, ontological, and scientific themes.

The Primacy of the Postulate

In striving to achieve an architectonic unification of systematic knowledge, Kant asserts that the "ideas" that reason possesses of God, the soul, and freedom are necessary for the ultimate unification of philosophy, which would include scientific and ethical knowledge. Schelling of course follows Kant in this aspiration, and will ground such a unification of knowledge in the unified practical aims of reason itself. The genuine significance of Kant's assertion of the "primacy of the practical" (the legitimate, practical employment by reason of the ideas) requires a brief sketch of the nature of transcendental philosophy and of the problem of synthesis. Kant's formulation of the problem is familiar: "How are *a priori* synthetic judgments possible?" he asks, and his response entails the renowned "transcendental deduction of the categories of the understanding" in the first *Critique*. Universal and necessary judgments, such as those of mathematics, are possible according to Kant's analysis because the structure of the human cognitive apparatus is such that (in a judgment of experience) it synthesizes a concept and an intuition, subordinating the part to the whole, the conditioned to the condition. In theoretical or speculative philosophy the categories, such as cause and effect, are "deduced" and their necessity justified. More relevant for practical philosophy, even if often overlooked, is Kant's deduction of the "pure concepts of reason," or the ideas mentioned above. In speculative reason, scientific ideas like that of "unity" receive a justification and objective validity insofar as they are indirectly related to experience by means of the understanding's categories; even formal knowledge remains disorganized unless an idea of the "system" regulates the organization of knowledge. The case is somewhat different with an idea such as "freedom." It turns out not only to have a certain status within theoretical philosophy in its "practical employment," but it becomes the very synthetic anchor of practical reason itself, underlying both the theoretical and practical use of reason (as the very possibility of "critique"), and is justified by means of the deduction of the "fact of reason" in the second *Critique*. It was this implicit assertion of the unity of theoretical and practical philosophy that J. G. Fichte (interpreter of Kant and Schelling's philosophical mentor) and Schelling endeavored to make the basis of systematic philosophy.[6]

For Fichte and Schelling, it was obvious that Kant had successfully legitimated the claim to a priori synthetic knowledge; what Kant failed to do epistemologically was provide an account of synthesis in general, that is, an account of the fundamental relation between thought and the world, or representations and objects. A brief look at much of Schelling's philosophical exertion in the years 1794 to 1796 reveals a more general statement of and characteristic solution to the problem of synthesis. He will conclude that the real solution to the problem of synthesis is to be found in practical, rather than theoretical, philosophy. Even for the philosophy of nature this solution retains its definitive significance.

The discussion of the "postulate" in Schelling is meant to emphasize the deliberate collapse of theoretical into practical philosophy, or the mediation of all theory by practice, typical of the post-Kantian tradition. This is critical with reference to Schelling's philosophy of nature, because unless it is seen as an attempt to reground science itself in the soil of practical philosophy, it will be (and has been) viewed at best as merely another narrative, myth, or story about nature, and at worst a collection of speculative, bizarre, and "unverifiable" theories. For Schelling, nature philosophy is not merely another "representation" of a nature to which human beings maintain only a distant and instrumental relation. For him, the first postulate of philosophy must express the dynamic synthesis of self and world, subject and object, as an ontological unity from which both terms are derived. Self and world are of one substance, and we will continue to misunderstand ourselves and undervalue the natural world unless this ontological identity is expressed philosophically.

In "On the Possibility of a Form of Philosophy,"[7] taking Kant as his point of departure, and Fichte's opposition between the self and the nonself as an indispensable principle, Schelling notes that the problem of synthetic judgments implies the opposition between the manifold of the "given" and the unity of the self.[8] For there to be synthesis at all, there must be two distinct terms. For synthesis to occur, the two terms must be unified by means of a common term or medium. Because the common term cannot stem from one of the two opposites, there must be a "prior" unity within which the understanding and the manifold of sense are themselves synthesized. This prior unity is called the "absolute I." The synthesis of an empirical I with an opposed empirical multiplicity, a synthetic judgment, is a function of the cognitive faculty or understanding. Both terms are conditioned by definition and cannot find the ground or reason for their opposition within themselves, so they must be "preceded" by an unconditioned. The empirical synthesis in a judgment of experience depends on a transcendental (logical and ontological) synthesis in the absolute. Kant shows by the transcendental deduction of the "concepts of reason" that the ideas of freedom and a divine being can be regulatively employed for practical purposes, that is, to keep us from mechanism and fatalism, but can provide no constitutive knowledge. Epistemology and ontology, or form and substance, are kept rigorously distinct—this is what keeps critique from becoming "dogmatic metaphysics." For Schelling, however, the simple fact that we have an idea of the absolute, the whole or totality, means that we are, in some sense to be determined, the Absolute. Likewise, if our experience is characterized in terms of a separation of subject and object, and knowledge consists in the reunion of these two terms, then we must contain the unity of these two terms in ourselves (given that nothing can act on the mind "from without").[9] Where Kant had only set out to show how the understanding determines an indeterminate "given" content, Schelling argues that this purely formal account of synthesis leaves the substance or material of the world out of account. It also means that we cannot eschew ontology or metaphysics, but must pursue it with renewed vigor. A new ontology becomes indispensable for the solution to the problem of synthesis in knowledge, as well as the problem of transcendental freedom.

If theoretical reason necessarily seeks the unconditioned (as Kant also held), then it must also admit that the endeavor which produces a synthesis in each act of knowing, a reunion of subject and object, ultimately demands the affirmation of such a unity as principle. That is, this endeavor is the symptom of a desire to achieve a state in which synthesis is no longer necessary. If synthetic activity is to eventuate in a "thesis" or affirmation of unity, the opposition between subject and object must be eliminated. Historically, Schelling remarks, either the object has been considered absolute and envelops the subject (which he considers Spinozism), or the subject is treated as absolute and is both content as well as form, a "*subject-in-itself*" (Fichte's idealism). Because theoretical philosophy deals exclusively with the relations between a subject and an object (conditioned factors), the annihilation of this opposition cannot come to pass through theoretical means. Therefore, this passage from synthesis to thesis is at the same time the transition from theoretical to practical philosophy. If theoretical philosophy seeks to apprehend the unconditioned, but is unable to *realize* it, that is, to know or prove it solely theoretically, it must become an object of action; the idea itself "*demands* the *act* through which it *ought* to be realized."[10] Kant showed that reason is compelled to go beyond its proper limits, ending up in dialectical paralogism and antinomy (the false problems of "dogmatic metaphysics"). The only way to avoid dialectical illusion is to resort to the doctrine of regulative or problematic ideas, and to turn the constraint felt in the positing of ideas to his advantage by situating it in the moral law or "fact of reason." Schelling, even more explicitly than Kant, submits theoretical to practical reason; there is not merely a "practical employment" of the ideas of theoretical reason, as for Kant, but practical reason is itself the primary form of reason. The first principle of philosophy—if all of philosophy is to be deduced from certain a priori principles, a Fichtean premise—must be both theoretical and practical at once.[11] This first principle must necessarily take the form of a postulate, because only the postulate expresses a theoretical principle with a practical force.

"The first issue of philosophy, to act freely upon oneself, seemed to [me] as necessary as the first postulate of geometry, to draw a straight line. Just as little as the geometer proves the straight line should the philosopher attempt to prove freedom."[12] The postulate is that which cannot be demonstrated or proved because it is that assumption or principle with which a science must begin. It differs, however, from the "theorem" or "axiom" in that it cannot merely be theoretical, since it must provide the foundation for all philosophy (theoretical *and* practical). Schelling urges that philosophy cannot be reduced to "abstract principles," and that he wants "to prove that true philosophy can start only from free actions, and that abstract theorems as the core of this science could lead only to the *death* of all philosophy."[13] The theory of the postulate must be more than an epistemological meditation on the first principle of philosophy, for he "believes that humanity was born to *act*, not to *speculate*, and therefore that his first step into philosophy must announce the arrival of a free human being."[14] This notion has obvious antideterministic implications, and is immediately relevant to the philosophy of nature and its critique of mechanism. "What is caught up in mere mechanism cannot step out of the mechanism and ask: how has all this become possible?"; "*That* I am capable of posing this question is proof enough that I am, as such, independent of external things; for how otherwise could I have asked how these things themselves are possible *for me*, in my consciousness?"[15] In this critique of mechanism Schelling implies a merely negative concept of freedom—freedom as independence and spontaneity—which he ultimately finds inadequate and reformulates in terms of his own uniquely positive conception.

If the idea of freedom is the sine qua non for philosophy, then the relationship of the postulate to the idea is revealed as a mutually dependent and constitutive one. The idea is sufficient, constitutive, reason for endeavor; it is not merely a regulative condition but a genetic, compulsory element. The idea will be expressed not merely "by a theoretical but by a practical capacity; not by a cognizing, but through a *productive, realizing* power; not through *knowledge* but through *action*."[16] Schelling capitalizes on Kant's theory of the ideas by making the idea the locus of intersection of ontology and ethics; ontological because the power through which reason postulates the ideas is none other than a mode of absolute causality as such; ethical because philosophy cannot begin unless, like the geometer, one draws the line of freedom in a postulate. Determinism in natural causality and in ethics is forestalled by this first postulate of philosophy, the exercise of "transcendental freedom."

In the treatise "On the I as Principle of Philosophy" Schelling then devotes more attention to this problem of freedom. The subject's freedom can no longer be thought of as a mere noumenal spontaneity, since the "thing-in-itself" has been rejected. The question of freedom necessarily involves the possibility of genuine causality through reason. Schelling insists that the "absolute freedom of the absolute I," or the pure self-affirmation of being, has never been a question, since being realizes itself by absolute causality—being is infinite power. The problem has always been to think of an empirical subject as free, which Kant called its "transcendental freedom": "[T]he freedom of the empirical I cannot possibly *realize* itself, because the empirical I as such does not exist through itself, through its own free causality."[17] As empirical, it could never theoretically affirm the sheer absolute reality of the I, but is under the standing obligation to (practically) produce it. Through the demand to produce this reality, it intuits the causality that is the genetic source of this very productivity.[18] But such a demand can be made only upon an empirical self that is itself not absolute freedom (because it is not a pure unity), but also "whose causality *does not differ from* the absolute causality in quality, but only in quantity."[19] The argument takes the typical form "not without":[20] no transcendental freedom without absolute freedom, no empirical realization without absolute positing or affirmation. Reason is obliged, if it postulates, to affirm the reality of the causality by means of which reason itself can determine the will. We might say that Schelling develops a critical ontology adequate to the Kantian practical philosophy.

From Postulate to Deduction

In subsequent studies Schelling takes this practical and ontological ground of philosophy and develops it along two lines: transcendental and natural philosophy. The practical demand of philosophy (that philosophy be defined by freedom, that is, absolute causality or potency) entails that the philosopher regard neither objects in Nature nor categories in the mind as ultimates, and that each domain receive a "genetic deduction" in its own right from the first principle of philosophy or absolute substance, understood as "pure activity" or "pure productivity." Inspired by Fichte's move to explain the categories by means of analyzing the activity of the thinking subject, Schelling insists that this activity is the real medium of both transcendental and natural philosophy. In transcendental philosophy the ideal necessary conditions of experience are seen to be engendered through the play of activities, both free and necessary, in the thinking subject. Transcendental philosophy begins with the ideal or subjective element and explains the objective on this basis. In the philosophy of nature the real necessary conditions of objects are seen to be engendered through essential forces in the material world. "Forces" are the empirical manifestation of nature's "productivity" or activity, and all matter, organic or inorganic, is composed of a play of forces, both free and constrained. Nature philosophy begins with the real or objective side, as Schelling will advance in the *Introduction*, and considers even reason itself to have "developed"[21] from nature. In the *Introduction* he says that

> what we call "reason" is a mere play of higher and necessarily unknown natural forces. For, inasmuch as all thinking is at last reducible to a producing and reproducing, there is nothing impossible in the thought

that the same activity by which Nature reproduces itself anew in each successive phase, is reproductive in thought through the medium of the organism.[22]

The subjective and objective "halves" of philosophy are shown to be two different expressions of the ontological ground of activity. "Activity" is univocal being itself. In the *System* of 1800 Schelling says that "it cannot be said of the I [or of *natura naturans*] that it 'exists,' precisely because it is *being itself*. The eternal act of self-consciousness, existing beyond time, which we call *I*, is that which gives existence to all things, and thus itself requires no other being to support it; bearing and supporting itself, rather, it appears objectively [in Nature] as *eternal becoming*, and subjectively as *infinite productivity*."[23] In the introduction to the *Ideas*, Schelling asserts that philosophy

> is nothing other than a *natural history of our mind*. From now on all dogmatism is overturned from its foundations. We consider the system of our representations, not in its *being*, but in its *becoming*. Philosophy becomes *genetic*; that is, it allows the whole necessary series of our representations to arise and take its course, as it were, before our eyes. From now on there is no longer any separation between experience and speculation. The system of Nature is at the same time the system of our mind, and only now, once the great synthesis has been accomplished, does our knowledge return to analysis (to *research* and *experiment*).[24]

This genetic philosophy of becoming will "deduce" the necessary structures of thought as well as the categories of nature; the former in transcendental philosophy, the latter in nature philosophy. The categories adduced in each case will have a logical and, in some cases, more than logical necessity. What we are compelled to assert derives from the constraint of the real. This notion of necessity or constraint is unfortunately one of the most obscure parts of Schelling's philosophy.

For example, there is a purposive interconnection of cause and effect exhibited to us by an organism. Schelling takes the "regulative idea" of the organism from Kant's third *Critique* and gives it a constitutive sense. The organism is an existence "for itself," it organizes itself, and every part presupposes the whole (in contrast to the "aggregate" which lacks such a systematic interconnection of parts, as well as a power of self-organization). Form is really inseparable from matter, and we have no choice but to think of an organism as an independent whole; we do not arbitrarily impose the form of purposiveness on individual units of matter disorganized in themselves—we are compelled to think of an organism as purposive. To be sure, purposiveness is conceivable only in relation to a judging intellect, and only in relation to an intelligence can anything be called purposive. But "at the same time, you are no less compelled to concede that the purposiveness of natural products dwells *in themselves*, that it is *objective* and *real*, hence that it does not belong to your *arbitrary*, but to your *necessary* representations."[25] Thus just as the separation between representation and object is resolved by establishing their mutual dependence and deducing them from a common source, there is a reciprocal dependence between the judgment of organization and purpose and the actual existence of organized beings—neither can exist without the other, and both derive from a common source. Our feeling of constraint informs us of this reality. Schelling argues that both mechanism (in causal series) and teleology depend implicitly on an idea of Nature as an organized and organizing whole. In the introduction to the *Outline* he argues that this idea is at the root of all investigation into Nature, and, as the first postulate of nature philosophy, it is an "involuntary" and necessary postulate. Materialism and mechanism occupy the standpoint of reflection, and cannot think life. Only a philosophy of nature that transcends the standpoint of separation, of mechanism, can think both the life in nature and the freedom in humanity:

> As long as I myself am *identical* with Nature, I understand what a living nature is as well as I understand my own life; I comprehend how this universal life of nature reveals itself in the most manifold forms, in measured developments, in gradual approximations to freedom. However, as soon as I separate myself, and with me everything ideal, from nature, nothing remains to me but a dead object, and I cease to comprehend how a *life outside me* could be possible.[26]

In the *Ideas*, although Schelling asserts that "this absolute purposiveness of the whole of Nature is an idea which we do not think arbitrarily, but *necessarily*," he does not explicitly define the "secret bond that couples our mind to Nature," or the nature of this necessity.[27] Not until the introduction to the *Outline* does he return explicitly to this idea. I will devote some attention to it since all of the problems that define the philosophy of nature, and that it sets out to solve, can be localized in the theory of the idea. With it Schelling is able to affirm the primacy of practical philosophy—since ideas must be postulated—and as a result dispense with determinism a priori; with the theory of the idea he relies on a new ontology, since both the postulating subject and the world "opposed" to the subject derive from the same source; with this theory he takes the Kantian doctrine of regulative ideas to a new level, and pushes Kantian "deduction" to its limit. Before its complete articulation in the *Introduction*, however, in the next text devoted to philosophy of nature—a book much praised by Goethe—Schelling arrives at a temporarily suitable conception of the original unity of real and ideal. He names this text *On the World-Soul, a Hypothesis of the Higher Physics towards the Explanation of the Universal Organism*. Hypothesis here clearly means "postulate," and the "*Universal Organism* is what constitutes nature as a whole. Key portions of this text are devoted to the question of the nature of organism and organization, and how the system of nature is an expression of the system of thought. In accord with precisely the same logic that provoked Schelling to postulate a prior unity from which representations and objects flow, here he argues that organism is prior to mechanism and explains it. In short, "the *world* is an *organism*, and a *universal organism* is the very *condition* (and to that extent the *positive* factor) of *mechanism*."[28] Schelling will only be able to demonstrate the dependence of mechanism on organism by explicitly executing a deduction of this necessary relation, in addition to the inductive evidence for such a relation provided in this text.

Schelling's understanding of "induction" and "deduction" is specific. He calls his approach in *On the World-Soul* "inductive," but it means more than the inference of a rule or concept from a case and comparative observation. Induction entails an ascent from the conditioned to the causal condition, from objects to the a priori principles of the production of objects, and it is precisely Schelling's aim to show that mechanism cannot give a complete account of these conditions, neither for organic phenomena nor for the inorganic. The mechanically simple cannot be shown to ground the dynamically complex, but the complex is the ground of the simple. In the first part of the text dealing with inorganic nature, Schelling successively discusses the material phenomena of light, combustion, air, electricity, and magnetism, showing that the allegedly more basic phenomenon would itself not be possible without the more complicated; for example, the polarity in electrical processes cannot be understood unless it is acknowledged that magnetic polarity is already operative throughout nature as a whole. Moreover, at a more fundamental level, a principle (cause, ground) of polarity reigns throughout nature, from which specifically polar materials (electrical or magnetic) derive. Part 2 of the text is an ascent to the conditions of organic life, since it has been acknowledged that the inorganic does not provide the principles of explanation for the organic. As he articulates it in the *Outline*, Schelling's aspiration is to show that the primary forces of nature, both in the inorganic and organic realms, although superficially diverse, are at bottom conceivable as substantially identical: "To be sure, there must be *one* force that reigns throughout the whole of Nature.... But this force may be capable of infinitely many modifications and may be as various as the conditions under which it operates."[29] In *On the World-Soul* he employs a classical formulation to describe the relation between this omnipresent power and the empirical phenomenon: "[T]he universal principle of life individualizes itself in every individual living being (as if in a unique world) according to the different degree of its receptivity."[30] Although with reference to "life" here specifically, Schelling conceives fields of virtual powers (like the electromagnetic) to be omnipresent throughout the cosmos, manifesting themselves where the conditions are suitable. For Schelling, even every movement implies some degree of sensibility, and therefore some form of sensibility must permeate the entire cosmos, making the world a single organic whole.[31]

Deduction, in contrast to induction, will begin with a priori conditions and descend to the conditioned. Schelling notes that the *Outline* adopts this procedure (these distinctions are already evident in the two parts of the *Ideas* as well). Since this method characterizes transcendental philosophy after Kant and reflects an athletic appropriation of Kantian principles, I will spend some time unpacking this notion of deduction or "construction" with reference to the "idea of Nature" which becomes thematic in the *Introduction*.

Transcendental Deductions and The Idea of Nature

It is typical of Kantian critical method to sidestep the question of the truth or falsity of a claim, and instead to examine the assumptions that support it. "Transcendental deductions," wherever they occur, aim at the demonstration of the "right to possession" of a claim or knowledge, or at demonstrating the legitimacy of this knowledge, vindicating its employment or possession in the face of the critical objection that such a claim rests on an arbitrary, illegitimate presupposition or "foundationalist" assumption. Deduction, as a method, is the complementary odd half of the critical project. If critique forces thought to reveal its own presuppositions, then deduction is there to show that we must begin with at least some assumptions to which we have a "right." In a juridical deduction (Kant's model), there are two questions that are to be answered. The question of right (*quid juris*) can be settled separately from the question of fact (*quid facti*), and questions in the form "How is x possible?" are questions of right. Dieter Henrich remarks that deduction is defined by "the process through which a possession or a usage is accounted for by explaining its origin, such that the rightfulness of the possession or usage becomes apparent.... In a state of doubt about the rightfulness of our claim to be in the possession of genuine knowledge, it seeks to discover and to examine the real origin of our claim and with that the source of its legitimacy."[32] The task of deduction is then precisely to show that the only assumptions that are made are those that are necessary, or "indispensably necessary," and not "arbitrary."

Without this complementary method one remains at the level of analysis, which indeed takes off from the conditioned to search for its conditions, but which cannot itself guarantee the truth or necessity of just those premises it finds and clings to when confronted by antinomy. Moreover, deduction not only shows that the revealed "presupposition" is necessary, but also that it has a "synthetic" character, earning it the moniker "transcendental" (i.e., related to objects, but not derived from experience of objects). Kant tells us that "the explanation of the manner in which concepts can thus relate *a priori* to objects I entitle their transcendental deduction; and from it I distinguish empirical deduction, which shows the manner in which a concept is acquired through experience and through reflection on experience, and which therefore concerns, not its legitimacy, but only its *de facto* mode of origination."[33] The *Critique* provides these transcendental deductions, and most often the "source" of the legitimacy of such claims is found in the "nature" of the understanding or of reason. Schelling uses this strategy to show that thought is compelled or constrained to think the reality of purposive organisms, as shown above, and to think of nature as an organized whole by means of the "idea of Nature," as shown below.

After Kant it was incumbent upon anyone following in his footsteps to justify the existence of the branches of science they pursue within the system of philosophy, since Kant argued that he had shown exhaustively which faculties and categories are necessary not just for thought and experience, but by extension, for science as well. In the definitive *Introduction* to the *Outline*, Schelling deduces the possibility of a "speculative physics," or Naturphilosophie. Speculative physics would be shown to be possible (i.e., legitimate) as a science if it could somehow be demonstrated that in our investigations into nature we already employ certain ideas or acts of the mind, and that without these conditions natural science itself would not have achieved anything thus far.[34] For speculative physics as a science to be possible, one must consider the idea that serves as a principle or a rule for organizing knowledge in that science—the idea of the unconditioned whole. For Schelling, mechanistic physics cannot claim to be a science at all because it treats nature as fragmentary, and does not seek the ground of phenomena. Only the unconditioned can be a final ground and sufficient reason. The idea of nature involves the philosophical postulate of nature's autonomy and autarchy (analogous to reason's own), and through it the turn from nature's "products"—conditioned things, objects of mechanistic physics—to the "final ground" of these products, an unconditioned "productivity," is accomplished.

The deduction proceeds in two stages: Schelling first shows that certain ideas are already employed in natural science, and then he argues that the ideas are not arbitrary or regulative, but constitutive and necessary. He suggests that the very nature of experimentation as employed in the natural sciences betrays the existence of necessary a priori conceptions in the minds of the experimenters, for example, the idea that nature is a whole is inevitably assumed. By showing that experimentation would not be possible without the employment of an idea of Nature, nor the idea itself without the actual existence of a self-organizing Nature, the possibility of natural science is deduced from the actual behavior of scientists and the acts of the mind involved in human knowledge. Ultimately the deduction will be successful if all of the "necessary" phenomena of nature can be deduced from a first principle, and in turn confirm the legitimacy of this first principle when the science is complete. The deduction of Schelling's natural science differs from Fichte's deductions in the *Wissenschaftslehre* in that the latter, according to Fichte, can achieve perfect completeness, while all other sciences (such as nature philosophy) must remain incomplete because a measure of freedom is involved in the first principles of these sciences. Schelling recognizes this incompleteness of his nature philosophy when he speaks of "intermediate links" ("necessary" phenomena) in development, and admits that it is the task of speculative physics only to show the need for them where they have not yet been discovered, not to enumerate them exhaustively. Thus, natural philosophy is, in a sense, incomplete, and relies on "experiment" to fill in the gaps; but this is a necessary incompleteness, since "the complete discovery of all the intermediate links in the chain of nature, and therefore also our science itself, is an infinite task."[35]

In his consideration of experimental practice, he says that to know is to know the "principles of possibility" of a thing, the conditions under which and by means of which it was produced, its genesis. It would be impossible to know natural objects if it were not possible for human beings to "invade" nature by means of freedom. Nature can be "compelled to act under certain conditions" that do not exist, or at least not in a pure form, in nature. Such an "invasion" is an "experiment." But experiments are not random: "every question [put to nature in an experiment] contains an implicit *a priori* judgment; every experiment that is an experiment, is a prophecy: experimenting itself is a production of phenomena. The first step, therefore, toward science, at least in the domain of physics, is taken when we ourselves begin to produce the objects of that science."[36] Experimentation produces the "necessary" in phenomena, and in effect transforms the a posteriori (some aspect of the experienced) into the a priori (a universal condition of experience). The natural philosopher learns to see all things as necessary insofar as they have their final ground in the principles according to which the unconditioned productivity of nature operates. Construction is the deduction of the unconditioned conditions of natural productivity, and also the reproduction of these conditions in thought. Therefore, "construction by means of experiment is, after all, an absolute self-production of the phenomena."[37]

What are "produced" then are a priori principles, such as the idea of polarity, that is, condition of the theories of electricity and magnetism. As universal and a priori, the nature of polarity can never be found in the objects of nature, and so our knowledge of this regularity is not originally produced by means of experimentation on things. We already know that things will behave in certain ways when submitted to certain experiments: this is the element of "prophecy." Schelling treats polarity, for example, as the "final *cause*" of phenomena, which itself cannot be phenomenal. We must then "put it into nature, endow nature with it" in our interpretations. This is the result of the first stage of the deduction. The postulation of these final causes is already part of experimental practice. As he moves to the second stage of the proof, he must ask whether these ideas are necessary.

He considers the objection that anything "put into nature" in this way is clearly hypothetical, and that a science founded on such a principle therefore must be hypothetical.[38] Schelling answers the objection by admitting that the hypothetical nature of the science "would be possible to avoid in only one case, i.e., if that presupposition [of Nature's wholeness] itself were involuntary, and as necessary as Nature itself."[39] Again, Schelling notes that one does not choose to view an organism as something purposive; one is forced to think this way, compelled; therefore it is not a merely "regulative" idea:

> You feel yourself constrained in your judgment; you must therefore confess that the unity with which you think it is not merely *logical* (in your thoughts), but *real* (actually outside you). . . . Or, if it rests with your choice whether or not to impose the idea of purposiveness on things outside you, how does it come about that you impose this idea only on *certain* things, and not on *all*, that further, in this representing of purposeful products, you feel yourself in no way *free*, but absolutely constrained?[40]

As already mentioned, this apparent fact of compulsion in thought proves that something real corresponds to it, for if our perceived reality is entirely constructed by ourselves it would follow that we would not be compelled to think of it in one way rather than another.[41] Therefore, the idea of purposiveness is one that legitimates the project of a philosophy of nature to investigate "nature as subject" (as autonomous productivity) and not only view "nature as object." Schelling believes himself to have demonstrated both the inevitability and the necessity of the idea of nature as an organic whole. It remains to be shown specifically what is implied in this idea.

I have mentioned that the idea of nature as a whole entails an essential duality. Nature is both conditioned and unconditioned at once, *natura naturata* and *natura naturans*, product and productivity. Thus a perpetual dialectic between the productivity of nature and its products is implicit within the idea of nature, an idea forced on us by the constraint to conceive nature in a certain way, even while acting (thinking) freely or spontaneously. One can never hold that the productivity of nature, the final cause of all things, is identical to the sum of its products, since this would amount to holding fast the dynamic movement or "oscillation" between nature insofar as it is productive and nature insofar as it is product. This universal duplicity of principles maintains nature in continual activity, and thus universal duality must also be a principle of the explanation of nature, "as necessary as the Idea of nature itself." But this "deduction of all natural phenomena from an absolute hypothesis" means that "our knowing is changed into a construction of nature itself, that is, into a science of nature *a priori*."[42] What in Kant was called "transcendental deduction" becomes (by way of Fichte) the method of construction, and what Schelling calls an "*absolute* hypothesis" no longer has the character of the speculative, Kantian regulative idea possessing "objective validity" and only an indirect "objective reality." The first postulate of nature philosophy expresses the reciprocal presupposition of an actually organized nature and its idea, or the identity of real and ideal.

All of these provisions culminate in the theory of construction. It emphasizes the element of "causality through freedom" involved in postulating or positing the idea, and reciprocally concerns the structures or principles of experience that are necessary in order for one to be able to think and experience nature at all. Nature produces (freely), but this production can occur only according to necessary regularities. The idea of the whole is "involuntary"—a necessary or "absolute" presupposition that does not spring from the relativity of our knowledge but from the real constitution of the cosmos. For Kant the notion of the regulative idea or of the reflective judgment entailed a deep tie to his theory of freedom: we must understand the realm of appearances as necessitous, but there is some amount of interpretation respecting the causes assigned to phenomena (an empirical or intelligible character). This indeterminacy is a space opened up by the regulative judgment, which is interpretive and merely guided by the idea of system. Schelling goes beyond this to show that there must be a real content that forces thought to think, otherwise our judgments would be arbitrary. For Schelling to think a whole is to think constitutive self-production, and to think self-production is to think a whole. Through experiment one is led to lawfulness, through lawfulness to the whole, through the whole to self-production. This means that we do not merely translate, through knowing, the experience of objects into the necessity of thought, but that nature is necessary "in itself": "It is not that *we know* nature [as a priori], but that nature *is a priori*, that is, everything individual in it is already determined through the whole or through the Idea of nature in general."[43] Thus the philosophy of nature is at once an epistemology, an ontology, and an eminently practical philosophy (in the Kantian sense, meaning "pertaining to freedom").

Logogenesis, Construction, and Potency in the Philosophy of Nature

From the dualism of productivity and product implicit in the idea of nature, Schelling attempts to derive an entire graduated series of increasingly complex stages in the evolution of nature. From simple qualities to inorganic forces, from light to organic sensibility, he shows that, as in transcendental philosophy, construction is not designed to trace an actual, empirical or experiential awareness of development, but is the extraction of the necessary from the contingent in a deductive development. It is the determination of necessary "conditions of possibility" of the experience of an objective world and a justification of their legitimacy. In natural philosophy and its graduated series of stages Schelling sketches the categories of natural ontology and epistemology, deploying the least number of concepts needed to provide an account of the world and experience. Hermann Krings calls this a "logogenesis."[44] Since construction reveals the "necessary," and the necessary is such because the real compels thought, then Schelling can also sometimes say that nature "constructs itself." For example, the "dynamical process" is nothing other than the "self-construction of matter."[45]

If, as has been mentioned many times, it is Schelling's objective to define the set of concepts necessary for us to be able to think experience and nature at all, and to furnish a genetic account of both the subject and the object from a single source, it is obvious that he cannot be satisfied with the inventory of concepts supplied by Kant because they are strictly subjective. He deduces such categories as the understanding possesses in the 1800 *System*, but a deduction of "objective" categories is needed as well. He calls the phenomena of gravity, magnetism, and electricity the "categories of physics," precisely in resonance with Kantian usage. To have a complete system of the categories of experience, we need to uncover not only the subjective but the "objective" categories that are also "conditions" of our experience of things or objects in nature. The simple Kantian dualism between the synthetic, form-bearing subject and the unknown content-in-itself to be informed must be superseded by an account of both a form and content of spirit as well as a form and content of nature. Just as we cannot experience an object that does not occupy space or time (as Kant asserted), we also cannot experience an object that is not involved in the operation of gravitational, electromagnetic, and chemical forces. If Kant's critique of Sir Isaac Newton is to have any force, then it is not enough to say (as does Kant) that space is a "form of intuition," but the specific nature of space and the matter that fills it (e.g., dynamical rather than atomic) must be tied to our deduction of categories. It is in Schelling's variously executed "construction of matter" that these categories are deduced.

Schelling's dialectic is driven by the persistent attempt of the (absolute) subject to become an object for itself, making its way to higher powers of subjectivity or inwardness in the process.[46] Thought is necessarily and inherently synthetic, and begins with a genuine opposition of factors; either something is opposed to thought itself or there are two factors contesting in thought. From these initial factors a dialectic ensues that necessitates a third synthetic moment, and this new whole can itself be treated as one factor or product at the next level or stage of development. We obtain the image of nested spheres of activity or "products." A product consisting of two simple "factors" can itself become one of two factors constituting another product or sphere of activity, and so what is a mere factor for one stage of development could itself be a product from the perspective of an earlier stage. Schelling sought not only to think organic life as a single unfolding continuum, but also Nature as a whole, including the inorganic realm. The organic and the inorganic could be unified only if at bottom both realms were constituted by the "same" forces. The dialectic of forces in the inorganic realm, specifically the "construction of matter" out of chemical, electrical, and magnetic forces, must in some sense be contained or be implicit in the organic realm. In the "General Deduction of the Dynamical Process, or on the Categories of Physics" (1800), written after the *System* and appearing in the first issue of his *Journal of Speculative Physics*, Schelling takes to its furthest development the basic construction of matter and force that he has approached in various ways since 1797. The "Dynamical Process" names the perpetually active "self-constructing of matter recapitulated at diverse levels," and a deduction of the dynamical process will be equivalent to the complete construction of matter itself. In it "we distinguish various moments in the construction of matter that we allow it to *pass through*," but we do not have to think that "Nature actually passes through these moments in time, but only that these moments are dynamically—or if this is more meaningful—metaphysically grounded in Nature."[47] Since "all genuine construction is *genetic*," it is necessary to think of "moments" analytically where in Nature itself (from an ontological standpoint) there is unity and no temporal sequence. From the dynamical extension of space itself into three dimensions (length, breadth, depth) as a consequence of the dialectical interaction of fundamental forces, Schelling deduces or constructs the categories of gravity, magnetism, electricity, and the chemical process as forms as fundamental to our apprehension of objects as is space itself. Magnetism, for example, because it must necessarily be conceived as a duality in identity and a unification of polar opposites, is the "form" or category of physics under which length itself must be thought, and magnetism is the conditioning factor of all length. This means that magnetism is a universal function or power of matter, and is not bound to any particular substance in nature (in contrast to the representation of a "magnetic fluid," a notion current at the time). Magnetism is an expression of the synthesis of repulsive and attractive force in one and the same body, and Schelling even holds that every magnet, because it is a synthesis of opposites, is "a symbol of Nature as a whole."[48] Magnetism is itself a product of force relations, and constitutes matter as such (not any specific material); only the addition of a further determination through "negative conditions" makes this matter a specific body or object.

According to the central tenet of natural philosophy's critical epistemology, in contrast to mechanical and atomistic philosophy, the main objects of investigation are dynamical forces or productivity, and not static objects or products. The static object is always secondary with respect to the forces and powers ("functions" or "factors") that generate and maintain it. Thus there is unity at the level of production and diversity at the level of products in nature. Nature is conceived to be in perpetual *becoming*, while *being* is "becoming suspended." A major problem for nature philosophy emerging from these ontological theses is then the nature of permanence. It is no longer the task of *philosophia naturalis* to solve the problem of hylomorphic "substantial change," but to explain how products appear to be permanent at all in this continual flux. Schelling's preferred figure to characterize the product of nature is the whirlpool:

> A stream flows in a straight line forward as long as it encounters no resistance. Where there is resistance—a whirlpool forms. Every original product of Nature is such a vortex, every organized being. E.g., the whirlpool is not something immobilized, it is rather something constantly transforming—but reproduced anew at each moment. Thus no product in Nature is *fixed*, but it is reproduced at each instant through the force of Nature entire. (We do not really see the subsistence of Nature's products, just their continually being-reproduced.) Nature as a whole co-operates in every product.[49]

"Products" are specifically the result of a production through a relation of forces, and will constitute the ground of an existing body (matter), from which a determinate body is formed.[50] If each product is a relatively "permanent process," the continual reproduction of its own substance and processes, and yet all products are suffused by the powers of nature as a whole, what are the factors that distinguish one product from another? One product is specifically different from another not only in its material composition, but in the relation and proportions of its constitutive forces among themselves (and the former depends on the latter). As in his earlier essays on transcendental philosophy where the self, in striving to preserve its identity, "imitates" the absolute as far as it can by exercising "absolute causality," all of the vortical existents are preserved in their existence by their striving to express the whole, in increasing individualization, to approximate a single perfect organism, a single archetype.[51] On the other hand, Schelling speaks often of the tendency of nature to return to a state of indifference, where no strife exists, where all individuality, therefore, is eliminated. He repeatedly insists that the "individual exists, as it were, against the will of nature."[52] The perfect balance of these two tendencies is expressed in the preservation of the genus or species at the expense of the individual. On a larger scale, what a species expresses of the whole is not a particular mixture of materials and forms, but a certain proportion and intensity of primary organic functions. It is through the continuity of organic functions that the whole diversity of the natural world is connected and forms a single whole organism.

In contrast to the school of comparative anatomy and the old natural history, Schelling establishes a "comparative physiology" of organic functions. Just as "speculative physics" is a science of the fundamental powers of matter, he endeavors to establish a science of the various degrees and proportions of essential powers that belong to all organic beings. Through these diverse expressions in various proportions, nature as a whole achieves the realization of its ideal of unity in plurality. Just as Jean-Baptiste Lamarck oriented his system of zoology with attention to the basic functions of reproduction, circulation, and sensation, rather than to structure or external form, Schelling defines every organism as a permanent process by attending to its specific proportion of reproductive force, irritability, and sensibility. Every organism is defined, not primarily by its external form (although its form and organs will follow from the disposition of its powers), but by the particular proportion of forces active within it. All organic beings are suffused with all three powers, and yet plants, for example, have a preponderance of reproductive force while sensibility in them approaches zero. Mammals, in contrast, have a preponderance of sensibility but produce few offspring, their reproductive power narrows to a capacity to reproduce only the organism itself through growth, assimilation, and maintenance.

The order of the powers of matter I have presented is also one of decreasing extensity and inversely proportional intensity. Light (or chemical process) reaches everywhere, electricity is less widely distributed, and only a few bodies are magnets (although magnetic force permeates the entire cosmos). The organic functions are isomorphic with the inorganic. Force of reproduction is positive, expansive, and the most widely distributed. Irritability is reactive and designates the juxtaposition of inner and outer worlds, and is less widely distributed. Sensibility is the synthesis of the two because it is the source of movement and cause of life's reproduction of itself and of its reactions to the world. Because these powers are dialectically related however, and because they are all modifications of a single omnipresent power of nature, neither the second nor the third in each group is absent from the the manifestation of the first. In a sense, the third is there from the beginning. This is the reason why Schelling analyzes these powers in the *Introduction* in the opposite order, elucidating their dialectical genesis more explicitly. Magnetism is the first power of inorganic nature because its factors are in a state of unity: in one and the same being opposite factors are expressed, a unity in difference. In electricity these poles are separated into different bodies, independent products in their own right, and here two products are opposed in the electrical process; they are in a state of explicit difference. In the chemical process two initially separate products are seen to recombine, to achieve an "intussusception" and unification, or return to a state of "indifference." The dialectical formula is repeated in the organic. Sensibility is, therefore, the stage of unity in difference, irritability that of explicit differentiation, and reproductive drive the stage of return to unity or indifference. Provocative statements based on these parallels (or "analogies"), such as "sensibility in the organic is the higher power of magnetism in matter," always attract the most ridicule in negative receptions of the philosophy of nature. While some followers of Schelling abused the method of analogy, Schelling was committed to its employment in a determinate form. It is the notion of "potency" that allows Schelling to present these structural repetitions in a determinate concept.

    To achieve a unified view of nature means to reduce the domains of inorganic and organic nature to a single expression, and Schelling can do this by employing his ontological and dialectical methods. The distinctions between products are drawn on the basis of their limited expression of universal powers. One could argue that univocal being expresses itself as a whole in each being, but each being receives univocal being only to the degree that it is able. Being is not unevenly distributed into substance and property or quality; this substance-predicate model is rejected along with the primacy of static objects. Being is expressed in processes whose material conditions facilitate the expression of greater or lesser degrees of intensity. Schelling therefore understands the powers of reproductive force, irritability, and sensibility as isomorphic with chemical process or light, electricity, and magnetism, respectively, and the former are nothing other than the latter "raised to a higher power."[53] In his much later lectures *On the History of Modern Philosophy* Schelling notes that the idea of "potency" is more determinate than the idea of "analogy." In speaking of light, he says it "is obviously an analogy in the extended world for spirit or thought, and if we reduce this indeterminate concept of an analogy to a determinate concept, then light is nothing but spirit or thought at a lower level or at a lower potency."[54] Here analogy is to be distinguished from potency. As Kant held, for example, analogy does not entail a similarity between things, but only a likeness in the relations between things.[55] But just as Schelling goes beyond Kant's regulative idea to a positive ontology, he uses the indeterminate concept of analogy as a means to indicate where a determinate concept of potentiation may be thought. There is a likeness of relation or proportion of one power to another (e.g., reproductive drive is most extensive in the organic as chemical process is in the inorganic), but there is also an ontological community that the concept of analogy does not imply. It is the determinate conception of a difference in intensive quantity of being that produces a qualitative difference within the structure of dialectical development, driven by the endeavor of Nature as subject to become object to itself. He uses this concept to show that organism and the inorganic are not essentially opposed. One cannot simply oppose mechanism by developing a "philosophy of organism," but must seek the common expression of both. For Schelling the opposition between the inorganic and organic realms of nature is merely apparent, and in closing the *Introduction to the Outline* he notes that the attempt to reduce one to the other is futile, and a false problem. They are not opposed at all; the organic is nothing but a "raising to a higher power" of the inorganic forces. The word *Potenzierung* names Schelling's original and powerful concept for conceiving this identity in diversity.

Conclusion

For the philosophy of nature, universal nature is a whole, a living organism, and every individual in nature is an expression of this whole. All things in nature are conjoined by virtue of a universally omnipresent but virtual power that manifests itself in various modes, depending on the material conditions of its manifestation. Every individual is defined not as a static formed substance, but as an enduring process or contest of forces restricted to a particular sphere of activity. The play of forces within every limited sphere obeys a regularity and logic that is necessarily dialectical, and in accord with which their expressions in products, of whatever type, can be situated in a graduated scale of development that indicates the intensity or degree of evolution (emanation) of the powers of nature manifest in a particular being. Schelling's philosophy of nature is dynamic and genetic, but the processes it describes are "static geneses," the expression of virtual powers in actual materials, and not the historical description of a genesis from actual term to actual term. As Krings notes, Schelling's philosophy of nature may be thought of as a logogenesis, not a real genesis. Construction, or the deduction of the categories necessary to think and experience the natural world, is the method employed to span the depths and heights of the graduated series in nature. It is often overlooked that the method of construction I have described also involves, in the act of postulating, the engagement of human freedom in transcending mechanism from the start. Not only epistemological and ontological, the philosophy of nature is an expressly ethical project.

It is hoped that the appearance of the following text in English for the first time will not only contribute to the current Schelling revival that has been gradually gaining momentum among philosophers and theorists in the English-speaking world, but will also provide a valuable resource for those interested in holistic metaphysics, environmental philosophy, ecology, and the sciences of complexity and self-organization.

Works Cited

Emerson, R. W. *Ralph Waldo Emerson: Essays and Lectures*. Ed. J. Porte. New York: Library of America, 1983.

Esposito, Joseph J. *Schelling's Idealism and Philosophy of Nature*. Lewisburg: Bucknell UP, 1977.

Fichte, J. G. "Concerning the Concept of the Wissenschaflslehre." *Fichte: Early Philosophical Writings*, ed. Daniel Breazeale. Ithaca: Cornell UP, 1988. 94–136.

Gould, Stephen Jay. *Ontogeny and Phylogeny*. Cambridge: Harvard UP, 1977.

Gower, Barry. "Speculation in Physics: The History and Practice of Naturphilosophie." *Studies in the History and Philosophy of Science* 3.4 (1973): 301–56.

Harding, Sandra G. *Is Science Multicultural?: Postcolonialisms, Feminisms, and Epistemologies*. Bloomington: Indiana UP, 1998.

Hegel, G. W. F. *The Difference between Fichte's and Schelling's System of Philosophy*. Trans. H. S. Harris and Walter Cerf. Albany: State U of New York P, 1977.

———. *Differenz des Fichte'schen und Schelling'schen Systems der Philosophie (1801)*. Leipzig: Philipp Reclam Verlag, 1981.

Henrich, Dieter. "Kant's Notion of a Deduction and the Methodological Background of the First *Critique*." *Kant's Transcendental Deductions*. Ed. E. Förster. Stanford: Stanford UP, 1989. 20–46.

Hoffmeyer, Jesper. *Signs of Meaning in the Universe*. Trans. Barbara J. Haveland. Bloomington: Indiana UP, 1996.

Holz, Harald. "Perspektive Natur." *Schelling: Einführung in seine Philosophie*, ed. Hans Michael Baumgartner. Freiburg/München: Verlag Karl Alber, 1975. 58–74.

Jantsch, Erich. *The Self-Organizing Universe: Scientific and Human Implications of the Emerging Paradigm of Evolution*. Oxford: Pergamon, 1980.

Kant, Immanuel. *Kritik der reinen Vernunft*. Ed. W. Weischedel. Vol. 3–4, *Werkausgabe*. Frankfurt am Main: Suhrkamp Taschenbuch Wissenschaft, 1974.

———. *Opus Postumum*. Trans. Eckhart Förster and Michael Rosen. Ed. Paul Guyer and Allen W. Wood. *The Cambridge Edition of the Works of Immanuel Kant*. Cambridge: Cambridge UP, 1993.

———. *Prolegomena to Any Future Metaphysics*. Trans. James W. Ellington. Indianapolis: Hackett, 1977.

Kauffman, Stuart A. *Investigations*. Oxford: Oxford UP, 2000.

Krings, Hermann. "Die Konstruktion in der Philosophie. Ein Beitrag zu Schellings Logik der Natur." *Aspekte der Kultursoziologie*. Ed. J. Stagl. Berlin: D. Reimer, 1982. 338–352.

———. "Natur als Subjekt: Ein Grundzug der Spekulativen Physik Schellings." *Natur und Subjektivität: Zur Auseinandersetzung mit der Naturphilosophie des jungen Schelling*. Ed. R. Heckmann, Hermann Krings, and R. W. Meyer. Stuttgart: Frommann-Holzboog, 1983. 111–28.

Lewontin, Richard C. *The Triple Helix: Gene, Organism, and Environment*. Cambridge: Harvard UP, 2000.

Lovelock, J. E. *Gaia: A New Look at Life on Earth*. Oxford: Oxford UP, 1987.

Margulis, Lynn. *Symbiotic Planet: A New Look at Evolution*. New York: Basic, 1998.

Poser, Hans. "Spekulative Physik und Erfahrung. Zum Verhältnis von Experiment und Theorie in Schellings Naturphilosophie." *Schelling: Seine Bedeutung für eine Philosophie der Natur und der Geschichte*. Ed. L. Hasler. Stuttgart/Bad Cannstatt: Frommann-Holzboog, 1981. 129–38.

Prigogine, I., and Isabelle Stengers. *The End of Certainty: Time, Chaos, and the New Laws of Nature*. New York: Free Press, 1997.

Rudolphi, Michael. *Produktion und Konstruktion: Zur Genese der Naturphilosophie in Schellings Frühwerk*. Ed. Walter E. Eherhardt. Vol. 7, *Schellingiana*. Stuttgart/Bad Cannstatt: Frommann-Holzboog, 2001.

Salthe, Stanley N. *Development and Evolution: Complexity and Change in Biology*. Cambridge: MIT Press, 1993.

Schelling, Friedrich Wilhelm Joseph. *Ideas for a Philosophy of Nature*. Trans. E. Harris and P. Heath. Cambridge: Cambridge UP, 1988.

———. *On the History of Modern Philosophy*. Trans. A. Bowie. Ed. R. Geuss. *Texts in German Philosophy*. Cambridge: Cambridge UP, 1994.

———. *Sämmtliche Werke*. Ed. K. F. A. Schelling. 14 vols. Stuttgart: J. G. Cotta'scher Verlag, 1856ff.

———. *System des transzendentalen Idealismus*. Hamburg: Felix Meiner Verlag, 1992.

———. *The Unconditional in Human Knowledge: Four Early Essays (1794–1796)*. Trans. Fritz Marti. Lewisburg: Bucknell UP, 1979.

———. *Werke: Historisch-kritische Ausgabe*. Ed. Hans Michael Baumgartner, Wilhelm G. Jacobs, and Hermann Krings. Stuttgart: Frommann-Holzboog, 1976ff.

———. *Zur Geschichte der neueren Philosophie*. Ed. M. Buhr. Leipzig: Reclam, 1975.

Schelling, Friedrich Wilhelm Joseph von. *System of Transcendental Idealism (1800)*. Trans. Peter Heath. Charlottesville: UP of Virginia, 1978.

Smolin, Lee. *The Life of the Cosmos*. New York: Oxford UP, 1997.

Treder, H.-J. "Zum Einfluβ von Schellings Naturphilosophie auf die Entwicklung der Physik." *Natur und geschichtlicher Prozeβ: Studien zur Naturphilosophie F. W. J. Schellings*. Ed. H. J. Sandkühler. Frankfurt a. M.: Suhrkamp, 1984. 326–34.

TRANSLATOR'S NOTE

This translation is based on volume 7 (2001) of the historical-critical edition published by the Schelling Commission in affiliation with the Bayern Academy of Sciences (1976ff.). The translation preserves the sometimes abrupt and unpolished quality of the original lecture text and manuscript notes, while it also endeavors to be acceptable to the English reader's sensibility. A perfect compromise was not always possible. I have preserved Schelling's use of emphasis and liberal employment of the em dash. To add clarity to an often obscure course of argumentation, the chapter headings have been supplied by the translator and are derived in all cases directly or indirectly from the introductory "outline of the whole" provided by Schelling himself. Numbers in brackets indicate the critical edition (AA) pages; for corresponding SW page numbers the reader may consult the page concordance.

All endnotes are my own. Footnotes to the main text, unless marked "Original note," derive from a handwritten manuscript used by Schelling in the Jena lectures, unfortunately destroyed during the Second World War. However, they were included as footnotes in the *Sämmtliche Werke* edition. On the grounds of a directive provided by Schelling himself regarding the fate of this manuscript ("Hardly to be used, best if eliminated," AA I,7 11), the editors of the critical edition of Schelling's works decided to distance these remarks from the first edition text and have placed them in a separate section of their edition. Nevertheless, I believe these explanatory notes, supplements, elucidations, and clarifications are indispensible for a greater understanding of the text and of enormous benefit to the reader, and I have chosen to follow the earlier edition and include them as footnotes to the main text, which otherwise follows the critical edition.

I have benefited from the 1867 translation of the *Introduction* to the *Outline* by Thomas Davidson, which appeared in the *Journal of Speculative Philosophy* (vol. 1, no. 4). It has provided me with generous resources both here and in the translation of the *Outline* itself. The translation of the *Introduction* presented here (following the translation of the *Outline*) may be considered a thoroughly revised version of the Davidson translation. Bracketed page numbers in that text refer to the SW edition of Schelling's works, since the critical edition of this text is not yet available.

An appendix containing a few biographical details about the numerous now-famous or now-forgotten scientific figures and philosophers to whom Schelling refers throughout the two texts and notes has been added for the reader's convenience.

Erster Entwurf

eines Systems

der

Naturphilosophie

Zum Behuf seiner Vorlesungen

von

F. W. J. Schelling

Jena und Leipzig
bey Christian Ernst Gabler
1799

FOREWORD

The same demands cannot rightfully be made upon a treatise that has been written solely and exclusively to serve as a guide for lectures—like the one before you—as upon a text which was primarily intended for the public at large.

This treatise may surely be called a *first* outline, because no attempt of its kind has previously existed—for no one has yet *ventured* for *dynamic* philosophy what has been done for the *mechanistic* philosophy by Lesage.[1]—But the title has another sense as well.

The author has too lofty a notion of the magnitude of his undertaking to announce in the present treatise anything *more* than the *first* outline, let alone to erect the *system itself*.

Thus, I ask but one thing: that the reader remember, in levelling a judgment, that all of the facts are not yet in. The reader should pass judgment least of all on what "philosophy of nature" or "speculative physics" means for the author (for those who do not know already); rather, if he must pass judgment, let him await my explanation, which will follow shortly in a certain treatise *On the Foundation and the Inner Organization of a System of Speculative Physics*.—For now, the following outline may take the place of an introduction.

March 20, 1799
Jena

OUTLINE OF THE WHOLE.

First Division.
Proof that nature is *organic* in its most original products.

I. Because to philosophize about nature means as much as to create it, we must first of all find the point from which nature can be posited into *becoming* (pp. 13–15).

In order for a real activity to come to be out of an infinite (and to that extent ideal) productive activity, that activity must be inhibited, *retarded*. But because the activity is originally an infinite one, it cannot result in finite products, even when it is inhibited; and even if it should result in finite products, these can only be merely *apparent products*; i.e., the tendency to infinite development must lie once again in every individual; every product must be capable of being articulated into products (pp. 15–19).

II. III. Thus the analysis can not be permitted to stop at any one thing that is a *product*; it can only cease with the purely *productive*. But this *absolutely* productive character (which no longer has a substrate, but is rather the cause of every substrate) is that which absolutely blocks all analysis; precisely for that reason, it is the point at which our analysis (experience) can never arrive. It must be *simply posited* into Nature, and it is the first *postulate* of all philosophy of nature.—It must be that which is *insurmountable* in Nature (mechanically and chemically); such a thing is thought to be nothing other than the cause of all *original* quality (p. 19). Such an absolutely productive character is designated by the concept of the *simple actant*.[1]—(Principle of a dynamic atomism)—(pp. 20ff.).

If the absolute analysis were to be thought as actual, then, because an infinite product evolves in Nature as object, there would have to be an [68] infinite multiplicity of simple actants, thought as the elements of nature and of all construction of matter (pp. 20–22).

(Here we must at once recall that this absolute analysis in nature can never be reached, that those simple actants are therefore only the ideal factors of matter.)

Yet these simple actants cannot be distinguished from one another in any other way than by the original *figure* that they produce (a point we owe to the atomists). Yet because absolute evolution does not eventuate due to the universal compulsion toward combination that holds Nature together as *product* (pp. 28–29), these fundamental configurations cannot be thought as existent (*contra* the atomists).* Therefore they have to be thought as self-canceling, as *interpenetrating* (cohesion, pp. 25ff.). The most original product of this interpenetration is the *most primal fluid*—the absolute noncomposite, and for that reason the absolute decomposite. (A look at caloric, electrical, and luminous phenomena from this point of view (pp. 29ff.)).—Such a principle would entail the cancellation of all individuality—hence also of every *product*—in Nature. This is impossible. Hence there must be a counterweight in *Nature*, by means of which matter disappears from the other side into the absolutely *indecomposable*. However, this in turn cannot exist except by being at the same time the absolutely *composable*.² —Nature cannot lose itself in either extreme. Nature in its originality is therefore a mediator on the basis of both (pp. 32–33).

The state of *configuration* is therefore the most original in which Nature is viewed.—Nature = a product which passes from figure to figure—in a certain order to be sure, through which, however, [69] it cannot result in determinate products without absolute *inhibition* of the *formation*.—I demonstrate that this is conceivable only if the formative drive splits in opposite directions, which on a lower level will appear as *differentiation of the sexes* (pp. 35–36).

Proof that by this means the permanence of various stages of development in Nature is assured (pp. 39ff.).

Yet all these various products = *one product that is inhibited at sundry stages*. They are deviations from *one* original ideal. *Proof* on the basis of the continuity of the dynamic graduated series of stages in Nature (pp. 48ff.); on that basis we discern the fundamental task of all nature philosophy: TO DERIVE THE DYNAMIC GRADUATED SEQUENCE OF STAGES IN NATURE.

IV. Individual products have been posited in Nature, but Nature implies a *universal* organism.—Nature's struggle against everything individual.

---

*If one considers nature as object to be *real*, and as having originated not by evolution but by *synthesis* (and we have no alternative from the empirical standpoint), atomism is necessary, whether it be mechanical or dynamic.—In the transcendental view to which speculative physics ultimately elevates itself, everything is entirely transposed.

Deduction of the necessary *reciprocity of receptivity* and *activity in everything organic* (which will be presented below as "excitability") (p. 56), and the cancellation of this reciprocity in the opposed systems

- a) of chemical physiology, which posits mere receptivity (no subject) in the organism, and
- b) of the system that posits in the organism an absolute activity (mediated by no receptivity)—an absolute life force (p. 60).

Unification of the two systems in a third (pp. 61ff.).

If, however, receptivity is necessarily posited in the organism as the mediator of its activity, there lies within the organism itself the presupposition of a world that stands opposed to it—an *anorganic*[3] world that has a determinate effect on the organism.—This world, however, precisely because it is a determinate (unalterable) world, has to be subject itself to an external effect (it must be, as it were, in a state of compulsion), in order that it may form, together with its organic world, once again through some kind of commonality, something *interior*. [70] This would have to be derived from the conditions of an anorganic world in general.

Second Division.
Deduction of the conditions of an anorganic nature.

Deduction of the possibility of sheer contiguity and exteriority (p. 71). Because such a thing is conceivable only as a tendency toward penetration, a cause is postulated that sustains this tendency.

- a) Deduction of universal gravity (pp. 71–72). Opposed systems,
    the mechanical and
    the metaphysical system of attraction (pp. 73–78).
A third system on the basis of the other two: a system of physical attraction derived from the theory of universal cosmic formation (pp. 78–93).
- b) With universal gravitation, the *tendency* toward universal intussusception[4] in Nature is founded. Accepted as a hypothesis, namely, that there is *real* intussusception, the action of gravity would be only the first impulse toward it; thus another, different action would be adduced to it in order to make it actual.—We are required to demonstrate such a thing in Nature (p. 194).

Proof that the principle of all chemical process of a determinate sphere is not in turn a product of the same sphere, but of a higher sphere. (Deduction of oxygen) (pp. 94–96).—Conclusion that the positive action in every chemical process within the lower sphere must take its point of departure from the higher one.

Proof that, in the part of the universe that is known to us, *light* is a phenomenon of such a dynamic action, exercised by celestial bodies of a higher order on subaltern bodies. (Combustion = a transition of opposed spheres of affinity into one another) (pp. 96–100).

[71] c) Deduction of a relation in all terrestrial substances that is opposed to that action—*electrical* relations of bodies.

Distinction between the electrical and chemical processes. The principle that immediately intervenes in the one is the mediately determining principle of the other (pp. 102–104).

d) Relation of the action of gravity to chemical action (p. 104).

Third Division.
Reciprocal determination of organic and anorganic nature.

I. The supreme concept by which the interconnection of the organism with an anorganic world is expressed is the concept of *excitability*.—Duplicity, which is thereby posited in the organism, and a derivation of the same from the general organization of the universe (pp. 105–108).

Complete unification of the opposed systems wherein the organism is posited either as mere object or mere subject in a third system, which posits the organism as *excitable* (pp. 108ff.).—Derivation of a *cause* of excitability, the condition of which is duplicity, a cause which in its tendency is chemical, and which precisely for that reason cannot originally be chemical; thereupon a grounded and complete determination of the *possibility of a higher dynamic process* (the same as the life process), *which, although not itself chemical, nevertheless has the same cause and the same conditions as the chemical process* (p. 113).

[72] II. *Derivation of individual organic functions from the concept of excitability.*

a) Because excitability presupposes duplicity—the cause of the former cannot be the cause of the latter. Thus a cause is postulated that no longer *presupposes* duplicity—a cause of *sensibility*, as the source of organic activity (pp. 116–117).

b) Determination of the activity whose source is sensibility, and the conditions of this activity (in Galvanism)—*irritability* (p. 125).

c) Extinction of this activity in the product—*force of production*, with all its offshoots (nutrition, p. 125, secretion, pp. 126–130, growth, p. 130, technical drive (animal instincts in general), pp. 130–139. —Metamorphosis, reproductive drive, pp. 138–140).

III. Consequences of the preceding.

a) That the organic functions are subordinated, one to another, in such a way that they are *opposed* with regard to their *appearance* (their coming to the fore), both in the individual and in the whole of organic nature.

b) That by this opposition (because the higher function is repressed by the surfeit of subordinate functions) a *dynamic* sequence of stages in Nature is founded.

c) Demonstration of this dynamic sequence of states (p. 141f.) on the basis of:

    aa) a reciprocal determination of sensibility and irritability (pp. 142–147);

    bb) a reciprocal determination of sensibility and force of production (p. 147f.);

    cc) a reciprocal determination of irritability and force of production (pp. 147–148), throughout organic nature.

Conclusion: that it is one and the same product that, beginning from the highest stage of sensibility, ultimately dissipates in the reproductive force of plants.

d) Demonstration that *the same dynamic sequence of stages prevails in universal and anorganic nature as in organic nature* (pp. 149–159).

[73]   *General schematic of this sequence of stages*

Organic	Universal	Anorganic Nature
Formative drive	Light	Chemical Process
Irritability	Electricity	Electrical process
Sensibility	Cause of magnetism?	Magnetism?*

---

*Since the subordinate forces in universal Nature, as in the organic, already presuppose an original heterogeneity, then a cause that *brings forth* heterogeneity (from homogeneity) is postulated, in whose place is situated, as merely hypothetical, the cause of universal magnetism.

e) *Supreme problem of the philosophy of nature: What cause brought forth the first duplicity* (of which all other opposites are the mere progeny) *out of the universal identity of Nature?* (p. 158).

(APPENDIX TO III: Theory of illness, derived from the dynamic sequence of stages in Nature, pp. 158–172.)

IV. Not only the subordinate functions of the organism but also the general forces corresponding to them (electricity, chemical process) presuppose an original heterogeneity—the solution of that problem (What is the cause of the original heterogeneity?) is thus at the same time a theory of chemical process, and vice versa.

Universal theory of the chemical process (pp. 172–187).

a) *Concept of the chemical process* (pp. 172–174).
b) *Material conditions of the chemical process.*—Demonstration that in the chemical as well as in the electrical process only *one* opposition prevails (pp. 174–179).
c) Inasmuch as all chemical (and electrical) process is mediated by a *first* heterogeneity, the latter has for universal nature the same function that sensibility has for organic nature.—Complete demonstration that MAGNETISM is for universal [74] Nature what sensibility is for organic nature, that all *dynamic* forces of the universe, such as sensibility, are subordinate to it—that magnetism, like sensibility in organic nature, is *universal* in anorganic nature (and *canceled*, wherever it is canceled, only for *appearance*).—Conclusion: the identity of the ultimate cause of sensibility and magnetism (p. 184).
d) *Complete construction of the chemical process and of all dynamic process* (pp. 184–187).
    aa) inasmuch as an intussusception between heterogeneous bodies is possible only insofar as the *homogeneous* is itself sundered *in itself*, no homogeneous state can be *absolute*; rather, it can only be a *state of indifference*. In order to explain this, we must suppose that in the universe there is a universal effect that propagates itself from product to product by means of (magnetic) distribution, which would be the universal determinant of all quality (and we must therefore suppose that magnetism is universal) (p. 186).
    bb) Further, in order to bring heterogeneity into the particular dynamic sphere, and thereby the possibility of canceling the dynamic state of indifference, we must suppose a *communication* between the higher and the lower spheres of affinity

(through the medium of light, p. 186). By means of the lower sphere, the *external* condition of the dynamic process (heterogeneity) is given; by means of the higher sphere, the *inner* condition (the diremption[5] in the homogeneous itself) is given.

V. The *dynamic organization* that we have now derived presupposes the universe as its *scaffolding*.

Deduction of the forces by which (presupposing an original duplicity in Nature) *the evolution of the universe is conditioned—*

Deduction of the expansive force,
of the retarding force,
and of the force of gravity, which alone (in their independence from one another) make Nature possible as a determinate product for every moment of time and space, and which alone make possible a real *construction of matter* (pp. 186–192).

[75–76 blank]

[77]

FIRST DIVISION

I.
The Unconditioned in Nature

The subject which is to be the object of philosophy in a given instance must be viewed, in a word, as *unconditioned*. The question arises as to what extent *unconditionedness* might be ascribed to Nature.

1) First of all, we must try to secure the concept of the unconditioned. To this end, however, we are in need of a few principles that are assumed as well known from transcendental philosophy.

FIRST PRINCIPLE. *The unconditioned cannot be sought in any individual "thing" nor in anything of which one can say that it "is." For what "is" only partakes of being, and is only an individual form or kind of being.—Conversely, one can never say of the unconditioned that it "is." For it is* BEING ITSELF, *and as such, it does not exhibit itself entirely in any finite product, and every individual is, as it were, a particular expression of it.*

ELUCIDATION. What is asserted by this principle obtains universally overall and for the unconditioned in every science. For although only transcendental philosophy raises itself to the Absolute Unconditioned in human knowledge, it must nevertheless demonstrate that every science that is *science* at all has its unconditioned. The above principle thus obtains also for the philosophy of nature: "the unconditioned of Nature *as such* cannot be sought in any individual natural object;" rather a *principle*[1] of being, that itself "is" not, manifests itself in each natural object.—Now, since the unconditioned cannot be thought under the predicate of being, it obviously follows that as principle of all being, it can participate in no higher being. For if everything that "is" is only, as it were, the color of the unconditioned, then the unconditioned itself must everywhere become manifest through itself—like light that requires no higher light in order to be visible.

[78] Now, what is this *being itself* for transcendental philosophy, of which every individual being is only a particular form? If, according to these very principles, everything that exists is a construction of the spirit, then *being itself* is nothing other than *the constructing itself,* or since construction is thinkable at all

13

only as activity, *being itself* is nothing other than the *highest constructing activity*, which, although never itself an object, is the principle of everything objective.

Accordingly, transcendental philosophy knows of no originary being.* For if *being itself* is only *activity*, then the individual being can only be viewed as a determinate form or limitation of the originary activity.—Now *being* ought to be something just as little primary in the *philosophy of nature*; "*the concept of being as an originary substratum should be absolutely eliminated from the philosophy of nature*, just as it has been from transcendental philosophy." The above proposition says this and nothing else: "Nature should be viewed as unconditioned."†

Now Nature itself is, according to general consensus, nothing other than the *sum total of existence*;‡ it would therefore be impossible to view Nature as an unconditioned, if the concealed trace of freedom could not be discovered in the concept of being itself.§ *Therefore* we assert: every individual (in Nature) is only a form of being itself; *being itself* however = absolute activity. For, if being itself is = to activity, then the individual being cannot be an absolute *negation* of activity. Nevertheless, we must think the natural product itself under the predicate of being. However, viewed from a higher standpoint, this being itself is nothing other than a *continually operative*‖ *natural activity* that is extinguished in its product.—Originally, no *individual being* at all (as an accomplished fact) is present for us in Nature, for otherwise our project is not philosophy, but empirical investigation.—We must observe what an *object* is in its *first origin*. First of all, everything that is in Nature, and Nature considered as sum total of *existence*, is not even present for us. To philosophize about Nature means *to create* Nature. Every [79] activity perishes in its product, because it reaches only to this product. Thus we do not know *nature as product*. We know Nature only as *active*—for it is impossible to philosophize about any subject which cannot be engaged in activity. To philosophize about nature means to heave it out of the dead mechanism to which it seems predisposed, to quicken it with freedom and to set it into its own free development—to philosophize about nature means, in other words, to tear *yourself* away from the common view which discerns in nature only what

---

*of no being *in itself*.

†The philosopher of nature treats nature as the transcendental philosopher treats the self. Thus Nature itself is an unconditioned to him. This is not possible, however, if we proceed from objective being in Nature. In philosophy of nature objective being is as little something originary as in transcendental philosophy.

‡and to that extent Nature would be understood as *object*.

§if the trace of a loftier concept, the concept of activity, did not lie in the concept of being itself.

‖uniformly operative

"happens"—and which, at most, views the act as a *factum, not the action itself* in its acting.*

2) We have answered the first question (how unconditionedness may be ascribed to Nature) through the assertion that Nature has to be viewed as *absolutely active*. This answer itself drives us to the new question: how can Nature be observed as absolutely active, or more clearly expressed: *in what light must the totality of Nature appear to us, if it is absolutely active?*†

The following principle must serve us in answering this question.

SECOND PRINCIPLE. *Absolute activity cannot be exhibited by a finite product, but only by an* INFINITE *one*.

ELUCIDATION. The Philosophy of Nature, so that it does not degrade into an empty play with concepts, must demonstrate a corresponding *intuition* for all of its concepts. Therefore, the question arises how an absolute activity (if there is such a thing in Nature) will present itself empirically, i.e., in the finite.

—Possibility of the exhibition of the infinite in the finite—is the highest problem of all systematic science. The subordinate sciences solve this problem in *particular cases*. Transcendental philosophy has to solve the problem in its greatest *universality*.—This solution will doubtless eventuate in the following result.

The illusion that surrounds the entire investigation concerning the infinite in all sciences issues from an amphiboly in this concept itself.—The *empirically infinite* is only the external intuition [80] of an *absolute (intellectual) infinity* whose intuition is originally in us, but which could never come to consciousness without external, empirical exhibition. The proof of this is that this intuition comes to the fore precisely when the empirically infinite series lying before the imagination is obliterated ("I blot it out, and you lie fully before me"²). If, that is, the finite can be intuited only externally, then the infinite can not even be presented in external intuition otherwise than through a *finitude* which is never complete, i.e., which is *itself infinite*. In other words, it can only be presented by *infinite becoming*,‡ where the intuition of the infinite lies in no individual moment, but is only *to be produced* in an endless progression—in a progression, however, which no power of imagination can sustain. Therefore, reason determines either to obliterate the series,§ or to assume an ideal limit to

---

*In the usual view, the original productivity of nature disappears behind the product. For us the product must disappear behind the productivity.

†productive

‡by *letting-become*

§When the series is obliterated, nothing remains except the feeling of an infinite tendency in ourselves—this tendency now emerges in intuition, and the above expression of the poet should be considered in this regard. From this it becomes clear that originally all infinity lies *in ourselves*.

the series which is so far removed that in practical employment one can never be compelled to go beyond it (as the mathematician does when he assumes an infinitely large or small magnitude).

But now, how must one represent an infinite series if it is only the external exhibition of an *original* infinity? Are we to believe that the infinite is produced in the series through *aggregation*, or rather ought we to represent any such series in *continuity*, as one function running to infinity?—The fact that in mathematics, infinite series are composed of magnitudes, proves nothing on behalf of that assumption. The *originally infinite* series, of which every individual series in mathematics is an imitation, does not arise through *aggregation*, but through *evolution*, through evolution of a magnitude already *infinite in its point of origination* which runs through the entire series. The whole infinity is originally concentrated in this one magnitude. The succession in the series signifies only, as it were, the individual *inhibitions** which continually set bounds to the expansion of that magnitude into an infinite series (an infinite space), and which moreover happens with an infinite velocity and would permit no real intuition.

[81] The genuine concept of an *empirical infinity* is the concept of an *activity*† that *is infinitely inhibited*. But how could it be inhibited to infinity if it did not flow into infinity and if it did not deposit its whole infinity in every individual point of the line that it describes?

CONSEQUENCES FOR THE PHILOSOPHY OF NATURE
(which are at once to be seen as the response
to our second question above).

FIRST CONSEQUENCE. *If Nature is absolute activity, then this activity must appear as inhibited ad infinitum.*‡ *(The original cause of this inhibition must again only be sought* IN ITSELF, *since Nature is* ABSOLUTELY *active).*

SECOND CONSEQUENCE. *Nature* EXISTS *nowhere as product; all individual productions in Nature are merely apparent products, not the absolute product that always* BECOMES *and never* IS, *and in which the absolute activity exhausts itself.*§

According to the first principle, an *original duality* must simply be presupposed in Nature. For it permits of no further derivation, because it is the only

---

*through reflection

†tendency

‡otherwise no empirical presentation of it is possible.

§Productivity is originally infinite; thus even when a product comes to be, this product is only an apparent product. Each product is a point of inhibition, but the infinite still "is" in each point of inhibition.

condition under which an infinite is finitely presentable at all, i.e., the condition under which a Nature is at all possible. Through this original antithesis in itself, Nature will now be for the first time truly whole and complete in itself.*

Since Nature gives itself its sphere of activity, no foreign power can interfere with it; all of its laws are immanent, or *Nature is its own legislator* (autonomy of Nature).

Whatever happens in Nature must also be explained from the active and motive principles which lie in it, or *Nature suffices for itself* (autarchy of Nature).

They are both contained in the proposition: *Nature has unconditioned reality*,† a proposition which is precisely the principle of a philosophy of nature.

[82] The absolute activity of Nature should appear as inhibited to infinity. This inhibition of the universal activity of Nature (without which "apparent products" would never once come to be) may be represented, at any rate, as the work of opposed tendencies in Nature. (Let one force be thought, originally infinite in itself, streaming out in all directions from one central point; then this force will not linger in any point of space for a moment (thus leaving space empty), unless an energetic activity opposing (retarding) its expansion did not give it a finite velocity.‡) However, as soon as one undertakes to carry out the construction of a finite product from these opposed tendencies, one encounters an irresolvable difficulty. For if we let both coincide at one and the same point, then their effects toward one another will reciprocally be canceled, and the product will be = to 0. Precisely for this reason, it must be assumed that no product in nature can be the product in which those opposed activities absolutely coincide, i.e., in which Nature itself attained rest. One must, in a word, simply *deny* all *permanence* in Nature itself. One has to assume that all *permanence* only occurs in Nature as *object*, while the activity of Nature as subject continues irresistably, and while it continually labors in opposition to all permanence. The chief problem of the philosophy of nature is not to explain the *active* in Nature (for, because it is its first supposition, this is quite conceivable to it), but the *resting, permanent*. Nature philosophy arrives at this explanation simply by virtue of the presupposition that for Nature the permanent is a limitation of its own activity.§ So, if this is the case, then impetuous Nature will struggle against every limitation; thereby the points of inhibition of its activity

---

*and so it should be.

† = Nature has its reality by virtue of itself—it is its own product—a whole, self-organizing, and organized by itself.

‡ = Kant's repulsive and attractive forces—which is merely the mechanical expression for something higher.

§Or rather, it becomes permanent only in that it is a limit for the productivity of Nature.

in nature as object will attain *permanence*.* For the philosopher, the points of inhibition will be signified by products; every product of this kind will represent a determinate sphere which Nature always fills anew, and into which the stream of its force incessantly gushes.

[83] However, when one asks (and this is the principal question), "how is it at all possible to view all of these individual products in nature as only apparent products?" we find the following answer. Evidently every (finite) product is only a *seeming* product, *if again infinity lies in it*, i.e., if it is itself again capable of an infinite development. If it engages in this development, then it would have no permanent existence at all; every product that now appears *fixed* in Nature would exist only for a moment, gripped in continuous evolution, always changeable, appearing only to fade away again. The answer given above to the question, "how could Nature be viewed as absolutely active?", is now reduced to the following PRINCIPLE:

Nature is absolutely active if the drive to an infinite development lies in each of its products.

With this the course of our further investigations is marked out. That is, to begin with, the question arises, "How must a product that is capable of an infinite development be constituted, and is such a product really found in Nature?"—Let it be noted that with this question we respond at the same time to another which must definitely be answered, namely: Why is the tendency to infinite development in such a product just maintained, and why, as fixed, does it seem oblivious to this tendency and not lose itself in the infinite?

REMARK. The proposition that the *whole*—the infinite—mirrors itself in each individual being in Nature, has been heard in transcendental philosophy more than in the philosophy of nature. The former science also has exactly the

---

*Example: a stream flows in a straight line forward as long as it encounters no resistance. Where there is resistance—a whirlpool forms. Every original product of nature is such a whirlpool, every organism. The whirlpool is not something immobilized, it is rather something constantly transforming—but reproduced anew at each moment. Thus no product in nature is *fixed*, but it is reproduced at each instant through the force of nature entire. (We do not really see the subsistence of Nature's products, just their continual being-reproduced.) Nature as a whole co-operates in every product. Certain points of inhibition in Nature are originally set up—consequently, perhaps there is only *one* point of inhibition from which the whole of Nature develops itself—first of all, however, we can think infinitely many points of inhibition—at each such point, the stream of Nature's activity will be broken, as it were, its productivity annihilated. But at each moment comes a new impulse, as it were, a new wave, which fills this sphere afresh. In short, Nature is originally pure identity—nothing to be distinguished in it. Now, points of inhibition appear, against which, as limitations to its productivity, Nature constantly struggles. While it struggles against them, however, it fills this sphere again with its productivity.

same difficulty to explain: How opposed activities coincide in the intuition of the finite without reciprocally canceling each other. It will have to be denied that they coincide in any product absolutely; one will assume that spirit does not have an intuition of itself in any individual product—that it has no intuition of itself in unity, but rather in the infinite *keeping apart* of its opposed activities *from one another* (which are only unified at all by virtue of this holding apart). It must be assumed that just for this reason each *individual* intuition is only *apparently individual*, and that actually the intuition of the whole universe is contained in every individual. The originary strife of self-consciousness—which is for transcendental creation [84] precisely what the strife of the elements is for physical creation—must, like self-consciousness itself, be infinite; therefore, it cannot end in any individual product, but only in a product that always becomes and never is, and is created anew in each moment of self-consciousness.—In order to unify absolute opposites, the productive imagination enlarges their reciprocal cancellation into an infinite series; the finite is brought into being only through this infinite extension—this infinite nudging back of absolute negation.

II.
The Original Qualities and Actants in Nature

A product is only an *apparent* product if infinity lies in it once more, i.e., if it bears the capacity for an infinite development. This capacity cannot occur in it, however, without there originally being an infinite multiplicity of unified tendencies in it.

A. The question arises, by what means do these tendencies manifest themselves in Nature at all?

THEOREM. *The most originary points of inhibition of Nature's activity are to be found in the* ORIGINAL QUALITIES.

PROOF.—Our science has an ineluctable demand to fulfill: that it accompany its *a priori* constructions with corresponding external intuitions, since otherwise these constructions would not have meaning for us anymore; no more than the theory of color for those born sightless. Now, it has been asserted in the preceding that an absolute activity can appear empirically only under infinite negations. Infinite negations of one and the same original activity must be discovered in Nature through analysis.

An *unconditioned* would have to reveal itself in these negations. No *positive* external intuition of the unconditioned is possible, however. Thus, at least a *negative* presentation of it has to be sought in external experience.

Now, we have determined the unconditioned as that which, although it is principle of all being, yet itself *never "is."* Every external being is a being in space. Therefore, something has to come to the fore in experience [85] which, although itself not in space, is yet principle of all occupation of space.*

1) *It should not itself be in space.*—What is in space can also be affected by physical force; it is mechanically† or chemically destructible. Thus a principle that is not itself in space must, admittedly, not be liable to being overpowered either mechanically or chemically. Nothing of the kind is discovered in experience except for the *original* elements (principles) of all *quality*.

2) *It should be principle of all occupation of space.*—Accordingly, it must be that which, if the (mechanical) division of matter proceeds to infinity, preserves every little piece of matter, no matter how small, for further division; in short, it must be that which *makes* the infinite divisibility of matter *possible*.‡

Now, if the infinite divisibility of matter were impossible, then one would, finally, have to reach a part in the division of any material which one could not recognize any more as a *part* of that material, i.e., no longer as *homogeneous* with the material itself. Since, therefore, the divisibility of matter proceeds to infinity, then every material must remain infinitely *homogeneous* as far as it is divided. Infinite homogeneity, however, is recognized solely in the permanence of qualities, thus the permanence of qualities is the condition of the possibility of mechanical division to infinity; accordingly, the principles of the qualities are also the principles of the occupation of space itself.

The originary qualities are thus the most originary negative presentations of the unconditioned in Nature. Now, since the unconditioned is everywhere = to absolute activity, but absolute activity can only appear empirically as an infinitely inhibited activity, then the most original points of inhibition of the activity of Nature are determined for us by the original *qualities*.

CONCLUSIONS. 1. *The divisibility of matter must be finite in one respect, simply due to the fact that it is infinite in the other.*

The atomist is mistaken only in that he assumes *mechanical* atoms, i.e., the finitude of *mechanical* divisibility. [86] In every concrete space no part must be the absolute smallest, just as in mathematical space. What IS *in space* is in space by means of a continually *active* filling-up of space; therefore, in every part of space there is moving force, so also *mobility*, and thus infinite *divisibility* of each part of matter, no matter how small, from all the remaining ones. The original actants, however, ARE not themselves *in space*; they cannot be

---

*Is, nevertheless, principle of all being in space or of all occupation of space.

†mechanically infinitely divisible.

‡The concept of infinite divisibility is necessarily contained in the concept of matter or the concept of the occupation of space.—How does it happen that matter, although divided to infinity, does not disappear for us but still remains a substrate? What is the substrate of matter supported by, and through what does divisibility become possible?

viewed as *parts* of matter.* Accordingly, our claim can be called the principle of *dynamic atomism*. For us, every original actant is just like the atom for the corpuscular philosopher; truly *singular*, each is in itself whole and sealed-off, and represents, as it were, a *natural monad*.†

2. *Each quality is an actant of determinate degree, for which there is no other measure than its product.*

a. It is *action* in general, thus not *itself* matter. If it were itself matter—*stuff*, as the popular chemistry expresses it, then it would even have to be

---

*They are the constituent factors of matter. So, if "atomism" designates a theory which assumes something simple as constituent *of* matter, then the true philosophy is nonetheless atomism. However, since it only asserts a *dynamic* simple constituent of matter, it is dynamic atomism. Each original quality is for us an actant of a determinate degree, and every such actant is—truly singular.— No individuality is to be attributed to matter without such original *unities*, which are not the unities of a product, but of *productivity*.

†In brief, our opinion is this: If the evolution of Nature were ever complete (which is impossible), then after the general decomposition of each product into its factors nothing would be left other than *simple* factors, i.e. factors which are no longer themselves products. Therefore, these simple factors can only be thought as *originary* actants, or—if it is permissible to express it this way—as originary productivities.

Our opinion is thus not that *there are* such simple actants in Nature, but only that they are the *ideal* grounds of the explanation of quality. These simple actants do not really allow of demonstration—they do not *exist*; they are what one must posit in Nature, what one must think in Nature, in order to explain the originary qualities. Then we need only prove as much as we assert, namely, that such simple actions must be *thought* as ideal grounds of explanation of all quality, and we have provided this proof.

"What is *indivisible* cannot be material, and conversely; it must lie beyond matter. But beyond matter is pure *intensity*—and this concept of pure intensity is expressed through the concept of action.—It is not the *product* of this action that is simple, but the *actant itself* abstracted from the product, and it must be simple in order that the product may be divisible." (Cf., the "Introduction to the Outline" [below p. 208; SW III 292—Trans.])

The philosophy of nature assumes, 1) with *atomism* that there is an original multiplicity of individual principles in Nature—it brings multiplicity and individuality into Nature with it.— Each quality in Nature is a fixed point for it, a seed around which Nature can begin to form itself. However, our atomism does not assume these principles as actual material parts, but as original and simple activities. 2) The philosophy of nature is in *agreement* with *dynamic* physics in that the ground of qualities does not itself consist again in material bits—every actant is pure activity, not itself matter once again; it is *not in agreement* with dynamism in that it does not allow all diversity of matter merely to consist in a variable relation of attractive and repulsive force (through which mere difference of density originates).

The philosophy of nature is therefore neither dynamic in the accepted sense of the word, nor atomistic, but is a *dynamic atomism*.

(We have posited *simple actants* of indeterminate, i.e., of infinite multiplicity in matter, as *ideal* ground of explanation. This basis of explanation is *ideal* because it presupposes something ideal, namely, that Nature has unfolded itself into simple factors.—If we proceed further down this path we shall arrive at an *atomistic* system. However, this system, on account of its insufficiency, will finally just drive us back to the *dynamic* system.)

presentable in space. However, only its effect is presentable in space, the action itself is prior to space (*extensione prior*).—(Why has chemistry not presented any of its *substrates* purely—isolated from all material?)—Action is just as little something merely inhering in original matter (the atoms, as the atomist teaches) as is figure, nor is it something that results from the collective action of atoms. For, if they do not have any qualities themselves, how is such a thing produced through their collective action?

b. *Quality is action, for which one has no measure other than its own product.* This means that the actant itself, abstracted from its product, is nothing. Indeed, it is nothing other than the product itself viewed from a higher perspective. One cannot expect to be able to take a look into the interior of that actant itself and determine the magnitude (the degree) of the action, as if by means of mathematical formulas. All attempts to do this have until now led to nothing real. Our knowledge does not reach *beyond* the product, and no other expression [87] for the magnitude of the action can be given than the *product itself*. The philosophy of nature has nothing further to do than recognize the unconditionally empirical in these actants. Empiricism extended to include unconditionedness is precisely philosophy of nature.*

**Quality is originally absolutely *inconstructible*, and it must be, because it is the *limit* of all construction by virtue of which every construction is a determinate one. All previous attempts to construct qualities have been incapable of leading to anything real for this reason. The atomist believed himself able to express qualities through figures, and assumed, therefore, an actual shape in Nature for each quality.—We have moved beyond this mode of construction.—With so-called dynamic philosophy the attempt is made to reduce qualities to analytical formulas, and to express them by means of the variable relations of repulsive and attractive force. Indeed, Kant has nowhere genuinely ventured to construct the specific (qualitative) diversity of matter out of his two basic forces. A few who wished to apply his dynamic principles have gone further. I will name only *Eschenmayer* here, rather than all of them. (One ill-conceived attempt to construct the qualities and series of degrees of qualities according to Kantian principles is to be found in his "Principles from the Metaphysics of Nature" and his "Investigation," which try to derive the magnetic phenomena *a priori*. In any case, it is to be recommended for the sake of understanding the first principles of Kant's dynamics).

Very diverse—and in part strange—manners of thinking concerning the concept of dynamical philosophy still generally prevail, and I think it necessary, therefore, to say something in general here regarding the concept of dynamical philosophy.

Many believe that dynamical philosophy consists in the fact that one assumes no particular materials for the explanation of natural appearances; e.g. who denies the materiality of light, or the existence of a galvanic fluid?—a dynamical philosopher. Only there is a bit more to it than that—one cannot get off the hook so easily. Others believe that dynamical philosophy consists in tracing everything back to the basic forces (repulsive and attractive force). The latter are, at any rate, closer to the matter at hand. All original, i.e. all dynamic natural phenomena, must be explained from forces which reside in *matter* also at rest (for Nature is movement while also at rest; this is the foremost fundamental principle of dynamic philosophy)—therefore, those appearances, e.g. the electrical, are not *appearances* or effects of determinate individual materials, but rather alterations of the *subsistence* of matter itself; and, if one lets matter consist in repulsive and attractive force—(as one

NOTE. With the preceding we have brought the construction of matter *in general* to completion. Since the identity of a material is ascertained only by the permanence of its qualities, its identity in no way differs from the latter; every material is thus nothing other than a *determinate degree of action*, no material is originally *mechanically aggregated*; for were it so, then—presupposing infinite divisibility—it would have to be dissoluble into *nothing* and originally constructed from *nothing*. Therefore—(*ne res ad nihilum redigantur funditus omnes* [3])—if matter should, for anyone, arise mechanically, it must aggregate out of *atoms* (an assumption which envelops one in a slew of other troublesome consequences).

However, let no one believe, on this account, that we have already deduced the *specific difference* of matter, or that we wanted to deduce it. Although every material is a determinate degree of action, this action can nevertheless be *highly composite*, as, according to Newton, white light is composed of seven simple ones, and these seven perhaps of other simpler actants. It is, in fact, truly nonsense to want to explain the infinite multiplicity of material in the world through various degrees of one and the same—simple—action. Does it follow from this that the original qualities are to be viewed as simple actants, that even every—also derived—quality is likewise a simple actant? If this were to be demonstrated, how is it that in experience not one original quality is found nor can be found?—Yet what are philosophical reasons for, where experience speaks

---

does at the standpoint where Nature is viewed only as *product* and not as *productivity*, i.e. as I call it, at the standpoint of mechanism, which must let matter so originate)—if one generally lets matter consist in repulsive and attractive force, then those appearances are, at any rate, only alterations in the relation of these basic forces.

All these effects also appear at the lowest level of their appearance (in the chemical process) as, at any rate, alterations—of cohesive force, of density, of specific gravity, i.e. as alterations of those basic forces. However, this is only the farthest, lowest level of their appearance—and those alterations in the relation of the basic forces cannot again be explained *from* such alterations. For appearance, every dynamic process is to its farthest extent an alteration in the relation of the basic forces—but the question is by what means these alterations have been produced, and this has not been answered by any previous research; and that question lies far higher—and yet deeper, and ultimately in the construction of matter. I want to make another remark regarding the impossibility of constructing qualities mathematically, or of submitting them to calculus.

One has transferred the familiar laws of mechanics to the dynamic appearances and wished to give them a higher, dynamical meaning. For example, it is a well-known law of mechanics that the single force *does equal work* in doubled time as with doubled force in a single unit of time. However, this law, applied dynamically, does not hold true. Let us take, e.g., two completely equal pieces of iron, the one in the focal point of a concave mirror, the other in unconcentrated sunlight. Let us say the force of light in the focal point is = to a thousandfold of that outside of the focal point, and that the time in which the metal melts in the focal point = one minute. Then, according to that law, the single force will do equal work in 1000fold time to 1000fold force in a single unit of time, i.e., if the iron in the focal point will melt in one minute, *outside* the focal point it will melt in 1000 minutes, which is absurd.

loudly against them? If that opinion were grounded in truth, then the difference of qualities would have to run completely parallel to the difference of specific gravities and densities; the inspection of a table of the latter will convince one of the contrary. And how does one ultimately wish to explain those entirely peculiar—not by virtue of specific gravity and density, but peculiar through their most intimate mixture [88]—products of nature in their organic operations? Or do we believe that here too Nature does nothing other than decrease and increase density and specific gravity?

Finally, it must be remarked here that since our science takes off from an unconditioned empiricism as principle, one can by no means speak of a transcendental construction, but solely of an *empirical construction of matter. How is matter in general originally produced*? Precisely this will become clear through our following investigations.

B. Qualities = Actants, this proposition is demonstrated. *In all of these individual actants one and the same* original activity of Nature *is* inhibited. This is not thinkable unless these actions, presented collectively, *strive toward one and the same product*; for all natural activity aims toward an absolute product. For this to happen, it is required that various actants are able to combine themselves in one and the same collective product; in short, that there should be composite actants. They cannot combine themselves, however, without having reciprocal *receptivity* for one another. One actant must be able to *prehend* the other. For two different actants, there must be one common point in which they unite—(this point will be named—at a much lower level to be sure—the chemical product).

Since an infinite multiplicity of actants together ought to exhibit one absolute product, the PROBLEM is presented: *find the point in which this infinite multiplicity of diverse actants can be unified in Nature.**

The qualification must necessarily be added that in this circumstance the *individuality* of no actant would cease to exist. Otherwise, the multiplicity would be annihilated. The unity should not be achieved at the cost of the multiplicity. *The multiplicity should remain, and yet a collective product result*, which holds that infinite multiplicity together.

(It may be noted that if *one* such product actually arises in Nature,† in this respect matter is also *dynamically* [89]—although not *divisible*—actually divided

---

*The dynamical philosophy cannot even arrive at this problem, and we can discern here the difference between dynamical and atomistic philosophy quite clearly. Nature is given as product to the atomist only through its *constituents*; to the dynamical philosophy, in contrast, the constituents are given through the product. The dynamist, therefore, does not ask how the product originates from these constituents; for the product *precedes* the constituents; the atomist on the contrary asks how the product emerges from these constituents (because *to him* the constituents precede the product).

†i.e. if nature is such a product

to infinity, since no individuality is to be extinguished in that whole. The importance of presupposing the endurance of *each* individuality in this product will be shown below.)

SOLUTION. The two actions restrict themselves reciprocally, through interaction, to the *mutual effect*. (Only this mutual effect is the *tertium* in which they are able to rest.* There is, again, no other expression for the interaction of the two *than* this effect.)

Now, the striving of all original tendencies aims generally

> a) toward the *filling-up of space*; their prehension of one another is thus a striving toward the filling of a *collective* space, such that in every part of a given material, no matter how small, all tendencies would still be met with. (From this one sees in passing what dynamic divisibility is really like. That is, the *quantity* of material is completely unimportant; in the largest as in the smallest part of the same material the same tendencies must be met with. Therefore, even through a mechnical division carried to infinity, universal homogeneity can never be reached. It can also be seen here that a composite actant does not come into being in Nature originally, but already through particular natural operations, the likes of which we perceive in chemical suffusions.†) Through this striving toward the occupation of a common space, such a space would have to be actually continually filled anew.—Therefore, rest‡ is not an absolute negation of movement, but rather an uniform tendency toward the filling-up of space, and the perseverance of matter itself = to a constant being-reproduced.— Further, the filled space is only the appearance of a striving whose principle is not itself in space. Space is thus filled, as it were, *from inside out*, a very important concept. (The inner in contrast to the outer is always called that which is *principle* of the occupation of space.) If that striving toward the filling-up of a common space were heralded in experience [90] by a resistance to the cancellation of the shared occupation of space, this would give the phenomenon of connection—

---

*In and for itself each actant, as highly individual, excludes the other from its sphere. They are thus only able to meet in a third.

†But *how* the actants unify themselves—permeate each other, is still unexplained here, and is a special problem. (The dynamical philosopher does not even have to inquire about that, as was said; because he never has to allow the actants to *separate* themselves. He does not need to explain *how* they penetrate each other, but only by what means they are held together, how the absolute separation— the absolute evolution—may be hindered.)

‡of matter

cohesion. The force with which that cancellation would be resisted would be called the *cohesive force*.*

REMARK. The cohesive force is thus a composite force, not a simple one like the attractive force.—There are difficulties to the customary explanation of cohesion through mere attractive force, since, in the majority of the materials we know, the relation of the cohesive force of their smallest parts to the square of their distance would have to be completely different than it should be according to the law of universal attraction. This is not to mention the fact that this hypothesis presupposes atomistic conceptions, and the diversity of cohesive forces under this presupposition would be nearly inexplicable.—Further, in relation to the universal attractive force, all matter spread throughout infinite space, balled-up into planets is = to one material; that universal attraction extends to infinity, and with respect to it no space can be thought of as *empty*.† Conversely, cohesion strives against the universality of the attractive force, for it constantly *individualizes* and leaves the space outside the sphere within which it alone works *empty* (unoccupied by its force). Genuine cohesion occurs only within an *individual body*. Therefore, it must be strictly distinguished from adhesion, and from that special kind of attraction which occurs between *different* materials, e.g., in the contact of water and glass.

> b) Further, each tendency is a completely individual and determinate one, i.e., a striving to fill space in a *determinate way*. This is betrayed through determinacy (individuality) of *figure*. In Nature there is a continual determination of figure from the crystal to the leaf, from the leaf to the human form. Therefore, we consider the atomist correct in that he attributes to the elements originary figure (leaving aside the fact that he needs originary figures of atoms for the possible construction of specifically different materials); we assert only that for the original actants it is never a matter of the production of these original [91] figures, nor can it be; we assert that, therefore, these original shapes nowhere exist in Nature, because no simple actant is to be met with in Nature (which, to be sure, we are here not yet able to prove).

Now if, however, every actant is limited through the inifinity of all the remaining ones, then all together they mutually derange each other in their pro-

---

*With this said, what the *cause* of the force of cohesion would be remains unexplained. It will be *the* force through which the actants in Nature bind themselves.

†Space, emptied of matter, is at least filled by that force.

ductions, and none is allowed by the others to achieve the production of the originary figure, i.e., they reduce themselves reciprocally to *formlessness*.*

The shapeless = the fluid. The fluid (at least of the second order, which owes its fluidity to a higher principle) is—not the absolutely formless (= the μὴ ὄν of the ancient Greek physicists), but rather that which is *receptive to every form*, formless (ἄμορφον) for just this reason. The fluid must generally be defined as a mass wherein *no part is distinguished from another by figure*. From this definition, at least, all others previously sought can be derived, insofar as they are correct. Absolute continuity, the complete absence of friction in all fluids, and the fundamental laws of hydrostatics can be deduced in this way. The major principle is *the equivalence of actions* (accordingly also the attractions) *in the fluid in all directions*.†

Accordingly, the most original and most absolute combination of opposed actions in Nature must generate the *most original fluidity*, which, because that combination constantly runs ahead of itself (the *actus* of organization is constantly underway), presents itself as a universally extended entity that simply works against nonfluidity (solidity), and continually endeavors to *liquefy* everything in Nature.

(This principle is called the *principle of heat*, which is, consequently, no simple substance, no material at all, but always only the phenomenon of constantly diminished capacity (of original actants for one another), and is therefore

---

*The most original product of Nature is, therefore, the formless or the *fluid*.

†That is, because the original actants in the fluid nullify one another reciprocally.—For the dynamical philosopher the formless is the original, because it is that which comes nearest to pure productivity. In the pure productivity of Nature there is yet no determination, thus also no form. The nearer Nature is to pure productivity the more formless, the nearer to the product, the more formed.

The atomist distinguishes the fluid of the first and the second order, or the absolute and the relative fluid. The fluidic *in general* will be explained *here* as that wherein *no part is distinguished from another by means of figure*. A few of Kant's disciples explain the fluid as that wherein the attractions in all directions are equal. Let's consider that 1) if an individual particle is drawn to direction A, then it will be just as strongly drawn in the opposed direction—these opposing attractions therefore cancel themselves; there is thus *within* this space no attraction to overcome, and each individual particle within this whole can be moved in all directions without resistance. Hence the *relative mobility of parts*.—Further 2) with equal force of attraction in all directions the spherical shape is necessary because this produces the greatest contact of particles among one another and the smallest amount of empty space. 3) If all attractions cancel themselves among one another, no figure can be produced—which is our definition; but if there is no figure, then there is also no rigidity, no friction, which is necessary according to the laws of hydrostatics. If there were friction in a fluid mass, then an impulse could not propagate itself in all directions equally, which is a fundamental law of hydrostatics. Therefore, we understand the equal height of water in both channels of a bent pipe having unequal masses. Enough concerning the concept of the fluid in general. The concept of the *absolutely fluidic*, of the most original product of Nature, primarily concerns us here.

proof of the steadily enduring process of organization in Nature.—New theory of heat according to these basic principles.)

[92] Now, if there were nothing in Nature that preserved the balance with the fluidizing principle, then the whole of Nature would resolve itself into a universal continuity. The *individuality* of the original actants, however, strives against this *universalization*. The individuality of all actants *ought to* be maintained in the absolute product together with the most complete combination.

Since everything in Nature—or rather, here just that absolute product—is conceived continually *in becoming*, then it will neither be able to achieve absolute fluidity nor absolute nonfluidity (solidity). This will furnish the drama of a struggle *between form and the formless*. That product always in becoming will be conceived continually in the leap from the fluid to the solid, and conversely, in the return from solid to fluid.

It will run through all possible forms within the sphere that it comprehends since that struggle (between form and the formless) is endless, and it will transform itself into all forms like an ever-changing Proteus.

This Proteus will draw all qualities into his circle, gradually assimilating them, as infinitely manifold as they may be, and, as it were, throughout infinitely many attempts, seek the proportion in which the universal unification of all individual actants of Nature in one collective product is attainable. However, through this drive to unite everything *individual* in Nature in itself, a certain circle of possible forms will also be determined for it in advance. One will therefore be tempted to believe that with all the various forms through which it metamorphoses, creative Nature has in mind a common ideal operative in it to which the product gradually approximates itself; the various forms to which it commits itself will themselves appear only as *various stages of development of one and the same absolute organism*.

[93]
III.
Actants and Their Combinations

1) The *whole* of Nature, not just a *part* of it, should be equalivalent to an ever-*becoming* product. Nature as a whole must be conceived in constant formation, and everything must engage in that universal process of formation.

Everything that *is* in Nature must be viewed as something having *already become. No material in Nature is primitive*, for an infinite multiplicity of original actants is in existence (how these arise will be precisely the ultimate problem of the philosophy of nature).—These actants should together represent only *one* absolute product. In that case, Nature must combine them. Therefore, a *universal compulsion toward combination* must occur throughout the whole of Nature, for one cannot see how and why it should have limits; it is unconditional. So there is combination in every material, and no substance is *primitive*.

However, since every material differs from the others, *each material is the product of a particular natural operation*. These various natural operations must be deduced *a priori* in order to ascertain the possibility of a specific variety of material.

2) *No material in Nature is simple*. Since a universal compulsion toward the combination of elementary actants prevails in Nature, no actant can produce a form or shape for itself; every material has arisen by means of combination. There can be no objections to this from experience, since we will even deduce necessarily that there are *indecomposable* materials.*

3) *All diversity of natural products can only derive from the various proportions of actants*. All multiplicity of Nature is to be sought in the elementary actants alone; matter is everywhere *one*, only the proportions of the original combination are different. Since the compulsion toward combination occurs throughout the whole of Nature, the whole of Nature must originally suffuse each product. In each material all original actants are contained *originally*. But all original actants can only unite into the *absolutely fluid*, their individuality notwithstanding. However, the *absolutely fluid can reveal its existence in no other way than through decomposition*. It is indecomposable for sensation = 0, for [94] all actants mutually cancel themselves in the fluid, such that none allow the others to come to any sensible effect. But the *absolutely fluid is by its very nature the most decomposable*, for there prevails in it the most complete equilibrium of actants that, consequently, is disturbed by the merest alteration.—It is further evident from this that the absolutely fluid is only *decomposable*, but is not *composable*.

Fire or heat-matter is familiar to us as the original phenomenon of absolute fluidity.† Heat-matter seems to originate or to disappear where a merely

---

*Thus there is no primal substance in Nature at all out of which everything has become—somewhat like the Ancients thought of the elements. The single genuine primal substance is the individual actant. Thus, there are also no originally indecomposable factors in Nature, i.e. really *simple materials*. No material in Nature is *simple* (the actants are not material). So if there are indecomposable materials, then these materials cannot really be simple *materials*; their indecomposability cannot be explained on the basis of their simplicity. If they are to be indecomposable, then some other reason for their indecomposability ought to be evident. We find this reason when we reflect on the fact that the absolute indecomposable is established as the antithesis of the absolute incomposable. The indecomposable is opposed to the absolute incomposable. This is only possible if it is *itself* the *absolute composable*. Indecomposability and absolute composability must thus always coexist if there is an indecomposable factor without there being a simple one.

†this being inimical to all shape, and for this reason the favorite being for shaping—the universal *liquefying* principle, and therefore the mainspring of all formation and of all productivity in Nature.

quantitative decrease or increase of capacity takes place (enlargement or diminishing of the volume). The heat-matter appears as *simple*, and no duality has yet been perceived in it, or a decomposition into opposed actants, as e.g., with electricity. This is even the proof that in this most original of all fluidities the most complete combination appears yet *unperturbed*.

In contrast, the lightest contact of heterogeneous bodies produces phenomena of *electricity* (in galvanism, and in other recently presented experiments), and since heat as well as electricity is excited through friction (constantly repeated and intensified contact), it appears that in every repulsion of different bodies the absolute fluidity which permeates them all—(because it strives to liquefy everything)—is posited, both mechanically from equilibrium and dynamically from their original combination. The former furnishes the phenomenon of heat diffusion, the latter the phenomenon of excited electricity. Actually there is virtually no chemical process in which heat originates or vanishes that does not show traces of excited electricity; more exact analysis will here teach us much. This is not to speak of the fact that electricity expresses in many cases the same effects as heat, and that bodies are considered the same for both materials with respect to their power of conduction.

Meanwhile, it should particularly be taken into account that electrical experiments are conducted under highly complex circumstances; [95] therefore, in the electrical phenomena much can come to the fore that is not originally essential to electricity. Thus, for example, the Torricellian Vacuum does not glow, and electrical experiments conducted in a vacuum and in different media will consistently demonstrate different phenomena. Nevertheless, the galvanic experiments succeeded in nearly all media in which they have already been tried, and just as perfectly in a vacuum as in air itself.

Finally, what should be said of *light?*—Whether light is originally already split-up into a bunch of simple actants different from one another, whose total impression is white light (as according to *Newton*)—or whether light is *originally* simple (as according to *Goethe*), in any case the polarity of the colors in every solar image is proof of a duality prevailing in the phenomenon of light, whose cause is yet to be investigated.*

---

*What more than anything proves the affinity of light with electricity are the *prismatic* phenomena, such as Goethe has established in his contributions. From this I have concluded, and perhaps soon others will realize, that the Newtonian theory of white *light* as a composite of seven colored rays that become separated in the prism is wrong; that the prismatic phenomena have to do with something far higher than a merely mechanical or chemical decomposition of light.

That is, the colors of the prism do not show themselves in *continuity* when the experiment is precisely conducted; they are shown in continuity only under *particular circumstances*. Where these circumstances are lacking, i.e. as a rule, the colors of the prism are shown as opposed to one another—and distributed in opposite poles. The true structure of color formation is the following. In

4) No material can abandon the state of absolute fluidity unless some *actant* achieves preponderance. But no actant can achieve dominance unless another is subordinated to it, or is completely dissolved. Therefore, the greater the condition of fixity (solidity), the more *apparently simple* the substance (ores, metals, etc.). But no substance is simple. Every apparently simple (i.e., indecomposable) substance is the *residuum* of the universal process of formation, and although we lack the means to set its elements again in mutual independence, and to set free the actants subordinated to them, Nature might still have the means to accomplish this feat, and thus to take up these dead materials once more into the universal process of organization. Nevertheless, it is *a priori* demonstrable that there must be *indecomposable* substances in Nature, because the universal process of formation is only *infinite* to the extent that it continually turns back *into itself*. Even so, we must arrive at *final products* in this process, which Nature cannot further develop in the original direction, and with which, therefore, Nature is constrained to strike out on another path, and to cultivate them in the opposite direction.

[96] Here alone are the genuinely indecomposable substances recognized. They are materials that are only *composable*. It can already be concluded that, e.g., it is impossible for the *soils* to be indecomposable, and that the supposition will be confirmed that they are the debris of the great and universal process of combustion, which even now persists to some extent in the Sun and even on the surface of the Earth.[†]

No composition of indecomposable materials takes place unless bound actants in them become free. Just as Nature makes the absolutely indecomposable substances composable through decomposition, so the absolutely indecomposable substances, conversely, are inserted once again into the universal circulation of matter through composition. The composition cannot proceed unless the original combination of constituent actants is again altered in such substances; and since all actions originally permeate every individual

---

the middle, in the Indifference-point, as it were, the glimmer of white light is shown, and now on the margins of this glimmer—as it were, on the poles—the colors appear, and indeed just those colors that the eye has already distinguished as opposed, which, e.g., the eye of the artist has long since differentiated. Thus, here it seems that something far higher is at work. There is a manifest duality and polarity in the prismatic phenomena; therefore, the prismatic phenomena seem to belong in the class of electrical and autological phenomena.

[†]This supposition has been confirmed more strikingly since this was written.—There is no reason to consider, e.g. nitrogen, carbon, or phosphorus, to be *absolutely* indecomposable, that is, as actually simple. All of these substances are irreducible only on account of their great composability. Oxygen is doubtless the single really irreducible element—not as if it were *simple*, but for another reason that will be developed below. But even this material is also the most composable that we are aware of.

substance, Nature will possess the means wherewith to generate everything from everything.

Therefore, it is likely that in Nature the same antithesis exists in the great as is noted in the small, that is, that Nature on the one hand makes the indecomposable formative through composition, and the incomposable formative through decomposition. It is possible that on the stars as a whole, for example, the reverse process is underway, in contrast to that which takes place on the planets. If, according to universal experience, the indecomposable substances are those with greatest specific gravity, then it is to be expected that the most indecomposable substance lies at the center of every individual system. The illumination of the Sun betrays a continual process of combination; conversely, the same light that is developed in the solar atmosphere through such a process sustains a persistent process of decombination upon the dark planetary bodies; for neither vegetation nor Life is anything other than the constant awakening of slumbering activities, a continual decombination of bound actants.

[98] 6)[4] We are now aware of two classes of natural products, including on one hand the absolutely incomposable, and on the other the absolutely indecomposable substances. But Nature can tolerate neither the former nor the latter, for Nature does not at all tolerate any *final product*, nothing permanent, fixed once and for all. The direction of all natural activity will aim toward *mean products* (from each of the two opposed classes), toward materials which are absolutely composable and absolutely indecomposable at once, and *permanent processes* will appear in Nature (as object), through which the incomposable is constantly decomposed, and the indecomposable constantly composed. These processes, because they are *permanent*, and also because their *conditions* constantly exist, will have the appearance of *products*. The question arises of what sort these products shall be.

7) These products should lie in the middle between the two extremes, the absolutely decomposable and the absolutely indecomposable. In order to be absolutely *decomposable*, such a product would have to approach the *absolutely fluid*, i.e., unify in itself all constituent actants in the most complete *combination*. In order to be absolutely *composable*, the actants in it would have to be continually pushed out of their combination; a constantly disturbed equilibrium of actants would have to exist, that is, it would have to approach the SOLID. But it cannot achieve *either state*.

There must be the greatest *freedom* (mutual independence) and the greatest *linkage* (reciprocal dependence) of actants to one another in this product. The question arises as to what the result of this will be.

First of all, every actant will inhibit the other from producing its original figure. But only various degrees of intensity of every actant are possible.

Every actant will therefore be *a different actant* at every stage. At every stage, too, it finds its antagonists. The product will thus generally be equivalent to a series in which positive and negative magnitudes constantly succeed each other. But within this series the product cannot be inhibited, for it would be either $= 1-1+1-1$, i.e., $= 0$, or some positive actant [99] would have to gain preponderance. Neither of these alternatives can come to pass. Thus, the product cannot be at all *inhibited*, it must always only be conceived as in *becoming*.

(Thus, we have deduced here what type of product that *product always in becoming* would have to be, whose necessity we have deduced from the concept of an infinite activity of Nature. In it, that continual alternation of combining and decomposing processes will take place which we have demonstrated in Nature as universal and necessary.)

While the actants are *decombined*, left to itself each one will produce what it must produce according to its nature. To that extent, in every product there will be a constant drive toward free transformation. While the actants are continually *combined* anew, none of them will remain free with respect to its production. Thus, there will be compulsion and freedom in the product at once.

Since actants are constantly set free and recaptured, and since infinitely various combinations of them are possible (and in every combination a slew of various proportions are possible), then continually *new* and *singular* materials will be *originally* produced in this product. It is indeed possible to find the *elements* of these materials through the art of chemistry, but not the *combination itself*, that is, the proportion of the combination.

Since each actant is highly individual, and since each strives to produce what it must produce according to its nature, this will furnish the drama of a struggle in which no force entirely conquers the other nor completely submits to the other. The egotism of each individual actant must join itself to that of all the others; what is produced is a product of the subordination of all under one and one under all, i.e., the most complete *mutual* subordination. No individual potency could produce the whole for itself, but all together can produce it. The product does not lie in the *individual*, but in *all together*, for it is indeed itself nothing other than the external phenomenon or the visible expression of that constantly operating combination and decomposition of elements.

[100] The product, since it is a joint product composed of many different activities acting in concert, has the appearance of the *accidental*, and is a *blind* natural product, since only such a thing can come forth through this determinate original intensity of every individual actant, and with this determinate proportion of their unification. Thus, the *contingent* and the *necessary* are originally unified in it.

In every individual actant is an activity that strives to *freely develop*—according to its nature. Its *receptivity* for or restrictedness by all others, really lies in this tendency to *free development* of its own nature, because it cannot achieve the latter without the expulsion of all the others from its sphere. As foreign actants reach into its sphere, it is at the same time constrained to prehend the sphere of every other. Thus, a universal prehension by every actant of the others will take place. No actant can enter into this antagonism, for the development of them occurs *according to its nature*. The elements of such a whole will appear to have dressed up in another nature, so to speak, and their effect will appear to be completely different from that which they show outside of this antagonism. Yet the tendency lies in each one toward a development according to its nature, which will appear in this antagonism only as a *drive*. This drive will not be *free* in its direction; its direction is determined for it by the universal hierarchy; there is, as it were, a sphere circumscribed for it in advance, beyond whose limits it can never step and into which it constantly returns.

This sphere will again itself be infinite, however. Since it cannot at all end up in a product unless the actants maintain themselves in a state of mutual compulsion, and every individual actant strives against this compulsion, only after infinitely many attempts will the proportion of actants be found in which the greatest freedom of actants at the same time as the most perfect mutual bond is possible.

On the whole, we have no other expression for the proportion of actants than the configuration produced. Now, if the product produces all possible configurations by means of continual transitions, [101] and shuttles from proportion to proportion by way of imperceptible nuances, then a constant flowing of one form or configuration into another would exist in Nature; however, just for this reason nothing would be determinate or fixed, not for a minute something that would be a phenomenal product.*

However, the infinite activity of Nature that stirs in all individual actants should present itself empirically. Thus, it is necessary that that infinite product become fixed *at every stage of becoming*.

The product, however, is nothing other than productive Nature itself determined in a certain way; the inhibition of the product is, therefore, simultaneously an inhibition of Nature itself; but Nature itself is *solely active*. Therefore, it cannot be inhibited, unless this becoming-inhibited is itself, from another perspective, again = to *activity*.

---

*But an *apparent* product ought at least to present itself through the productivity of Nature.

IV.
Inhibition and Stages of Development

The PROBLEM arises: *to specify how Nature could inhibit its product at particular stages of development, without ceasing to be active itself.*

Solution.

1) The development of the absolute product in which the activity of Nature would exhaust itself is nothing other than an infinite process of *formation*. The process of formation is nothing other than a configuring. The various stages of development are nothing other than various stages of formation or of *configuring*. Every individual natural product runs through all possible forms up to the point at which it is inhibited (this must be accepted); however, it does not achieve an actual production at any stage. Each formation is itself only the phenomenal appearance of a determinate proportion which Nature achieves between opposed, mutually limiting actants. For as many proportions of these actions as are possible, there are as many diverse shapes, and just as many stages of development.*

[102] Every stage of development has a peculiar character.† *At every stage of development formative Nature is restricted to a determinate—sole possible—form*; it is completely bound with respect to this form, and in the production of this form it will show no freedom at all.

2) Now it may be asked how Nature, infinitely active, could be restricted to such a determinate shape. Nature contests the Individual; it longs for the Absolute and continually endeavors to represent it. It seeks the most universal‡ proportion in which all actants, without prejudice to their individuality, can be unified. Individual products, therefore, in which Nature's activity is at a standstill, can only be seen as *misbegotten attempts* to achieve such a proportion. The question is whether something may be found in Nature that might justify us in such a supposition.

A) If Nature had found or come across the true proportion for the unification of a multiplicity of actants, then it would have to be able to present them in a *joint* product, no matter how antithetical their natures may be. The proof that Nature has not struck upon such a proportion would be if a *diremption* of

---

*Each formation is only the phenomenon of a determinate proportion of original actants. If evolution were complete, then this would be = to the universal *dissolution* into simple actants. Every product, therefore, is = to a determinate *synthesis* of actants.

†a peculiar inner quality.

‡most complete.

actants came to pass in the product, as soon as it has arrived at a particular stage of formation; or, since the joint activity of the actants reveals itself as *formative drive*, the proof would be that if, at a certain stage of formation, the formative drive stirring in the product separates into opposing tendencies, such that Nature would be constrained to develop its product in opposing directions.*

Remark.

Throughout the whole of Nature absolute sexlessness is nowhere demonstrable, and an *a priori* regulative principle requires that sexual difference be taken as point of departure everywhere in organic nature.

[103] a) It is first of all a mere supposition that the so-called cryptogamic vegetables, such as fungus, conifers, tremella, etc., are merely budding plants and consequently absolutely sexless; the impossibility of demonstrating *sexual parts* in these plants is not at all a proof of this hypothesis.

b) Sexlessness is equally as little demonstrated in the animal realm, for even in polyps, since the discovery of *Pallas*, one cannot doubt the sexual functions. Where there actually is sexlessness, there is yet an *other*, *specific* direction of the formative drive. The sexual drive and the *technical drive* are equivalent for most of the insects before they have passed through their metamorphoses. The sexless bees are also the only productive ones, and yet without doubt they are only the mediators through which the formation of the one queen bee is achieved (in which the formative drive of all the remaining bees seems to be concentrated). Most insects lose all technical drive after sexual development.

As for sexual difference itself, though a great multiplicity of types seems to prevail, in the end it is reducible to a few varieties. The separation into different sexes happens for different organisms at different stages of formation, and this is itself proof for the assumption that each organism has a level of formation at which that separation is *necessary*. Nature has either unified the opposing sexes in one and the same product and has developed simultaneously in different directions (as in many species of worms, where mating is always dou-

---

*Neither of those opposed directions can fall outside of the general character of the stage of development, and yet neither completely expresses this character. If this were so, the product would not be able to separate into antithetical directions.—We know Nature first of all simply as organic or as productive. All of productive Nature is originally nothing other than infinite metamorphosis. It can never achieve determined and fixed shapes, i.e. fixed products, if the productive drive is not split into individual stages of development, or if the product does not separate into opposed directions just when it has reached a certain stage of formation. Now, if the separation into opposite sexes is just that separation which we assume as the reason for the inhibition of organic productions at a certain stage of development, then there must be no single product in Nature without an opposition of the sexes.

bled, as well as in most plants), or Nature has distributed the opposed sexes into different stocks (individuals). Here the one-sided sexual division is again distinguished only at different stages of development.

Plants generally attain sexual development through metamorphosis, like the insects (even those whose flowers unify both sexes). The development of the sexes is merely the highest zenith of the process of formation, for it occurs by means of the same mechanism through which progressive growth gradually takes place.

[104] The same law reigns with the insects: in the first stage of their formation no sexual difference shows itself (e.g., in the condition of the pupa), and the metamorphoses through which they pass are determined almost exclusively for the sake of developing the sex in them, or the revolutions of their metamorphosis are only phenomena of sexual development itself. This is because as soon as their metamorphosis is completed, sexual difference appears, and with it the sexual drive.—In addition, with both flowering plants and insects this is the highest summit of formation that they are able to reach; for the flowers fall to the ground and the transformed insect dies as soon as fertilization is accomplished, without having expressed any other drive.*

---

*In earlier times the metamorphosis of the insects was taken to be a kind of miraculous event and the symbol of something higher. The contemporary study of nature seeks to explain this phenomenon, and in order to be able to explain it more easily it first strips it of the breadth that it actually has. They have transferred the "involution" or "preformation system" to this phenomenon of organic nature as well. Already in the worm every part of the butterfly is supposed to be there, imperceptibly small, and yet individually preformed.

I do not yet want to invoke here the general principle that no individual preformation, but only *dynamic* preformation exists in organic nature, and that organic formation is not evolution, but the epigenesis of individual parts.—Various organs, parts, etc., signify nothing but different *directions* of the formative drive; these directions are predetermined, but the individual parts themselves are not. Let me just remain with the present phenomenon and ask whether *this* phenomenon is explicable in terms of Preformation theory. It is alleged that this individual preformation has even actually been proved by a specimen from Schwammerdam, by means of which he showed that, already in the pupa, a few parts of the future butterfly are distinguishable. But it is quite conceivable that when one opens the cocoon immediately before the final metamorphosis, after everything is already prepared for it, one can find everything that would shortly come to light on its own. If this example is supposed to prove something, then one ought to be able to show already in the pupa in its first moment of formation—one even ought to be able to show already in the caterpillar, those parts as individually preformed. At the moment when those parts can be seen the metamorphosis is for the most part already accomplished. Therefore, that specimen proves absolutely nothing about the preexistence of parts *before* the metamorphosis.

One therefore has no proof at all *for* that assumption, but surely there are proofs *against* it. When the *emergence of new parts* is explained as an individual preformation, how does one explain the *disappearance* of the parts that were there before? Nothing is lost from the pupa, and yet one does not find in the butterfly the organs that were in the pupa. One would have to represent the

The universal separation into opposed sexes must occur according to a determinate law, and indeed neither sex should be able to originate without the other simultaneously originating with it. We see that where both sexes are unified in one individual, they originate through one and the same formation. Therefore, the law which is observed in the latter must be extended over Nature as a whole.

---

caterpillar then by the burst shell—but then where is this shell?—Why is it not said as well that the *blossoms* of the tree have been *individually* preformed in it? What the blossom is in relation to the tree, the butterfly is in relation to the caterpillar. If need be, something can still be thought by "preformation" if an organ can be preformed in a *seed*, but how an organ could be contained in another cannot be conceived. I want to mention just a few things in order to exhibit this inconceivability. For example, the caterpillar nourishes itself through crude nourishment (from the hardest leaves). The butterfly sucks in ethereal nourishment—from the nectar of plants. Therefore, the organs which are designed to ingest the nourishment of the caterpillar must be totally different from those which conduct the fluid nourishment of the butterfly. Are the nutrient-channels of the butterfly supposed to have been preformed in the cruder ones of the caterpillar?

Another example. In the first days of its existence the pupa still needs the caterpillar's organs of respiration—(air passages, openings over the whole surface); the pupa soon learns to do without these organs, and when the butterfly is developed up to a certain point one finds no more trace of them—but in place of them a completely different organ of respiration, unlike the former and differently constructed. Now was this one also preformed, and where was it?

That transition from one phase of the metamorphosis to the other is not at all just a *partial* alteration, but a total one. For example, in the butterfly the direction of circulation is the reverse of that in the caterpillar. In the latter, the main artery (which runs along the back) pumps the liquid to the head, while in the pupa and the butterfly it runs away from the head.—The unfolding of the wings, which follows soon after the final development of the butterfly, happens by means of a rapid and forceful development of the vascular system in the center, by an influx of liquid from the interior—not, as others have believed, through the mere expansion of the folded wings or through the pressure of the air forcing its way in from outside.[5]

All of these phenomena prove that the metamorphosis of insects does not occur by virtue of the *mere* evolution of already preformed parts, but through actual epigenesis and total transformation. Now, how should these phenomena be explained? They are at any rate not explicable other than in terms of the theory of the graduated series in all organic formation that has been presented; these phenomena prove *a posteriori* what we have proved *a priori*. They demonstrate

a) that every organic individual must run through all *intermediate forms* (s.vv.) up to that stage of development at which it becomes inhibited.

b) that the ground of all permanence and all fixity in organic nature is to be sought in the separation of the sexes.—For the insects *before* they are transformed are sexless, or rather *just because* they are sexless, they are transformed. If the sex in them was decided, then they would have already arrived at the stage of development to which they are determined. Conversely, as soon as the metamorphosis of the insects is through, the sex is developed, or rather the reverse; as soon as the sex is developed the metamorphoses stand still. The butterfly has no sooner left its final shell behind than it begins to exercise the sexual functions. It seems to have accepted this final stage of development only in order to propagate its race.—The drive which expresses itself in the metamorphoses tends toward the *race*, as the highest to which an organic being can accede. The same law that holds for the metamorphosis of insects holds also for plants.

Thus if, according to our principles, the production of various genera and species in Nature is only *one* production captured at different stages, then the formations of the opposite sexes in the *same* genus and species must be only *one* formation, *one* natural operation, such that the different individuals of the same genus amount to only *one* individual, but developed in opposite directions. The universally evident proportion which Nature maintains between both sexes corresponds with this, at least in the animal kingdom (for in the plant kingdom the observations are wanting), not, indeed, as if the individuals of both sexes were equal in *number*, but that Nature substitutes, for the smaller number of individuals of one sex, a higher intensity of the formative drive, and conversely, substitutes the lesser intensity of the formative drive in one sex through the number of its individuals.*

[105] B) It must be demonstrated that the separation into different sexes is just the separation which we have furnished as the ground of inhibition in the productions of Nature. That is, it must be shown *that Nature is actually inhibited in its productions by means of this separation, without on that account ceasing to be active.*

1) From the moment of the diremption onward, the product no longer completely expresses the character of the stage of development at which it stood. It will not be a *finished* product, not a product upon which Nature could *cease* to work, although of course its further development is deranged by that separation and is then inhibited at this stage.† Now, what kind of activity will Nature exercise in this product?

First of all, if the product either separates itself into opposite directions or strikes out in a one-sided direction, Nature, which can never cease being active, will pursue the formation of the product toward either both or one of these directions to the furthest point, such that the product distances itself in the direction away from the universal character of its stage of development as far as possible. In other words, Nature will drive the individualization of the product to the extreme in both directions. Therefore, the most acute moment of individualization in each organism is also the most intense moment of natural activity in it.

---

*The formation of opposed sexes in most animal species is actually an interdependent one—also where they are distributed among different individuals; for example, the formation of the three types of bee is always *one* phenomenon, and precisely here that remarkable coexistence just mentioned enters, where the sexlessness of the productive bees is replaced by the intensity of the formative drive in the one queen bee.

†It will be shown below that the condition for an enduring activity is provided precisely in that opposition, since every condition for activity in Nature is a duality.

2) If the highest stage of individuality in both directions were achieved, then the organism could admittedly no longer be the *object* of natural activity, but instead means and *instrument*.*

If the highest stage is reached, then both directions are to be viewed as opposed; they are related to one another as positive and negative magnitudes. However, neither the one nor the other of these directions could be the one wherein the activity of Nature exhausts itself,† for the individual is everywhere opposed to the latter.‡

The opposing natural activities that are operative in the product toward opposite directions always become more independent from one another; the more independent from one another they become, the more the equilibrium is disturbed within the determinate sphere of Nature [106] that they describe.§ If they arrive at the maximum point of mutual independence, then the greatest moment of disturbed equilibrium is also reached.

However, in Nature the highest point of disturbed equilibrium is one and the same with the moment of the reestablishment of equilibrium. Between the two no time elapses. Those antithetical activities must, therefore, combine‖ themselves according to a necessary and universal law of nature. The product will be a *mutual* one, constructed from both of the opposed directions (of the formative drive); Nature will in this way return by a circular course to

---

*This moment of acute individualization is really the first moment of complete sexual development—the complete separation of the product. But just at this instant, Nature is shown in its greatest activity. Vegetative nature glistens with the brightest and most distinct colors, and this interval is the genuine moment of culmination for the animal as well. Nature has now completed its work. The product has become what it could become, within its limitations. It has been driven to the highest point of its existence. It can therefore no longer be an *object* of Nature.—For what is the *real objective* of Nature?

While Nature does develop individuality, it is not really concerned with the individual—it is rather occupied with the annihilation of the individual. Nature constantly strives to cancel out duality and to return into its original identity. This striving is precisely the basis for all *activity* in Nature.—The duality which imposes upon Nature the compulsion to constant activity is, where it exists, as it were, *against the will* of Nature—as it is here. Nature did not intend the separation.—Nature leads the product in both directions only for the sake of letting it sink (back) into indifference, as soon as it reaches the apex of development. For Nature was not concerned with either the one or the other of those directions, its concern was for the sake of the common product that was divided into them. Therefore, as soon as the product has reached the highest point in both directions, it fosters the *universal* striving of Nature toward indifference.

†can be that point toward which Nature tends.

‡to Nature

§to which they are restricted.

‖unify

that point from which it had departed; the product will, as it were, turn back on itself and will have adopted once more the general character of its stage of development.*

From this moment forward, since the *joint product* is secured, Nature will abandon the *individual*, will cease to be active in it, or rather† it will then begin to exercise an antithetical effect upon it; from now on the individual will be a *limit* to its activity, which Nature labors to destroy.

The genus must appear as end of Nature, the individual as means—the individual expire and the genus remain—if it is true that individual products in Nature ought to be seen as unsuccessful attempts to represent the Absolute.‡

3) The joint product will§ again run through the same stages of development from the fluidic forward,‖ up to that stage at which it must decide once more on one determinate direction, or split into two opposed directions, from

---

*We have departed from the assumption that all individual products of Nature can only be seen as abortive attempts to represent the Absolute. If the individual is only an abortive attempt, and Nature has developed it only under *compulsion* in order that by means of its construction Nature can achieve the collective product, then Nature does not have to tolerate it any longer as soon as it has ceased to serve as means. But as soon as the collective product is posited, the individual also ceases to be a means.

†since it can never stop being active

‡But is this really the case? This unimpugnable law of nature is most conspicuous, again, in the organisms which succeed to sexual development through perceptible metamorphoses. Flowers wilt, the transformed insect dies, just as soon as the genus is secured. The individual seems here almost to serve merely as a medium, only as a conduit, through which that organic vibration, the formative force (the spark of life) propagates itself.—But is this law of nature not also just as operative in the higher organisms, and does not the individual here too deceive us, seeming as if it were Nature's end and not merely a means? We do not perceive as strongly in higher creatures that demise of organisms, after the point at which that peak of opposition is achieved, partly because it happens with very attenuated speed, and because the product that was a longer task for formative Nature is also a longer task for destructive Nature; partly because here the sexes are much more separated than at the lower stages. If one makes a general comparison of the proximity and distance between the sexes of various organisms, one finds that for the most long-lived organisms the sexes are the most separated, and on the other hand, the more ephemeral the product the closer the sexes are to each other. Where Nature seems to want to preserve the individual longer in one species, it breaks the sexes further asunder from one another, as it were, makes them flee from one another. How separated the sexes are in the higher animal species, how near in the flowering plants, where they are gathered in a single calyx (as in a bridal bed)!

In conclusion we may suggest that the separation of the sexes happens against the will of Nature, as it were, and that because individual products originate only by means of this separation, these products are abortive experiments of Nature.

§wholly necessarily

‖for all formation takes the fluid as point of departure

which point forward Nature again adopts its previous mode of action.—(Let it be noted that there might be for each natural product a stage of formation at which—when the product has reached it (for many do not reach it)—*opposed* directions of the formative drive become unavoidable. This is an assumption to which we saw ourselves driven without initially being able to justify it.* It is enough that the supposition is *necessary* [107] in the context of our preceding investigations, although it is itself again a problem which we must subsequently solve. We must for the time being hold fast to the leading thread of our rationale and expect that in a decisive investigation of every problem remaining unresolved they will, at last, find their solution.)

We were only concerned for the moment to demonstrate inhibition as something *necessary* in the production of Nature. However, it would not be necessary if opposed directions of the formative drive were not *necessary* at every stage of development.

The difference of the sexes then, we assume, is the genuine and sole reason why (organic) natural products appear fixed. (But they *are* not in the least fixed. The individual passes away, only the *species* remains, but Nature never ceases to be active. However, since Nature is *infinitely* active, and since this infinite activity must present itself by means of finite products, Nature must return into itself through an endless *circulation*.) We cannot leave our last proposition without mentioning the consequences that flow from it. The most important CONCLUSION that proceeds from it is this: *the variety of organisms is finally reducible just to the variety of the stages at which they separate themselves into opposed sexes*.†

---

*That is, that such a diremption is necessary at each stage of development if the production is to be inhibited—this has already been proven. But we have not explained that diremption itself. It is thus a necessary supposition for us and is necessary in the context of our present investigation, although we are unable to explain it in itself. This explanation must be given subsequently, when our science is completely developed.

Still more cases of this kind will arise where we have to leave many things that must be postulated, for the time being, unexplained. It is to be expected at the outset that there will be, in the end, one universal solution for all of the problems that remain still unresolved.—There is no doubt only *one* opposition that splits into all of the individual oppositions of Nature. We have even already postulated this antithesis right at the start. But we still lack the intermediate steps in order to bring *this* antithesis (which divides itself into both sexes) into connection with that original opposition and thus to deduce it as necessary in Nature.

†Apparently paradoxical—but necessary. Nature is only one activity—therefore its product only one as well. Through the individual products it seeks to present just *one*—the absolute product. Thus its products can be distinguished only through the variety of stages. But many are already inhibited at the lowest level. The ones that stand at the higher stages must have had to pass necessarily through the lower, in order to succeed to the higher.

For since organisms overall are to be seen as only *one* organism inhibited at various stages of development* (this inhibition, however, being effected only through that separation), then all variety of organisms depends upon the various stages upon which that separation follows.—The formation of each organism will occur completely in step with the formation of all remaining organisms, up to the stage at which that separation occurs in it; the *individual* formation of every organism first begins with the development of the sex.

At which stage the separation happens can depend only upon the proportion of actants which is originally found in each organism.† Each organism thus expresses not only the character of a certain stage of development, but also a determinate proportion of original actants. It does not express this character completely, however, because [108] it could not be inhibited at that stage unless it divided itself into opposed directions. Now the *joint* entity that no *single* individual *completely* expresses, but all together express, is called the *species*. In organic natural products both *species* and *individual* are necessary.‡

A new CONCLUSION from the foregoing is that *organisms that are inhibited at the same stage of development must also be homogeneous with respect to their reproductive forces*.

Therefore, in empirical research one justifiably utilizes the shared fertility of species thought to be distinct as a proof that they are merely *variations* of the same genus or species, and can even for the first time raise that unity of the reproductive force to a principle of the system of Nature.

It is assumed that each inhibited product is restricted to a determinate sphere of formation. But Nature organizes to infinity, i.e., each sphere to

---

*One must not allow oneself to be led astray by the appearance of a lack of continuity. These interruptions of Nature's stages only exist with respect to the products, for reflection, not with respect to the productivity for intuition. The productivity of Nature is absolute continuity. For this reason we will present that graduated series of organisms not mechanically, but rather dynamically; that is, not as a graduated series of products, but as a graduated series of productivity. *It is but one product that lives in all products*. The leap from polyp to man appears gargantuan to be sure, and the transition from the former to the latter would be inexplicable if intermediate members did not step in between them. The polyp is the simplest animal, and the stalk, as it were, out of which all other organisms have sprouted. Other reasons why the graduated series of organisms not only *seems* interrupted, but actually *is*, will be suggested below.

†Until now it has been assumed that each organism designates a determinate stage of development. I can now propose, conversely, that the variety of stages alone constitutes the variety of organisms. But what then is the "stage of development" itself? It is indicated by a certain shape. But this determinate shape is itself only a phenomenon. The real, which is its foundation, is the inner proportion of forces which is originally found in each organism.

‡First of all, this is really just a consequence of the necessity of opposite sexes—but ultimately because in each organism an absolute product ought to be fixed, i.e., that each product be simultaneously fixed and not fixed, but fixed as species (as stage of development), not as individual.

which Nature is limited must again contain an infinity in itself. Within every sphere other spheres are again formed, and in these spheres others, and so on to infinity.*

This will give the impression of free directions of the formative drive within the general sphere of the species.† Since in *natural history* (in the authentic sense of the word), one must ascend to the individual as it immediately came from the hand of Nature, one must assume that in the first individuals of each species those directions of the formative drive were not yet indicated, for otherwise they would not have been *free*. Then, every first individual of its species, although it would incompletely express the concept of its genus, would have been itself again *genus* in relation to the individuals produced later. (What Kant said very truly in his treatise on the races of humankind may serve to elucidate this: "[A]s to how the form of the first human stock may have been designed with respect to *color*, it is at this point impossible to guess; even the character of the *Caucasian* is only the *development* of one of the original natural predispositions which were to be met with among the rest of them‡"6).

[109] The formative drive was *free* with respect to those directions because they were *all equally possible*; not, however, as if which of these directions it would take in any one individual were dependent on chance. There must, therefore, be an external influence on the organism in order to determine the organism toward one of these directions. Now that which is *developed* (but not, on that account, *brought forth*) through external influence is called *germ* or *natural predisposition*. Those determinations of the formative drive within the sphere of the general concept of the species, therefore, are able to be presented as *original natural predispositions* or *germs*, which were all united in the primal individual—(but such that the prior development of the one makes the development of the other impossible).

(That insufferable superficiality of explanation is hereby banished from a foundational science of Nature; i.e., as if the taxonomical differences in organic

---

*The product is *fixed*. But to what extent? Each product of nature can split again into new products. Nature organizes, where it organizes, to infinity. The product is then limited to this determinate sphere of formation, to be sure, but within this sphere still narrower spheres can again be formed. Although the product is fixed as *species*, it is still not fixed in every respect.—If the productive drive does not any longer proceed from the center to the periphery, then it goes from the periphery to the center; i.e., when the sphere of formation is no longer to be expanded, then narrower spheres arise, in these still others, and so on to infinity.

†and therefore the multiplicity of species, or, more precisely expressed, of the variations in organic nature. In the concept of the "variation" something of chance is thought, a determination which is not already necessary through the general character of the stage of development.

‡in the original of the human species

beings of the same kind were gradually impressed upon them solely through influences of external Nature, or even by art; while it is proven that in the organization of the species the disposition for such a characteristic constitution is originally laid already, and only has to wait for the developing influence of external causes.)

The organism indeed steps into a narrower sphere with the development of that original organic natural predisposition, but not, for that reason, out of the sphere of the concept of the genus itself or out of the sphere of its original stage of development; and since organic beings that are like one another with respect to their stage of development are also homogeneous with respect to their reproductive forces (above p. 43), then individuals of the same stage of development can be fertile together, as much as they may otherwise be different from one another in terms of taxonomy.

Therefore, they can be seen only as different *variations* or *races* of the same stock, not as various *species*.* (These variations are the most widespread in the plant kingdom, where the productive interbreeding of apparently different species has been taken extraordinarily far,† and where, even [110] for many organisms now existing, the original genus is no longer to be found.‡—In the animal kingdom the variation goes no less far in some species.§ Variation extends not only to external characteristics, for example, the color of the skin,‖ as might at first glance seem to be the case with human beings (although that is itself again a product of a characteristic organization of this organ of secretion), but far more to the inner construction of the body, principally in the bony structure of the head, and also most likely to the construction of the brain itself.#)—

---

*For example, the variety of human races proves absolutely *nothing* about the variety of human stocks. Rather, that they are fertile together proves that they are only deviations from one primeval original.

†Through the interbreeding of differently classified species one has completely transformed one into another, although just this transition is a proof that those differently classified species were only different variations of the same species.

‡For example, the various species of grain are apparently varieties preserved through intermixing of various species of grass, whose original does not even exist anymore.

§For example, from the hyena at one extreme to the Bolognese dog at the other, there is a continuum of variation. The variation of the wolf, the fox, etc., fall within this long series.

‖Indeed, this diversity of skin color also is not possible without an inner diversity of organization. It is now well documented that the black skin of the Negro derives from the fact that his skin is organized as an organ of secretion for the carbon of his blood—if the carbon is to be precipitated by the vaporous drying of the skin, then the skin must be organized in a particular way, which makes itself known in the blacks already just through the mere feeling of it.

#namely, according to the analogy of the shelled animals. The brain is, as it were, a shelled animal whose shell is the skull.—As the snail constructs its shell, so does the brain, in whose structure are large variations, according to this view; and it is in *this* respect that interesting things are expected from the procedures of Gall.

Since, however, those taxonomic differences are developments of the formative drive's original tendencies dwelling in the organism itself, they will, once developed, also be continuously and unfailingly inherited through continuous generations within the same variety, without having need to develop all over again in each single individual of the same class. Individuals of different classes will produce a compromise formation which, only when it always mixes with the same class, finally passes over completely into the latter.*

What is unfailingly transmitted is either so determined that it excludes all variety (as, e.g., the black color), or it permits Nature a wider playground, as does the white color, which admits of greater varieties. If this is so, then the variety cannot already be *determined* by the racial difference itself (e.g., the blond hair by the white skin color), for otherwise it would cease to be a *variety*. Just for that reason, it will not be inherited at the same time as the racial difference, but will appear rather as a *sport* of Nature; therefore, varieties do not establish diverse races, but only diverse sporting types. (As Kant notes in the previously mentioned treatise and in the treatise on the use of teleological principles.[7])

Finally, the progressively narrowing limitations of organic formation (within the general sphere of the species concept) proceed primarily in the human species to infinity, and Nature appears to be truly inexhaustible in the multiplicity of always novel external as well as [111] internal characteristics which it packs into the same original form.†

Notes.

1) "The product is inhibited at a determinate stage of development" does not mean that it absolutely stops being active, but that it is limited with respect to its productions; it cannot reproduce anything to infinity except *itself*. Since

---

*With racial difference the product enters into a narrower sphere of formation. But can Nature here cease to form still further? Still smaller spheres are possible within the sphere of racial differences as well. To the crude eye that sees only the gross outlines, these finer nuances surely withdraw from view.

†Most prominently, certainly, in the human species, where each form has a certain originality. Therefore Shaftesbury said, e.g. one could immediately distinguish *ideal* portraits from copies made by nature, because in the latter dwells a truth, i.e. such a precise determination, the likes of which are never achieved by art left to its own devices.[8]

Thus Nature does not cease to be productive in the individual, even when the species is fixed, until the individual as individual is completely determined. But this happens first with complete sexual development. In this moment the organism first fully enters the *narrowest sphere* of formation, e.g. the physiognomy fixes itself, is unchangeably determined.—But as soon as the product is driven to the peak of individuality, Nature ceases to work productively; it begins to work counterproductively and now sustains the individual solely to the extent that it struggles against its existence.

it is now perpetually active, it will be active only *for itself*, i.e., it will reproduce itself not only as individual but simultaneously as genus to infinity (growth and reproduction).

However, no organism can reproduce itself as *genus* which has not reached the stage of separation into opposed sexes. The propagation of plants and plantlike animals through buds or shoots is not reproduction, but only growth that can be driven to infinity by external influences.

Since each organism is limited to a determinate form, all of its activity must be directed toward the production and reproduction of this form. Therefore, the real reason why every organism reproduces only itself to infinity is to be sought in the *original* limitation of its formative drive, but not in some *preformed seeds*, for whose existence there is not a shadow of proof. The first* seeds of all organic formation† are themselves already products of the formative drive. There is also no reason to assume that in such a seed all parts of the individual—to the infinitely-small (individually preformed)—are present, but only that in it a multiplicity of tendencies is contained, which, as soon as they (every single one) are set into activity, must develop according to the antecedently determined directions. ("*Omnes corporis partes non actu quidem sed potentia insunt germini.*" Harvey's *De gen. an.*[9])‡ For all multiplicity of organs and parts signifies

---

*actually demonstrable

†e.g. the seed of the plant

‡It would lead me too far astray if I listed all of the reasons against the theory of individual preformation (reference to Blumenbach). Therefore, only a few of the principal reasons are listed:

1) Although as a rule in the production of the individual Nature expresses the original of the genus, it deviates from this as soon as it is compelled, e.g. as soon as any injury to the organism or any accidental lack is to be compensated.—So Nature here produces something whose production could not be calculated because it depends upon an accidental condition—something that also could not be individually preformed.

2) In particular, how are the reproductions of the lower animal species explicable?—Polyps are mutilated—dissected—rearranged—what is the life-principle of the seed here? Could it be the knife of the researcher?

3) Why, in all of the lower species, are the particular conditions of reproduction present only in young animals, with *higher* animals only those that are independent of the brain—or is there present a *particular* seed for every part?—Wild speculation. These reasons are already in themselves sufficient to defeat that system, not to mention the fact that preformation explains nothing.

I should say a few things about Blumenbach's system of the formative drive, which takes the place of the preformation theory, but this too can only be touched upon briefly here, since up to now the only genuine basis for a decision about it, i.e. the physical evidence, was lacking, and which we can expect to come to only later in the system. So just this much about it:

We are agreed with Blumenbach in that there is no individual preformation in organic Nature, but only a generic kind. Agreed, that there is no mechanical evolution, but only a dynamical one, and thus that there is only a dynamical preformation.

[112] nothing other than the multiplicity of *directions* in which the formative drive is compelled to operate at this determinate stage of development. *Therefore, all formation occurs through epigenesis.**,†

2) The hope which so many natural scientists seem to have cherished—to be able to present the origin of all organisms as successive, and indeed as the

---

*through metamorphosis, or through dynamical evolution

†As for the concept of the "formative drive," it is the most genuine designation that was possible for the state of physics at the time—although, at the same time, it is to be highly recommended in that it is an ultimate principle of explanation and not reducible to higher natural causes.

When we investigate *a priori* what sort of activity it might be that occurs in organic formation, it is immediately apparent that it cannot be *simple* productivity, like that through which the *product of the first potency*—and likewise dead matter—consists. Further investigations will show that it can just as little be a productivity of the *second* potency, which, e.g. is operative in the chemical process. It will then be productivity of a still higher kind than the merely *chemical*. It is this higher productivity that can at any rate be designated "formative drive."—The concept of the "formative drive" contains, 1) Freedom. Freedom is in the organic product because no simple productivity operates here, but a compound one, through which the appearance of freedom comes in the process of production. The individual action cannot produce anything in this antagonism that is in accordance with its nature; it is intensified to a higher productivity through the limitation in which it exists. However, 2) that freedom cannot be lawlessness. Although each individual action produces what it would *not* produce according to its nature alone, what it would necessarily produce if left to *itself*, then in this antagonism it cannot produce anything other than precisely that which it does produce.—To this extent, therefore, the product is again a *necessary* one. Thus, unification of freedom and necessity.

It is called "formative *drive*" to distinguish it from the concept of "formative force." This concept is completely justifiable, not insofar as it would name the cause *itself*, but insofar as it is a *name* for the cause. Among other thinkers, the Brownians should not object to this concept at all, since it actually expressed far in advance what Brown proposed only afterward—namely, that organic formation happens only through the mediation of the *process of excitability*. It is just by this process of excitability that the product is elevated, becoming a product of a potency higher than the merely chemical. Therefore, in the following we will make use of his concept, as long as we are able to lead this concept back to natural causes.

Recapitulation: We were concerned at the beginning of our investigation to explain how a fixed product was conceivable. We have fully satisfied this task; for where before it was as if the organism itself pursued an unorganic—not productive—world, Nature is for us only productive, i.e. organic.

It has now been deduced how Nature can be restricted to individual products without ceasing to be productive. For

1) within that sphere Nature organizes always narrower spheres of formations—variations—varieties—and so on, to infinity.

2) Just in the separation of the formative drive into opposed directions is furnished an enduring dualism, and with it the condition of an enduring activity, since dualism is condition of all activity of Nature. This activity cannot stand still until the identity of the genus proceeds again from the duplicity of the sexes; but this can never happen due to the same law according to which the sexes have originally separated themselves.

It has been further demonstrated through our investigation that, in organic Nature, we give an account of only *one* product inhibited at various stages of development, as different as the individual products may be. Only the diversity of the stages of development constitutes the diversity of the organisms.—This inhibited relation of production at individual levels of development happens only, and exclusively, through the separation of the sexes.

gradual development of one and the same original organism—disappears from our point of view, for the universal product could not be inhibited at various stages without at the same time dividing itself into opposite sexes.* As soon as there are opposed sexes in one organism all further formation is interrupted and it can only reproduce itself to infinity.†

Further, the diversity of stages at which we now observe the organisms fixed apparently presupposes a peculiar proportion of original actants‡ for each individual. It follows from this that Nature must have begun all over again with a totally new natural predisposition for each product that appears fixed to us. (It therefore remains a problem for the natural scientist to discover precisely these original natural predispositions, so that he does not reckon mere variations on an original plan as diverse species.)§

The assumption that different organisms have really formed themselves from one another through gradual development is a misunderstanding of an idea which actually does lie in reason. Namely, all individual organisms should together amount to one product. This would be thinkable only if Nature had had one and the same archetype for all of them, as it were, before its eyes.

This archetype would be the Absolute, that without internal difference in kind, in which individual and species coincide, which is now neither individual nor genus, but *both at once*. This absolute organism could not be presented through an individual product, but only through an infinity of individual

---

*All organisms, as different as they may be, are surely, in terms of their physical origin, only various stages of development of one and the same organism; they may be presented *as if* they had arisen through the inhibition of one and the same product at various stages of development. However, what holds for diverse organisms in terms of a *physical* origin, cannot hold good when transferred to the *historical* origin. For example, when one goes back to the original condition of the *Earth* and asks how and through what mechanism organic Nature has first arisen, then it would not suffice to accept only *one* original product and let this one product bring forth the various organisms through its gradual development. For, in order to bring forth a new product, Nature would have to begin again from the start.

†Once inhibited, a being can only infinitely reproduce *itself*.

‡forces

§Moreover, it does not follow from this that the productivity of organic nature cannot be seen as *one*. In the original productivity of Nature all products lay concealed. As soon as *determinate* points of inhibition in Nature were furnished, they emerged from identity. But in Nature there was originally only *one* point of inhibition—and so organic formation began, without doubt, from *one* product. In that Nature fought against this point, it raised it to *product*, cancelled it *as* point of inhibition. But just as certainly as Nature is limited originally and through itself, *just through* the cancellation of the one point of inhibition a new one must arise, and thus at any rate one product contained the *ground* of the subsequent one. The product C could not arise before B, and this not before A had arisen.—The *productivity* was thus one, but not the product. It was just not *one* already fixed and present *product* that developed itself into various organisms, for it could not be fixed at all unless it was always inhibited in its formation.

products which, seen *singly* depart infinitely from the ideal, but taken in the *whole* are congruent with it. Now, *that* Nature expresses such an [113] absolute original through all organisms together could only be proven if it is shown that all diversity of organisms is only a difference of approximation to that Absolute, which would then be the same for experience, as if they were originally only various developments of one and the same organism.

Now since that absolute product nowhere exists (but always only *becomes*, and so is nothing fixed), the greater or lesser distance of an organism to it (as to the ideal) cannot be determined by means of a comparison with it. However, since in experience such approximations to a common ideal must show the same phenomenal aspect (which would provide various developments of one and the same organism), the proof for the first point of view is provided when the proof for the *possibility* of the latter is given.*

This proof could be pursued either through comparison of the similarities and increasingly graduated divergences, partly in the *external construction* of organisms, partly in the *structure of their organs*, which is the job of a *comparative anatomy* (*anatomia comparata*). By this means one would have to gradually come to a far more natural ordering of the organic system of Nature than has been possible through the previous methods.† But since the *external shape* is itself only a phenomenon of the original inner proportion of organic functions,‡ one would have to provide a *comparative physiology* (*physiologia comparata*) for the discovery of these proportions, a science not yet attempted. It would furnish a far simpler principle of specification than that according to the diversity of shape and organic structure, although the latter can at least serve as guide to the discovery of the former.

Before we can follow this idea further, which promises to lead us by the shortest path to the goal, a few necessary preliminary elucidations are required.

---

*If it is *proven* that one can view organisms as various developments of one and the same organism, then it is demonstrated that Nature has expressed in all of them one and the same original. That is, *unity* in the productivity is proven at least. It has been attempted for ages to adduce that proof in various ways through the desire to prove a continuity of forms in Nature. That continuity of forms, namely, expresses none other than just the inner affinity of all organisms, as common descendants of one and the same stock.

†These differences that comparative anatomy discovers are really created by Nature itself. The conventional classifications do not exist in Nature and are only contrived as aids for thought. Severity of the Linnean method. Human being and bat, elephant and sloth in one class. This unnatural grouping-together is necessary as long as merely external characteristics count, e.g. whether the animal has teats, whether it has cloven or uncloven toes, how many teeth, etc.

‡How are these different functions related to that one principle with which we are familiar—to organic productivity? Are those functions perhaps themselves only different stages of productivity?

[114] a) Each organism is itself nothing other than the collective *expression* for a multiplicity of actants, which mutually limit themselves to a determinate sphere. This sphere is something perennially enduring—not something merely fading into the background as appearance—for it is that which *originates* in the conflict of actants, the monument, as it were, of those activities prehending one another; it is the *concept of that change itself*, which is the only enduring thing in the change. In all the lawlessness of the actants continuously jostling one another, there yet remains the *lawful aspect of the product itself*, which they (and no others) are constrained among themselves to produce; as a result, the perception of the organism as a product, in which what it is *it is through itself*—which is simultaneously cause and effect of itself, means and end—will be justified as in accordance with Nature.*

b) Now this conflict of actants by which each organic being comes into existence (as the permanent expression of them) will express itself in certain necessary actions, which, since they necessarily result from the organic conflict, must be seen as functions of the organism itself.

c) Since these functions proceed necessarily from the essence of the organism, they will be *common* to all organic natures.† All diversity in the organic realm of Nature could proceed from a *variation in proportion* of these functions in respect of their intensity.

d) But different proportions of these functions in terms of intensity could not occur if these functions generally stood in *direct* relation to one another, such that as the one increases in intensity the other must increase, and the converse;‡ for in this way only the absolute intensity of the functions could increase to infinity, but their proportion§ itself could not be altered. The functions then must stand in *inverse relationship of intensity* with one another, such

---

*The organism is 1) not mere appearance, not that which is known *merely* by its effects; 2) its activity is not at all directed to anything external, but is directed upon itself—it is its *own object* (new determination): it is, what it is, without any external *effect*.

†For example, if that alternation of expansion and contraction in the manifestations of irritability (pulsation) is a necessary condition of every natural product, every formation, then it cannot be lacking in any organism.

‡In the organism everything is cause and effect. None of those functions can exist without the others—none can overstep the other. *Distinction*: positive and negative relationship of causality.—If A is cause of B, the inactivity of A is cause of the activity of B. Employing the concept of negative causal relation here, the increase of the one can be cause of the sinking of the other, and conversely. This would not be possible if they stood in direct relation to one another.

§their relative intensity

that as the one is augmented in intensity the other diminishes, and conversely, as the one diminishes in intensity the other has to increase. In short, the functions must be *opposed* to one another [115] and reciprocally maintain each other in equilibrium, which in itself already corresponds with the concept of an organism.

e) In a single organism either one of these functions could be the *dominant* one; but in the degree that it is dominant, its opposed function must be supressed.* Or, these functions could maintain the *equilibrium* in one organism. However, since these functions are opposed to one another, such that the one excludes the other, then it is impossible that they be unified in *one* and the same individual. The *one* organism in which they were *all* unified would have to divide into many single individuals, as it were, and those various functions would have to be parceled out, so to speak, among these individuals. But these individuals must produce that organism once more through their collective action, and conversely,† the exercise of their functions must be possible only within this organism. They would relate simultaneously to the whole organism as cause and effect of its activity. That which so relates itself to the organism (as a whole)‡ is called an *organ*. Where opposed functions are united in *one* organism, these functions must be split up into various organs. Therefore, the more the multiplicity of the functions increases in the organic domain of nature, the more complexly must the system of the organs develop (in part called "system of vessels," which is completely wrong, for within the organism nothing is merely a *vessel*).§ Insofar as each organ exercises its special function, it would receive a *life of its own* (*vita propria*)—to the extent, however, that the exercise of this function is still possible only within the bounds of the whole organism, it would only receive a *borrowed life*, and it must be so in accordance with the concept of organization. If the manifold possible proportions of organic functions can be deduced *a priori*, the whole multiplicity of possible organisms would also be deduced, because the organic structure depends on this proportion.‖

[116] f) Now we can understand the problem: to determine the various organic functions and their various possible proportions *a priori*.—If this prob-

---

*The more the productivity has already passed into the product, or has materialized itself, the less the higher stages of productivity can be distinguished.

†because everything in the organism is reciprocal

‡and yet has its own individuality

§(For example, in polyps no organ is distinguishable.)—Therefore the affinity of comparative physiology with comparative anatomy.

‖This omnipresent, all-penetrating productivity is the invisible medium, as it were, that permeates every organism and binds them to one another.

lem is successfully solved, then a *dynamically graded series of stages* would not only be brought into Nature, but at the same time one will also have deduced the graded series of stages in Nature itself *a priori*, and what was *formerly* called *natural history* would be raised to a *system of Nature*.

Remark.

Natural history has been, until now, really the *description of Nature*, as Kant has very correctly remarked. He himself uses the name "natural history" for a particular branch of natural science, namely, the knowledge of the gradual alterations which the various organisms of the Earth have suffered through influence of external nature, through migrations from one climate into the other, and so forth. However, if the idea set out above were put into practice, then the name "natural history" would get a much higher meaning, for then there would actually be a *history* of Nature itself; namely, as it gradually brings forth the whole multiplicity of its products through continuous deviations from a common ideal—thus far freely—but not forming them *lawlessly* because of this (because it still remains constantly within the bounds of its ideal)—and so realizes the Ideal, not indeed in the individual, but in the whole.

With reference to this, it can be asked which principle of ordering should also guide the mere "description of nature" (which would be related to natural history in the sense just explained, approximately as anatomy is related to physiology). Since the continuity of species (*continuitas formarum*) is not met with in Nature as long as one investigates it merely according to external characteristics, it must present the chain of Nature either with continual interruptions as before, as through comparative anatomy; or it must, finally, as has just been attempted, use that *continuity of organic functions* as principle of organization. The latter is the objective of the following task, in which all of the problems of the philosophy of nature may be readily united, and for which, accordingly, the most universal expression has been chosen.

[117]

V.
Deduction of the Dynamic Series of Stages

Problem. To deduce a dynamic graded series of stages in Nature a priori.

Solution.

We have deduced in the preceding why it is necessary that the absolute product be inhibited at individual stages of development; we have also deduced how this inhibition itself occurs (III., IV.). However, it has not been shown how this inhibition could be *permanent*—how these individual natures which have torn themselves away from universal Nature, as it were, can maintain an individual existence, since all of Nature's activity is directed toward an *absolute* organism.

Now, the problem, to deduce a dynamic graduated series of stages in Nature, presupposes the permanence of individual natures. We cannot succeed in the solution to this problem before another *problem* is solved, namely: *how the individual is preserved in Nature at all.*

Solution. Assume that the whole of Nature = one organism, then within Nature nothing can come into existence which would not be joined to or subordinated to this universal organism, in short, nothing individual can remain in Nature.

Determined more precisely, our problem is this: *How can any individual nature hold its own against the universal organism?*

The universal organism operates absolutely by assimilation, i.e., it admits no production within its sphere that does not fit into it; it only allows that which joins itself to the absolute product to exist.*

[118] No individuality in Nature can, *as such*, maintain itself, unless it begins, just like the absolute organism, to assimilate everything for itself, to encompass everything within its sphere of activity. In order that it not *be* assimilated, it must *assimilate*; in order that it not *be* organized, it must *organize*.

In this act (of opposition) *inner* and *outer* are divided for it. It† is an activity *that works from the inner toward the outer*. But how could this direction be distinguished otherwise than in opposition to another activity that operates on it as on an external factor? And moreover, how could the latter operate on it as on an external factor, if it did not oppose itself to the inclusion into that activity (strive against the indentification with the universal activity of Nature)?

By the same action through which it excludes the whole of external nature from its sphere, it also makes itself into an external thing in relation to the whole of Nature.

External nature (for it) will struggle against it, but only insofar as it once more struggles against external nature. *Its* RECEPTIVITY *to external nature is conditioned by its* ACTIVITY *against it*. Only insofar as it strives against external nature can external nature act upon it as upon an inner factor.‡

The external world can hardly *be* taken up into it except to the extent that Nature *takes it in*. The external world is as good as not even there for it—

---

*Nevertheless, one could think that the individual had torn itself loose, as it were, from the universal organism. Each organism a unique, singular world—*status in statu*—.

†activity of the product

‡Dead matter has no external world—it is absolutely identical with its world.—The condition of an activity toward the outside is an influence from the outside. But conversely, the condition of an influence from outside is the activity of the product *toward* the outside. This reciprocal determination is of the highest importance for the construction of all living phenomena.

it has no reality for it, except to the degree that Nature directs its activity against it.

Its receptivity for the outer world is not only *generally* conditioned through its activity toward the outside, but the *way in which* the outer world acts upon it is conditioned through the *mode* of activity that it exercises toward the outside.

The external world does not act upon the inner factor as the external acts upon the external (dead thing upon dead thing). An external thing acts on an inner one only insofar as it engages negatively in its positive activity, or (what is the same) in the negative activity positively. But also conversely, the inner *takes* the outer *into* itself only because its activity in relation to it becomes positive or negative.

[119] Let us suppose that an external activity = X acts upon the inner factor. One abstracts from all mechanical efficacy, for such a thing has not yet been deduced here, and an inner factor as such cannot be mechanically acted upon at all. We are talking about a dynamic activity.

In general, let it be noted that we expressly hold that there is to be an influence upon the inner factor AS SUCH. The effect which that external activity exercises according to its nature is = A. But with A it cannot act upon the inner factor as such, unless the latter opposes to it an activity = $-A$. In this $-A$ lies the *receptivity* of the absolutely inner factor for the external activity = A.

(For example, let X be the activity of heat-matter. Its effect = A. In relation to this principle (the heat-matter) nothing is an *inner* factor other than what this principle produces *in itself*. The heat-matter also cannot exercise the effect = A on an inner factor *as such*, except insofar as the *proper* activity of the inner factor in relation to the heat-matter as an external factor is = $-A$. Both effects (A and $-A$) are *positive*. They are only positive and negative in relation to one another insofar as they reciprocally hold the equilibrium. Conversely also, the activity = $-A$ is extinguished without an external activity that is in relation to it = A, which keeps it in equilibrium and which is, as it were, its object.*)

The immediate effect which follows upon the effect = A in the inner factor, is the *negative* (i.e., not the negating, but the exact opposite of this effect = $-A$). (The heat-activity of its own body is = $-A$ in relation to the external influence of heat-matter.)

Indirectly, through this activity = $-A$ new transformations will be produced. If these transformations = Z, then Z will be the effect of both A and

---

*At any rate, the organic body generates heat in itself, but its own heat-activity is extinguished without an outer factor which it excites and which is its opposite, its object, as it were. If the inner factor brings forth activity in the outer, this means just as well that it produces its opposite.

$-A$.*—That is, X cannot act on the inner factor as such with the effect $= A$, except to the extent that the proper activity of the latter is in relation to the former $= -A$. Thus Z will also be determined in mode as well as in degree through the mode and degree of activity $= -A$.

[120] (To elucidate.—A *poison* acts upon the animal body. To what extent is it a poison, and why is it a poison? Is it a poison in itself? Hardly. For example, smallpox is a poison only once for each person; snake venom is not poisonous for the snake. Poison is not poison at all except to the extent that the body makes it so. For poison as poison the body has no receptivity, except to the degree that it is active against it. Poison does not attack the body, but the body attacks the poison.† The ultimate effect of the poison $= Z$ is determined in both mode and degree through the mode and degree of activity which the organism opposes to it. Therefore, it is really not an effect of the poison, but an effect of the activity $-A$.)

Conversely, however, the inner factor also exercises no activity $= -A$ except to the degree that it has a receptivity for an activity $= A$. The activity of the inner factor $= -A$ is thus itself again an effect of the activity of the external factor $= A$. Z will be indirectly determined in both mode and degree through the mode and degree of activity $= A$.

(The body will not be active against poison except to the extent that the poison is active against it. The form and the degree of its activity is determined through the form and degree of activity of the poison.)

Therefore A and $-A$ are reciprocally cause and effect of one another.‡

In the activity that the absolutely inner opposes to the external activity lies its receptivity for the outer, and conversely, its receptivity for the outer factor depends upon its activity. Neither the activity of the organism nor its receptivity can be purely known in themselves. The former is extinguished without an object against which it struggles, and conversely, nothing is an object for it except to the extent that it is active against it.

---

*It is the effect 1) of A, for through the activity of A the activity $-$(minus)A has first been excited; but 2) also of $-A$, for only by means of this could A produce transformations in the interior.

†The concept of poison only has a meaning for the organic product, like so many others, e.g., the concept of contagion, disease, medicine, etc.—Every body can become a poison, for it is so only through the activity of the organism.—Boundary between medicine and poison. *Kant*: what absolutely cannot be assimilated. But even every excretion is a poison. However, this much is true: poison is only poison by virtue of the fact that the organism directs its activity against it, strives to assimilate it.

‡conditioned through one another

Note.

In the synthetical principles just proposed two antithetical propositions are united.

a) *First principle. The activity of the organism is determined through its receptivity.** The organic activity is, therefore, through and through [121] dependent upon the[†] influence of external (material) causes. But matter can only act on matter, and only according to inexorable laws. Therefore, both the action of external causes on the organism and the functions maintained through it occur completely and entirely according to laws of matter. Matter acts on matter either through repulsive force (thrust) or through attractive force (gravity). The influence of external causes on the organism is explicable neither by the latter nor by the former type of effect alone, and neither is the activity of the organism animated by it—thus, it is explicable only from both taken together, or from the reciprocal action of both of those forces. This reciprocity produces what are called chemical phenomena.‡ The influence of external causes on the organism, as well as on the organic activity, is itself consequently *of a chemical sort*. All functions of the organism follow from chemical laws of matter, and life itself is a *chemical process*.

Remark.

(This theory appears to agree with experience, as made evident by the following.§)[10]

"Organization and life are entirely dependent upon chemical conditions. Already long ago, Nature made the first chemical sketches in the so-called inorganic world for the formations which it produces in the organic. The universal natural operations and those processes which are constantly underway must be seen as the first rudiments of all organization. Everything is swallowed up in a single chemical process. The preservation of the air circulation, e.g., in a uniform

---

*But not the reverse.

†direct

‡The merely chemical phenomena of matter already lie outside the merely mechanical, and are a dynamic source of movement in Nature.

§One will readily note that in the presentation the chemical system is idealized, but I found this necessary. (Up to this point original note.—Trans.) It would be quite natural to view the phenomena of life *entirely chemically*. With the great and important discoveries of chemistry which the chemical muse has spread through every brain, in particular the discoveries which have been made by means of chemistry in animal and vegetable nature, it is as if one would have come across them oneself without needing to come to this point of view in a scientific way. Least of all Reil, the chief defender of this chemical perspective, which he has presented in all of his writings, but without lending them the support of which this doctrine is capable.

proportion of mixture, is of the utmost importance for organic nature as a whole. Even the atmosphere, daily organized anew, already contains the first impulse to universal organization. The [122] meteorological phenomena are all without a doubt manifestations of processes through which they are always rejuvenated and replenished anew. That we do not know how to explain, for the time being, e.g., the aeration of water, and the disaeration, which seems to precede the rain, from our chemical knowledge, proves nothing against the assumption that both do not happen in a chemical way. Nature does not combine as the chemist combines. Nature and chemistry are related to one another like language and grammar.— Since the same substances in the atmosphere whose combination and decomposition also sustains animal and vegetable life are constantly combined and decomposed, the processes which always maintain the same chemical composition in the universal medium of life must be the first stirrings of universal organization. Even the perpetual maintenance of that proportion of factors in the whole through which the chemical bonding of both atmospheric substances never happens and is not permitted to happen, is not to be explained otherwise than from the perpetuity of constantly sustained chemical separation.

"Most of the indecomposable substances that are major components of organic matter also betray the strongest tendency toward combination in anorganic nature. None of these substances can be exhibited individually; one knows them either only in their combination with the absolutely fluid (as kinds of air), or in connection with solid substances. They thus stand in the middle between absolutely decomposable and indecomposable substances, and belong, like organic nature itself, to no one of the two types alone.

"The substances that are particularly active in organic nature are already conspicuous in anorganic nature, and conversely, the substances that in anorganic nature are the most efficacious are also the most active in organic nature. The heat-matter produced through a continually sustained phlogistical process in the animal body, extended everywhere, doubtless even in plants, flows through everything living. The electrical matter [123] gives the muscular system and the excitable plant fibers their elasticity. As a consequence of more recent observations, it is not impossible that a free development of light takes place in the eye. Plants draw the largest part of their substances from the ubiquitous water, the major components of animal matter are spread out in the atmospheric air. In the bones of animals the soils are hardened, and their veins conduct metallic content.[11]

The ground of all phenomena of organized bodies is, therefore, to be sought in organic matter, in the original diversity of its basic substances, in the peculiar proportion of their mixture—in the chemical alterations that are produced through external influences, also chemical, in them. The composition of organic matter proceeds to infinity because every organ organizes again into infinity, is again mixed and formed in a peculiar way, each distinguishing itself from the other by means of particular qualities.—But what is quality itself? If it were, according to the common notion, *dead stuff*, then the most complete composition of manifold substances would again require a new activity which sets everything in reciprocity and puts its dead forces into a free play. However, what appears to us as quality is itself already activity, and each particular quality is a particular degree of activity. Is it to be wondered at that a connection of such manifold qualities that is still, moreover, continually altered through the influence of alien actions (light, heat, etc.) brings forth such manifold and peculiar activities as we perceive in organic nature?

The explanation of organic *shape* only demands the unification of manifold activities which all lead toward production of an original figure.* Since the tendency to equilibrium dwells originally in each material, and this tendency in matter is unconditioned, it will take up every form in which it achieves equilibrium. Every individual organic material will commit itself to this [124] peculiar form freely, as it were, because this is the sole condition of a possible equilibrium of forces.

Accordingly, all difference of organisms will also be reduced solely to the diversity of substances that are united or separated in them, and the diversity of their functions solely to the diverse chemical influences to which they are receptive. The debated question about the difference between plants and animals is easily answered, naturally from a chemical point of view.

The two major opposing processes of Nature have prospered to the point of permanence in plants and animals. All manifoldness of material in the world is reducible to its relation to that substance which in our atmosphere enchains the element of light, and whose general possession seems to be the world system's luminous body. All materials are either burnt up,† or burning, or such as become combustible again. The major processes of Nature are combustion and decombustion processes, in the great—(therefore, the opposition between sun and planets)—as in the small. Organic nature is divided into both.

The animal destroys the atmosphere about itself, and preserves, increases and moves itself like the mobile, growing flame. The plant returns the power of combustion to the burnt, ubiquitous substance, and returns to the atmosphere that substance which makes combustion possible.—This difference between

---

*The explanation of organic shape only demands that peculiar chemical mixture that we presuppose in organic nature. A certain form is always inseparable from a certain mixture.—Proof in anorganic nature.—But even *a priori*. Matter cannot be *compelled* to take a determinate form, as in a determinate mixture, because that form is then the single condition under which an equilibrium of forces in that mixture is possible.

†e.g., the soils

plant and animal is the original one, grounded in Nature itself, from which stem all other differences between them. This difference itself, however, again derives solely from the different chemical composition of animal and vegetable matter. This is why the plant, therefore, at least for the most part, lacks the substance that makes the animal capable of retaining that principle in itself.

[125] Thus animals as well as plants are permanent chemical processes which are sustained through external chemical influence. The external condition of life for plants is *light*, for the animal *phlogistical* material. All of their functions engage in that chemical process and proceed from it."

The proposition: organic activity is determined through its receptivity, is consequently the principle of a *physiological materialism*.

b) *Second principle. The receptivity of the organism is conditioned through its activity*. If the receptivity of the organism is conditioned through its activity, then so is the action of matter upon it. No one can in any way experience the pure effect of any material as such, in—and on—the organism, for the effect is determined both in mode and degree through the activity of the organism. Matter cannot operate according to its forces freely and uninhibitedly in the organism. The connections of universal chemical affinity are dissolved by the organism, and new affinities are instituted. Whatever steps into the sphere of the organism adopts, from this moment forward, a new mode of action, alien to it, which it does not abandon until it is given back to anorganic nature.*

Remark.

(This system too appeals to experience.

"The organic preserves itself in a wholly peculiar mixture, without example in the rest of Nature. Chemistry indeed names the major constituents of this mixture. But if *these* substances only, and only these substances *in such a way*, are active in organic nature (as chemistry can demonstrate), how could the great multiplicity of organic products proceed from the different proportions of mixture of these simple substances? The organic body retains its proper degree of heat in every temperature. Out of mere air and water the plant kingdom [126] produces—and indirectly through the plant kingdom the animal too—the most disparate materials, the likes of which can be brought forth through no chemical art. The chemical forces of external nature have, instead of making organic matter like dead matter, exactly the opposite power as long as life endures. As soon as life

---

*In another respect, however, the problem is possible and solved, because the expression for the construction of the inorganic product is also the expression for the construction of the organic product, for we are permitted to take the categories of the inorganic, but in a higher potency, in order to transfer them to the organic. There is only one expression for the construction of a product; there are only products of different potency.

declines, the organic matter returns into the universal circulation from which it was previously withdrawn—it returns the more quickly, the less its elements were mixed according to the laws of affinity prevailing in dead nature, etc.")[12]

Now, the cause that in part cancels and in part alters the chemical forces and laws of matter in the organism cannot once again be a *material* one, since each material is itself subjected to the chemical process—therefore, it must be an *immaterial* principle, which is rightly called *vital force*.*

---

*That which is a law of nature is, just for that reason, an inviolable law. That it also appears as if Nature can again cancel its own laws might stem from the fact that, when soberly viewed, that which you call laws of nature are not actual laws of nature at all, but rather imaginary projections of your own. One only needs to take a look at most of the previous textbooks on medicine to hear (on almost every page in manifold forms, sometimes openly, sometimes concealed) that laws of nature suffer exceptions. However, this derives merely from the fact that the objects are regularly obstinate enough not to want to submit themselves to the received theory. For example, if a disease is found which is unable to be explained in terms of the received systems—no sooner is this disease an *ens sui generis* which follows completely singular and idiosyncratic laws.—One has that principle (that laws of nature have exceptions) to thank for the fact that the organic being has lain there for so long like a sealed book and has been whisked away from the region of natural explanation as if by a magic wand. It is this principle which until now has made all theory in medicine impossible and has reduced this science to the shallowest empiricism. This principle is at the same time so opposed to the first laws of the understanding that one has to give it another twist. This twist is: the laws of nature cannot, certainly, be canceled—this is conceded—except through forces of nature itself. Then, for example, the law of gravity cannot be canceled (e.g. the Moon cannot fall to the Earth); but now if there were in nature a force which acted counter to it (something like a negative gravity) then it would not be gravity itself but only its *effect* that is canceled—here no law of nature would be infringed, for the natural law of gravity only holds where no opposed force offers resistance to it.—Such is the case with the phenomenon of life. Nature cannot cancel the chemical and physical laws, to be sure, other than by the counteraction of another force, and just this force we call *vital force*, because it was completely unknown to us until now.

Already in this deduction of the vital force lies the admission:

1) that it is contrived solely as an expedient of ignorance and a genuine product of lazy reason;

2) that through this vital force we take no steps forward, neither in theory nor *in praxi*:

a) not in theory. For either α) one assumes that it is simple, like, e.g. repulsive force—or in accord with the usual idea of gravity; in other words, this means: it has *no empirical* condition: then one does not realize why it does not act just as universally as those forces. Or one assumes β) it is *composite*, i.e., dependent upon empirical conditions: then one must be able to provide these empirical conditions—until these are provided it is a mere word.—What the reference to gravity means here is, first, that it is not so constituted that it would have no empirical condition; secondly, gravity acts according to very simple laws. We would like to believe in vital force as soon as those simple laws are set out to us, and the existence and all appearances of organic nature are explained by means of them, just as the existence and the appearances of the universe are explained from the law of gravity. The concept of vital force helps just as little

b) in praxis, as in theory. The entire medical art is reducible to action upon this completely *unknown force*—naturally, to acting upon it not according to determinate laws which would be created for us from its nature, but according to a blind empiricism.

The proposition: *The receptivity of the organism is determined by its activity* is, therefore, the principle of a *physiological immaterialism*.

c) Neither one of these systems is true, for they reciprocally refute each other. Nevertheless there is in both something necessary; they are *both* true at once, or rather the true system is a third derived from both.

α) *Where* it expresses itself the principle of life is shown to be an activity that resists every infiltration of matter from outside, every impact of external force; but* this activity does not *express* itself unless it *is excited* by external impact. The negative condition of life is *excitation through external influences*.† Life, where it comes into existence, comes against the will of external nature (*invita natura externa*),‡ as it were, by a tearing-away from it. External nature will struggle against life; most external influences which one takes as life-promoting, are really destructive for life; for example, the influence of air, which is really a process of consumption—it is a continual attempt to subject living matter to chemical forces.

[127] β) But this struggle of external nature preserves life, because it always excites the organic activity anew, rekindles the flagging contest. In this way, every external influence on the living which threatens to subject it to chemical forces becomes an *irritant*, i.e., it actually brings forth exactly the opposite effect which, according to its nature, it should produce. That reciprocal determination of receptivity and activity is really that which must be expressed through the concept of *susceptibility*, a concept that is exactly the synthesis which unifies those opposed systems§ (in its greatest generality—one should completely forget the "susceptibility" of Haller[13]).

---

*in this lies the receptivity *for* external effect

†Here the organism submits itself to the laws of every other natural thing; no natural object is set in motion or activity other than through an external cause.

‡in contradiction to Nature

§It indeed sounds paradoxical, but is no less true, that through the influences which are contrary to life, life is sustained.—Life is nothing other than a productivity held back from the absolute transition into a product. The absolute transition into product is death. That which interrupts productivity, therefore, sustains life.

That proposition can be generally expressed in this way: the external influences on the organism bring forth in it precisely the opposite effects from those which it should produce according to its nature. The external influences tend toward the destruction of the product, but precisely in this, tend toward fanning the flames of productivity. Through those external influences, the activity through which the organ reproduces itself is fanned ever anew, such that the same influences which are *directly destructive* for the organism are, indirectly through productivity, *preservative* for the product.—By this means and by this means only does the outer become irritant, and *stimulant*, for the organism. By *irritant* we understand, for the time being, nothing other than an effect which sustains life as productivity, by never permitting the transition into a product.

The activity of life is extinguished* without an object, it can only be excited through external influence. But this external influence† is again determined through the organic activity.‡ Therefore, no external activity acts in the organic body chemically, according to its peculiar nature, which is why chemical forces seem canceled with respect to it.§ But no activity can be canceled otherwise than through an opposed activity. This opposite activity lies in the organic body as in a closed system. At each moment the organic system establishes an antagonism against every external effect that holds the former in equilibrium. (For example, the living body retains its proper degree of temperature in the highest temperatures, not because the universal law of the communication of heat is *canceled* with respect to it (this is impossible), but because it maintains equilibrium with the forces impinging from outside through opposed operations—(e.g., by increasing the capacity of the fluids circulating in it, by accelerating processes that absorb much heat).)[14] It is true that an external influence sustains organic activity, and that every such influence brings forth a determinate effect in the organic; but this *effect* is itself again a product of organic activity; e.g., of course opium acts as a narcotic, [128] but it has this effect not as *opium*; one would look in vain for the reason of this effect in its chemical constitution. The effect that it brings forth, it brings forth only *indirectly*, i.e., this effect is itself again an effect of organic activity.‖ Generally expressed: *Every external effect on the organism is an indirect effect.*

---

*would be extinguished

†on the organism

‡Because it acts directly only on the productivity—and only mediately and indirectly on the product. If the organic body were a product, without being productivity, then the outer would act on it exactly as it works on the *dead*. That it acts upon it *entirely differently* derives from the fact that it does not act directly on the product, but only on the productivity.

§Excitability = indirect affectability of the organism. It is immediately explained *from* this principle of *indirect affectability* why no external cause can act on the organism chemically unless one invokes a particular force that cancels the chemical force.

‖That opium acts as a *stimulant* is explained by its chemical, or what is the same, its electrical constitution (which is why it also acts in galvanism)—but its *mediate* effect, i.e. the effect mediated through the activity of the organism, is *narcotic*, and this effect is, to be sure, chemically inexplicable, since it is indirect. Thus it is shown on the whole, that just the same materials which cause the most intense excitability (which must be explained from their chemical and electrical constitution), indirectly exhaust excitability (which is now certainly no longer explicable in terms of its chemical constitution). It is no wonder that chemical explanations want to go no further than this. The ultimate effect of external causes on the organism *cannot* be chemically explained any longer. One does not require for the explanation of this appearance a fantasy like the *vital force*; it is not needed because it is a completely false assumption that the sublimity of life-processes over the chemical can only be explained in terms of an immaterial force.

(*Therefore*, no substance truly acts on the body chemically, but one does not at all need the fiction of a vital force on behalf of this idea; for either one understands by this a *simple*—original—force, like, e.g., attractive force: then it would have to operate just as universally as the latter. Or, if it is a *composite* force, then one must attempt a construction of it (e.g., if it proceeds from the antagonism that occurs in organic matter, then one would have to find a principle that continually sustains this antagonism and that keeps it from a chemical bonding of the elements, or that gives the chemical tendencies the peculiar direction which they take, e.g., in the animal body.) This could only be the function of a principle that does not enter into the chemical process itself, like, e.g., *absolute matter*, whose existence has been demonstrated in the foregoing, because this is absolutely incomposable, and because its conditions are present everywhere; where it is decomposed it must be composed anew in every moment.*

However, these presuppositions are not needed. The whole mystery rests on the opposition between *inner* and *outer*, which must be admitted if one admits anything individual in Nature at all.

[129] External nature will now struggle against every *inner* activity, i.e., against every activity which establishes *itself* at the central-point. The inner activity will itself be constrained to produce through this antagonism that which it would not have produced without it. The organic shape and structure is the single form in which the inner activity can assert itself against the outer, e.g., the form to which also belongs the manifoldness of individual organs, each of which adopts its particular function. Therefore, its formation is already itself an effect of that universal organic property of *susceptibility* (excitability through external influences) which is also found to be in agreement with experience. Conversely, the outer is, as it were, intensified to a higher kind of action through organic reaction, and precisely in this way the organic elevates itself over the dead.

---

*It was thus an overly hasty assumption which had been much too hastily interjected that there can be no matter that is inalterable through chemical life-processes, and that could not give to the chemical forces the particular direction that they take, e.g. in the animal body. For this reason I have proposed, in the treatise *On the World-Soul*, the hypothesis of an absolute matter (whose *necessary* existence in Nature is now proven) in opposition to the assumption that in order to explain that particular direction an *immaterial* principle is required. The hypothesis has been taken for an *assumption*—the *possibility* of such a substance has even been denied—for what reasons we will soon see. (Original note. See AA I,6 186–91.—Trans.)

†The activity of the organism = 0 without receptivity (for the organism should be neither *pure* productivity—activity *through itself*—nor *pure* product, but both at once)—but then receptivity is also only a minus of activity, thus not thinkable without activity.

Conclusions.

The activity of the organism is determined through its receptivity, and vice versa. Neither its activity nor its receptivity is in itself something real; both obtain reality only in this reciprocal determination.[†]

In addition, activity and receptivity are related to one another as opposed terms ($+$ and $-$). Thus, as the one factor increases, the other falls, and vice versa.

1) The beginning of life is activity; it is a tearing loose from universal Nature. But that activity is itself again receptivity, for receptivity is only the minus of activity.

Activity and receptivity arise simultaneously in one and the same indivisible moment, and precisely this simultaneity of activity and receptivity constitutes life.

Organic activity is not *activity* without external pressure. The external pressure against inner activity has precisely the *opposite* effect, however, in that it *enhances* activity, it *decreases* receptivity.[*] The *maximum of receptivity* (which one [130] can assume at life's beginning) *passes over first into a minus, and finally into a minimum of receptivity* (by virtue of the law of reciprocal determination).[†] In the degree that activity rises, receptivity must fall, until both enter into the most complete reciprocal determination, where they maintain equilibrium with one another—which is then the noon of life, as it were.

However, that complete reciprocal determination is only *momentary*, the organic activity is on the *increase*, receptivity on the *decrease*, and then the wheel of life rolls over toward the other side. The organic activity will increase more and more toward the minimum of receptivity, but since receptivity, as long as it has a *degree*, is itself only activity, it *passes over* by virtue of the inviolable law of reciprocal determination *from the minimum immediately into the maximum* (absolute receptivity), as soon as it sinks below all degree; the highest activity is $=$ to the negation of all activity, the maximum of activity $=$ to the maximum of capacity.[‡]

---

[*]It does not act on an organism as it does on a dead thing, it acts as an irritant.

[†]This happens, however, with retarded velocity.

[‡](In place of the last passage, the manuscript has the following.—Trans.) The organic activity increases, the receptivity sinks gradually always more toward the minimum. But receptivity is indeed itself the *mediator* of organic activity. Without receptivity no activity. So the law that the increase of activity $=$ the sinking of receptivity holds only to a certain limit. When this line is crossed, it is turned completely around. The minimum of receptivity passes over immediately into the maximum, by virtue of the inviolable law of reciprocal determination. This paradox is to be explained by reciprocal determination. A degree of receptivity is itself a condition of activity. Absolute negation of every degree of receptivity $=$ absolute negation of activity, and so the highest activity is immediately the limit of activity.—Maximum of activity $=$ Maximum of receptivity.

Life thus has two highest points between which it pulses, as it were, and from the one it passes over immediately into the other. The maximum of activity = the minimum of receptivity, but the minimum of receptivity = also the minimum of activity, that is, the maximum of receptivity, and so it is conceivable how each maximum in organic nature passes immediately into its opposite, the minimum, and the converse.

(Two observations can readily be made here.—*First*, concerning the transcendental significance of this natural law of the immediate transition from the minimum into the maximum, and conversely. For this is precisely the law *of all* activity, namely: that an activity which no longer has an *object* never reverts into itself, and likewise, that there is no longer an *object* for an *activity* that has ceased *to revert into itself*; that in this way the highest moment of all activity borders immediately on the dissolution of it.* Organic life begins in this way, with the reflection of an activity through an *object*, just as [131] the higher activity, and the *object itself*, falls within the *point of reflection*† for the organic as for the higher activity. If this point lies infinitely far away‡ then the activity is no longer reflected, it has no more intensity and dissipates into the infinite. If it lies infinitely near§ then it has no more extension and disappears into itself.‖,#

Secondly, this perspective offers analogies for use of a higher perspective on many other natural processes, e.g., through it the similarity of life with the process of combustion is first revealed. The effect of heat on the combustible

---

*The intensity of the activity is in inverse relation to its extension. Expansion of activity without any resistance = negation of all intensity.

†Only that which struggles against organic activity can be turned into an object—only *unruly* material can be formed.

‡ = absolute activity

§ = absolute receptivity

‖is a dead object

#Brown did not deduce the concept of excitability, but neither did he *construct* or explain it. He openly concedes: What excitability is, we do not know, and we also do not know how it is affected. However, if we do not know the latter, then our knowledge is empiricism through and through. When we do not know the *laws* of physics according to which excitability is affected, knowledge which is certainly impossible unless excitability is itself deduced from natural forces (i.e. to have constructed it physically), then our knowledge is—like all medical arts—only empiricism.—The fact that Brown knew nothing of how to set his theory in connection with physics (which is certainly excusable, since at that time the greater part of those physical discoveries were not yet made that have now been made)—was without doubt responsible for very many false conclusions of his system. That more, and very significant, false notions exist in his system will be proven in the following pages. I am not concerned here at all with the Brownian *system*: I speak here always only of the *principles* of this system, which Brown himself for the most part did not ground thoroughly and from which he did not always rightly reason.

body is the excitation of its activity, which one can think as repulsive force against heat—(heating)—and which, as soon as it has achieved the maximum, passes over immediately into the minimum. Therefore, the maximum of excitation or of activity in every phlogistical body is = to the maximum of capacity. This abrupt, sequential transition from the maximum of repulsive force (of activity) into the maximum of capacity (of receptivity) is really the phenomenon of combustion.)

2) From this follow a few fundamental laws of organic life.

a) it becomes evident that every *stimulant* only is a *stimulant* as far as it *minimizes* receptivity or *enhances* activity. It only *is* a stimulant because it produces its (real) opposite (activity).

b) however, since the function of the stimulant generally consists in the production of its opposite, it becomes evident that the stimulant can itself be of an antithetical kind, i.e., positive or negative, accordingly as it enhances or decreases the activity. A stimulant can only act *positively* at a certain degree of receptivity,[*] negatively only at a certain degree of activity,[†] because in the former case it ought to decrease the receptivity, in the latter decrease the activity. With a high degree of *capacity* for a negative stimulant the activity cannot be decreased by the stimulant, just as with a high degree of *activity* it cannot be increased through a positive stimulant. (Therefore, the phenomenon of desensitization to the stimulus through habit.)

[132] c) Let there be two individuals, the susceptibility of the one related to the other in a ratio of 1:2, and if both should be raised to the same height of activity, then the stimulus that acts upon both of them will have to have a ratio of 2:1 in terms of intensity; that is, the single susceptibility with double intensity of stimulus maintains equilibrium with the single intensity of stimulus with double susceptibility.

d) Finally, it becomes evident from this concept of stimulus (that it produces its opposite) why all stimulus[‡] finally ends with the absolute exhaustion of susceptibility, and how Nature achieves its aim with respect to every organism.

Nature achieves its aim in precisely the opposite way than the way in which it attempted to achieve it; the activity of life is the cause of its own dis-

---

[*]e.g. a lesser degree of heat only in a northerner.

[†]e.g. cold = negative stimulant only on a southerner.

[‡]also even the one that sustains life.

[§]Nature seeks to transform the receptivity of the organism to the external world, which is a determinate one, into an *absolute* one—but in doing so its receptivity is instead increasingly lessened, and in the same relation by which activity increases. In this way the organism achieves always greater independence from the influences of external nature—but the more it is independent of them, the less it is excited by them. Now, however, this excitability is, through external influences and the receptivity to them, a condition of life and organic activity: thus organic activity is extinguished along with organic receptivity. In this manner Nature achieves its aim, but in a completely roundabout way—and indirectly through organic activity itself.

Life comes into existence in opposition to Nature, but it would dissipate of itself if Nature did not struggle against it. Life, to be sure, ultimately subtends Nature, but it does not support the external pressure, only the lack of receptivity for the external. If, from the outside, the influence contrary to life serves precisely to sustain life, then in the same way, that which seems *most favorable* to life (absolute insusceptibility to this influence) becomes the cause of its demise. The phenomenon of life is paradoxical even in its cessation.

As long as it is organic the product can never sink into indifference. If it is to support the universal striving toward indifference, then it must first sink to a product of a lower potency. *As* an organic product it cannot die, and when it does die it is really already no longer organic. Death is the return into universal indifference. Just for that reason the organic product is absolute, immortal. For it is an *organic* product at all, because indifference can never be reached by it. Only at the moment when it has ceased to be organic does the product resolve itself into the universal indifference. The constituents that were drawn from the universal organism revert into it once again, and since life is nothing other than an intensified condition of common natural forces, as soon as this condition has passed, the product falls back into the dominion *of these* forces. The same forces which have for a time maintained life finally destroy it too, and so life is not anything in itself, it is only the phenomenon of a transition of certain forces from that intensified condition into the usual condition of the universal.

The system whose standpoint I have now just developed takes a stand between two opposed systems: the chemical system knows the organism merely as an object or product, and allows everything to act upon it as object upon object, i.e., chemically; the system of vital force knows the organism only as subject, as absolute activity, and allows everything to act upon it only as activity. The third system posits the organism as subject and object, activity and receptivity at once, and this reciprocal determination of receptivity and activity, grasped in one concept, is nothing other than what Brown called "excitability."

I have not only deduced the *necessity* of that reciprocal determination from the concept of the product (organic product), but I have also proven that the phenomena of life can only be completely constructed *from* this reciprocal determination. Thus I have to say *that Brown was the first to understand the only true and genuine principles of all theories of organic nature*, insofar as he posited the ground of life in excitability. Brown was the first who had had enough sense or fortitude to propound that paradox of living phenomena, at all times understood, but never *articulated*. He was the first who understood that life consists neither in an absolute passivity nor in an absolute activity, that life is a product of a potency higher than the merely chemical, but without being supernatural, i.e. a phenomenon submitted to no natural laws or natural forces.

It is a personal duty for everyone who realizes it to say this publicly, although, on the other hand, one must openly concede that the principle that Brown placed at the apex of his system was discovered more through a lucky groping than deduced in a scientific way, not to mention actually constructed.

a) Brown did not deduce the concept of excitability (as has already been said). It is at any rate to be deduced *a priori*, i.e. without any mediation by experience, in the most rigorous way from the concept of an organic product, and so it must be. Every science an *a priori* principle.

b) By far the fewest of Brown's disciples have understood the scientific seeds which lie in his principles. There is one exception, Mr. Röschlaub, whose writings no one can leave unread if they

solution. It is extinguished as soon as it begins to become independent of external nature, i.e., unreceptive to external stimulus, and so life itself is only the bridge to death.§

3) The task was to explain how the individual holds its own in Nature against the universal.* We struck upon the solution that the individual only exists through the pressure of an external nature. Inner and outer, however, are only differentiated in an act of opposition; therefore, there must be a mutual opposition between the individual and its outer nature; i.e., if the former is, in relation to the latter, *organic*, the latter must be *anorganic* in relation to the former. Therefore: no organic nature, no anorganic. No anorganic, no organic.

But if organic and anorganic necessarily coexist, *then the functions of the organism cannot be deduced otherwise* than *precisely in opposition to that anorganic realm.*

Conversely too, if the functions of the organism are possible only under the condition of a determinate external world, the organism and its external world must again be *of a common origin*, i.e., they must be just *one* product. (That is, popularly expressed, there must be between both a "relative purposiveness." Now, to explain this relative purposiveness [133] through a divine understanding† that has fit the one to the other is the demise of all sound philosophy. For example, "how wise it is that oxygen is not present in pure form in the atmosphere, because otherwise the vital air would consume the animals as quickly as a flame." However, if the atmosphere were pure oxygen, then the organisms of the Earth would have to be correlatively otherwise constituted (i.e., receptive to a purer air) *quite necessarily* and from *the same* cause which made the atmosphere pure oxygen. The reciprocal coming together of organic and anorganic nature can only be explained, therefore, from the *common physical origin* of both, that both are originally only *one* product.)

---

have any sense at all for medicine as a *science*, principally his investigations in pathology, and particularly a few essays in his journal of medicine (where he explains himself far more clearly and precisely about many things).—I hear that it has been said here and there concerning these writings that they are too philosophical, too scientific. For me just the reverse is true. I would like to say rather that they are not scientific enough, and also that Mr. Röschlaub has not yet thoroughly recognized the genuine depth and force of the principles which he defends—at least in his investigations on pathology.

I cannot enter into how well these principles agree with the dynamic mode of explanation—certainly not with the chemical or even mechanical modes of explanation, with which Mr. Röschlaub still seeks to reconcile these principles (if he has not long since left those modes of explanation behind). This will be further developed below.

*or: to deduce the graded series in productivity

†as a third thing

Nevertheless, they are *opposed* to one another. However, as opposed, they cannot unify themselves in any other way than by being in opposition to a third higher term *common* to both. In the act of opposition inner differentiates itself from outer. The organism and its outer world together have to be, in relation to an outer, once again an *inner*, i.e., again an organic being. This is thinkable only in the following way.—The organic presupposes an external world, and indeed an external world that exercises a determinate, permanent activity upon the organic. Now, this activity of the external world could indeed be an *incited* one, and the fact that it is *permanent* is not even *explicable* otherwise than by a continual excitation.*—Thus, the anorganic external world again presupposes another external world in relation to which *it* would be an *inner*. Now, since the activity of the original organic being is aroused only by the antithetic activity of its external world, this is again itself sustained through an *external* activity (in relation to it). Together with the external world that it immediately opposes to itself the original organic being would then be again jointly *opposed* to a third, i.e., again *mutually* an *inner* in relation to a third outer.

[134] The original organic being is immediately conditioned through *its* anorganic outer world; this does not bring us any closer to a third. It would have to be shown that the *anorganic as such*, according to its *nature*, cannot exist without an outer that has influence upon it, and the mode of this influence itself would also have to be determined. This is the object of the following investigation.

---

*By the fact that it is itself (the external world) held together by some force, it would be in a compelled condition. In the external world, which the organic presupposes, nothing can be *accidental*. This necessity in all alterations of the external world, this restriction to a determinate sphere of alterations, alone makes the the existence of the organic possible. Every activity that is not restricted loses itself in the infinite. The activity of the external world is also, therefore, restricted.

[134]

SECOND DIVISION

The nature of the anorganic must* be determinable through its opposition to the nature of the organic.† Now, if we attribute to the anorganic the opposite of everything that we have ascribed to the organic, then we have the following determinations.

If in organic nature‡ only the species is fixed, then in the anorganic exactly the contary must occur, the individual is fixed.§ However, the individual is itself only determinable in opposition to the species; nothing truly individual will exist in anorganic nature. No reproduction of the species would occur through the individual.‖ The extremes will not meet in it, as in organic nature, but will flee from one another. On the one hand, the matter in it will dissipate into the absolutely indecomposable, and on the other, into the absolutely incomposable. But an immediate contact should be possible between anorganic and organic nature. There will be certain mediating substances in it, in which the indecomposable is joined with the incomposable (with the *absolute fluid*). These materials must be without any *figure*, however, for only the formless (the formable) can immediately flow into the organic (types of air, fluids in general). There will, therefore, be a multiplicity of materials# in it,** but between these materials a mere contiguity and exteriority to one another will obtain. In short: anorganic nature is merely *mass*.

However, because no mutual interpenetration (no intussuception) is possible between them, these materials must [135] still be held together through

---

*merely for the time being

†1) the position of the first section was totally hypothetical—for organic nature is not completely explained as an *antithesis*. 2) Now the anorganic is to be deduced—but how? Merely from the opposition to organic nature (we were only driven to anorganic nature as explanation of the organic). We will be able to completely construct neither organic nor anorganic nature before we have brought their construction back to a common expression. (Compare the Outline of the whole, p. 6f.)

‡1)

§e.g. minerals are here not species, but only individuals.

‖2)

#which approach the mean between both extremes—the indecomposable and the incomposable—more to the one, less to the other. Both extremes therefore separated, between these materials . . .

**generally

some *external cause* or another; this would not be possible unless an external cause sustained a mutual tendency toward intussusception in these materials (where it would always remain only a *tendency*) down to their tiniest parts. It would have to be an *external* cause, because in these materials there cannot be any *proper* (organic) tendency to mutual intussusception.

This anorganic bulk would then again itself be an *inner* in relation to the outer which sustains that tendency, therefore, an *organic being*,[*] i.e., an—if not *actu* still *potentia*—organic being that always *becomes* organized and never *is* organized (because it remains sheer tendency).

However, that which is an *outer* in relation to the organic is an *anorganic thing*. Therefore, that external cause would have to be again *anorganic*, i.e., itself again only a *mass*.

In that it is *mass*, i.e., a contiguous aggregate of external parts without actual connection, it requires once again another external cause which maintains, by its influence, the tendency to mutual combination in all of its parts, yet without ever coming into combination itself, and so on to infinity.

A fragment of mass should have an influence on another to infinity,[†] such that its parts all have a common tendency toward one another; this influence must extend to the smallest parts of matter, or its intensity must throughout be proportional to the *mass*.[‡]

Every influence is also necessarily *determinate* with respect to its intensity, or (because the degree of intensity of a cause is measured through the extension in which it acts) it can be effective only in a *determinate space* with a certain *degree*. This space can be imagined as large or as small as one likes, but it can reach a degree of expansion if it is expanded more and more, where the degree of effect of that influence would be an evanescent one.

[136] If, therefore, a mass is to influence a mass with a certain degree, then the spatial relationship of these masses to one another must be determined, i.e., they must be maintained in a certain proximity or distance to one another.

Now, to explain this spatial relationship two opposite systems can be thought.[§]

---

[*]which through that outer would be preserved in a compelled condition.

[†](this is a condition of the anorganic world)

[‡]With respect to this tendency all materials of the Earth, e.g. are only *one* (we are not speaking at all about cohesion here).

[§]It remains until now still undecided whether that cause through which the coherence of matter—and that through which the proximity of masses to one another—is effected, is one and the same cause. This will at any rate be shown, since it is indeed one gravity which gives all materials of the Earth the tendency toward one another and maintains the whole, at the same time, at a determinate distance from the Sun.

First System.

1) Those masses are driven toward one another by an external impulse. That which drives them cannot itself belong to the anorganic mass. It would have to be the *first* in Nature. One would thus consider empty space to be filled originally with the simplest elements that no natural force has enough power to break down further.—These ultimate elements are in primal motion, and indeed they move in all *directions*, but only in *straight* unalterable directions (one is driven to this assumption through the analogy with visible substances, in which there is an original motion, e.g., of light, of positive electrical matter, among others).[1]

Now, cast one of those masses into any point of space that you like, let it be spherical; clearly it is infinitely larger than any of the atomic elements. The flow of primary particles meets with it, the flow is blocked. Since an infinite quantity strikes against the mass, it will achieve a certain velocity—but the elements are moved in all directions; to every flow another is opposed. Thus struck equally strongly from opposing sides the mass will come to rest. However, let the other large mass be set into space, then both serve reciprocally as a shield against the stream of atoms, each meets only *one* stream, each from the opposite side as the other; they will [138] be driven toward one another, and so *gravitate* to one another. Now, suppose that every mass originally has a unique motion by virtue of which it would move forward in a straight line; then out of both movements (the original and the transmitted) a third emerges, and the masses will move to a certain distance from one another in curved lines.

—In the second system we will speak of metaphysical objections to this doctrine. Here we will speak only of physical objections.—

"The atoms meet the masses like hailstones, i.e., only on the surface. But their gravitation toward one another should be directly proportional to the *mass*."—If, however, every single atom of the mass had its element among the atoms of gravitational matter that meets it, and would have to meet it—if so, would the gravitational matter be met by the stream all the way to its uttermost parts? The *possibility* cannot be denied, since *visible* substances* permeate bodies down to their smallest elements, like heat-matter (among others), and the hardest substances are transparent to many materials, e.g., light. Moreover, we are not suggesting that any body whatsoever is transparent to gravitational matter;

---

*Lesage can say that the visible (originally electrical) substances have in common with the invisible substances (the cause of gravity) that they act in all dimensions in straight lines. The point from which they depart is the central point of a periphery that is greater or lesser according to the relation of the center's intensity. For both, the law holds that its effect decreases inversely to the square of the distance.

rather, that every atom of the body is *nontransparent* to the gravitational matter; it is thus less a postulate than an objection.—"But then every body would have to finally increase in mass, and so become heavier."—With what the blocked (by the impenetrability of the body) gravitational particles have an affinity we do not know; on the whole surface of the Earth, which is magnetic at every point, perhaps they have an affinity with magnetism. Perhaps they give an electrical constitution to all bodies, just as it might seem that the gravitational stream returns from the surface of the Sun as a stream of light. Supposing, finally—but finally for whom!—that the Earth, e.g., grows in mass; then every other mass grows in correllation.—"But the intensity of gravity!"—The quantity of movement is just as much the product of the velocity to the [139] mass, as of the mass to the velocity. The velocity of the streams, however, can be accepted as nearly infinite.— "But that law has limits, e.g., the light, as fast as it moves, has no moment of impact." The velocity of the action of light, which can be expressed in determinate numbers, is incommensurable with the velocity of the action of gravity (as is proved in every lever).

Moreover, if the most primal affinity of all corporeal elements were the affinity to that principle, and if all other affinities were merely *derived*—and the final cause of *gravity* were also the final cause of all *chemical affinity*—then the gravitational stream would meet every single atom of every single body, which is not the case with light. (This grand thought is actually present in Lesage's system. Indeed, he says in one passage: "Universal gravitation could not completely explain the appearances of affinities, one must well distinguish, therefore, the true chemical affinities which are not dependent on laws nor on the cause of universal gravity from the inauthentic so-called affinities that are only particular cases of the universal phenomenon of attraction or at least follow the same laws as the latter." However—it only follows therefrom that the cause of gravity is not *immediately* the cause of chemical affinities. Lesage seeks the *latter* in a secondary fluid, the ether and its agitations, which are impressed on it through the gravitational principle.)

"But whence does this inexhaustible stream come, from what era does it derive, and what supports it continually?" Here the physicist must be permitted to complain about the general ignorance with respect to final causes—and so this system ends with the inexplicable, while within its bounds it explains as well as any other system and deduces as evidently as any other system all phenomena and the laws of universal gravitation.

[140] *Second System.*

2) A material principle of gravitation does not exist; the principle of gravity is an immaterial one, a basic force of all matter.

Since this theory of the Newtonians (for Newton himself was undecided) can have no physical grounds for itself, it must be *metaphysical*, which has just recently been maintained.*

We have the following:

For the construction of every material are required forces originally opposed. Matter's filling of space can only be conceived as a repulsive force extending in all directions. However, if another force does not put a stop to this repulsion, then the matter would scatter itself into infinity, such that in every given space only an infinitely small quantum of matter would exist, or, because the repulsive force decreases in inverse relation to its expansion, only an infinitely small resistance would be encountered. That check, however, cannot occur through a repulsion coming in the opposite direction. For where there is direction—where there is a *whence* and a *whither*, there is already a restricted force. A second force, one specifically different from the first, must be accepted which acts in the *absolutely* opposite direction in relation to the repulsive force and which makes infinite expansion impossible—attractive force.

The attractive force is, therefore, a necessary one for all matter as such, by virtue of the sheer construction of its concept.

Since it first makes all matter possible as determinate occupation of space, and so also something palpable, it also contains the ground of contact itself. It must then precede contact, be independent of it, i.e., its action does not depend on contact; rather, it is action through empty space.

Since the repulsive force also acts beyond the surface of contact, it is a *penetrating* force.

The effect of attractive force at a distance can infinitely decrease to be sure, but it can never completely disappear. Its effect stretches to every part of matter throughout the whole cosmos to infinity.

[141] The universal effect of attractive force which it exercises on every part of matter to infinity is gravitation. The action of attractive force in a determinate direction is called gravity.

Universal gravitation is an original phenomenon, and the attraction of all matter within its field is *real*, not merely apparent, as if it were mediated by the

---

*If the basic principle of this system can still be defended, then it must be defended as the principle of the construction of matter in general, in short, by a proof snatched from metaphysics. Here one can proceed in two ways, either, like Kant in his *Metaphysical Foundations of Natural Science* by an analysis of the concept of matter (I will present this proof below), or one can give a synthetic proof from the original construction of matter from the antithetical activities that come together in intuition and are unified. This is found in the *System of Transcendental Idealism*. This proof does not belong in the philosophy of nature, which gives no such transcendental proofs—and what it cannot prove physically it does not prove at all.

impact of another material. If one assumes that this matter is not itself *heavy*, then no force will offer resistance to its repulsive force and it will scatter itself into all eternity. Since it is, nevertheless, only different in degree from the other material, it could become as heavy as any other material through gradual descent to deeper levels of matter, and conversely, the specifically heaviest matter could ultimately pass into that negatively gravitational matter, which is contradictory. Or, if it is supposed that it is heavy in itself, then one requires for the explanation of the possibility of such a thing, again, an original attractive force.

The following principal theses are contained in this system:

a) For the original construction of matter one requires original basic forces.—I assert, however, that one can succeed with this construction from original basic forces only in mechanics (in the broad sense of the word, i.e., insofar as one views matter merely as occupation of space in general). However, it does not make the *formation* of a *single* material conceivable because, in this case, one abstracts from all specific difference of matter and does not bring any other difference into view than the various degrees of its *density* (i.e., its occupation of space), as is also the case in Kant's metaphysics of nature. In this work, Kant departs from the product, such as it is given in sheer occupation of space. Since the product as such offers no other multiplicity than the various *degrees* of extension, the product naturally also cannot be constructed otherwise than out of two forces whose variable relation gives *various degrees of density*. [142] Mechanics knows no other specific difference of matter, whose construction, then, may be just fine for explaining why a material has greater specific gravity than another, but not for making conceivable the *productive power* in matter. Therefore, these principles are in practice a dead weight for natural science.

(Incidentally, Kant handled the concept of matter solely analytically in his Dynamics, and limited himself to making conceivable the possibility of a construction of matter out of those two forces; it even seems that he held this to be impossible, according to a good many remarks.)

Our philosophy follows precisely the opposite course. It knows nothing of the *product*, it does not even exist for it. First and foremost, it knows only of the purely *productive* in Nature.—(The corpuscular philosopher has an infinitely great advantage over the so-called dynamical philosopher in that he brings something primally individual into Nature through his atoms, each of which has an original figure; it can be objected that it is impossible for these atoms, since they are themselves already *products*, to be *first* or *ultimate* things in Nature. Therefore, the philosophy of nature posits in their place *simple actants*, i.e., the *ultimate* in Nature, which are purely *productive* without being *products*—(hopefully, one has at least learned to think an activity *without substrate* and *before all substrate* by means of the transcendental manner of thinking). This pure productive power exists only *ideally* in Nature, to be sure, since simple factors can never be arrived at in the infinite evolution of Nature, and everything is still a product to infinity.)

Now, in order to explain how the productions of Nature are originally directed toward the development of a *determinate product*—to explain how every original actant is productive in a determinate way, which is revealed outwardly through the determinacy of figure—something negative [143] must be accepted in that infinitely productive activity. If, seen from the highest standpoint, all productive activity of Nature were only an infinite evolution from one original involution, it must be this negative factor (no longer a *product*) that *inhibits** the evolution of Nature, hinders it from reaching the end; in short, as we have shown above (p. 17), it is an originally *retarding force*.

To explain this *retarding* force—or, that Nature in general *evolves* with *finite velocity*, and so shows determinate products (of determinate synthesis) everywhere, seems to be the highest problem of the philosophy of nature. Only at the lowest standpoint, that of the perception of the product as sheer occupation of space, is that retarding force able to appear as attractive force. Moreover, this principle only serves to explain the finite, the determinateness in productions of Nature generally, not to explain how one natural object is finite in relation to another, e.g., the way the Earth gravitates toward the Sun. The former problem, to explain the finite in the productions of Nature *generally*, is already a transcendental problem (where one descends from the idea of Nature as a whole to the individual in it). But the latter, through which the Earth, for example, gravitates toward the Sun, is a purely physical problem, where one ascends from the individual in Nature to the whole; this ascent is, however, an infinite one, such that one is never constrained to force one's way to the final term that makes Nature finite *at all*. This is because the problem is always a *determinate* one, namely, to specify how this determinate number of bodies have organized themselves into a common system, which certainly would not be possible without a final principle that *inhibits* the evolution of Nature generally, or that gives it a finite velocity.†

---

*retards

†This retarding principle is that which Kant calls "attractive force" in his construction of matter. Now it becomes evident from the deduction of this retarding principle that it only serves to explain how determination and bounds arise in the original and indeterminate productivity of Nature; it serves to explain why the evolution of Nature occurs with finite velocity—but it does not explain how it becomes absolutely fixed, which is actually the effect of gravity. That which Kant called attractive force, and which we call retarding force, is a completely *intransitive* force, a force that is applied merely to the construction of individual products—and exhausts itself in them. On the other hand, gravitation is a *transitive* force, i.e. a force with which the product acts *outside* of itself.

I have two objections to Kant's construction of matter: 1) that it holds only from the standpoint of mechanics, where matter is already given as a product; 2) that it is incomplete, since that

Here we come upon the second thesis of the system, namely:

b) that the attractive force which belongs to the construction of every finite material is the same as the one that operates even outside of its sphere to infinity. For if it is thought that since this degree of [144] attractive force is applied in order to restrict the repulsive force to this determinate part of space, then it will exhaust itself in this repulsive force* and will not exercise attractive effects on other materials outside of its sphere; a difficulty of the system that is irresolvable.

(All difference of degree would have to be supposed to lie only in the repulsive force, and to assume attractive force to be *equal* in *every point* of space, such that it would not be absolutely exhausted by any degree of repulsive force. This mode of representation cannot be made comprehensible, at least from Kant's Dynamics, about which we will say more later.)

Third Possible System.

3) If everywhere antitheses unite themselves into a third true system, it must also be possible here.

A material principle that effectuates gravitation through impact cannot be thought, because one has no category in natural science for such a principle (since it would have to be heavy and not-heavy at once†). For example, that an immaterial force draws the Earth to the Sun; again no comprehensible concept of this is possible.—(That is, we are far from denying that ultimately something like attractive force exists in Nature. However, we assert that every attraction in experience is a *determinate* and empirically *determinable* one.)‡

Of course there could be something material, empirically determinable, in the phenomenon of gravity, if gravity—(we do not speak at all here of the *ultimate* factor, that which holds Nature together, its most interior principle), if gravity (e.g., of the Earth to the Sun) were conditioned by the reciprocal *specific* constitution of materials of both masses.

---

which Kant signifies by "attractive force" is a quite different force than gravitation; while the former is applied entirely to the construction of the product, the latter operates above and beyond the product. The attractive force for Kant still remains what it has always been—an unproven, and to that extent chimerical, principle.

*it will be applied merely to the construction of the product

†not heavy, because it first produces all weight; heavy, because otherwise one cannot conceive how a certain direction could originally belong to this material at all.

‡But to explain an individual attraction in Nature we cannot immediately pass to the ultimate term which holds Nature together overall. We also would not call this *ultimate* factor "attractive force" because this designation already presupposes false concepts and really only names the *outward aspect* of the matter, but not the thing itself. According to our system, attraction will also be something merely apparent—it is just that we do not allow it to act through impact. There is no "attractive force" in the usual sense; but that there exists in Nature something *like* attractive force, we do not deny.

[145] If there were, at the same time, something immaterial in this phenomenon, one would require no particular gravitational principle for its explanation aside from the *universal* specific constitution; rather, it could be said that all materials of the Earth gravitated to the Sun merely by virtue of a principle *common* to them—but in opposition to the materials of other planets *of a specific* constitution; although, perhaps this constitution itself would be *maintained* only through a material influence of the Sun, whose influence would then only be *indirectly* the cause of gravity.*

It has been ascertained above that that which keeps together a mass as a mere aggregate of materials external to and contiguous with one another must be just such an influence of a mass outside of it which gives all parts a mutual

---

*We have a determinate problem: to specify how a given number of bodies could organize themselves into a whole; thus the solution must also be not universal, but determinate.

To be sure, there must be *one* force that reigns throughout the whole of Nature and by which Nature is preserved in its identity; a force that we have not yet deduced, but to which we find ourselves driven for the first time. However, this force may be capable of infinitely many modifications and may be as various as the conditions under which it operates. The force—because we still lack the common expression for it—remains always a hypothesis. Nevertheless, to admit such a force does not entail that the phenomenon of gravity has no empirical meaning. This one force might admittedly be something immaterial indeed, but the *conditions* under which it operates could be material or *empirical*. For example (as mentioned above), if the gravitation of the Earth toward the Sun were conditioned through the specific reciprocal constitution of materials of both masses, then the condition of that force would indeed be material, but it could still be immaterial itself; i.e. a force that works immediately in matter just as soon as its conditions are furnished, without intervention of a particular material principle.

The empirical condition of gravitation of the two masses would, therefore, be the specific difference of both. But then what is to be thought under that "difference"? What is the condition of gravitation? Surely, no one will deny that between the superior and the subaltern planetary bodies (e.g. the Sun and the Earth) there is and must be a chemical difference. By what is this chemical difference conditioned? Without doubt by a *higher difference*—we speak here of the *higher* difference by which even the chemical is conditioned.

However, there is no other difference than that in relation to a higher third term, wherein the opposed terms are again unified. This will also be the case here. There will be a difference between the higher and the subaltern product, but this is a *reciprocal* one, as is said in the text. Both are reciprocally opposed to *one another*, but also opposed equally in relation to a higher third term—their common synthesis.

This explanation being assumed, the question is raised 2) how we come to the idea at all, or what reason do we have for positing precisely *difference* as the condition of gravitation—a question which we really ought to deal with first.

I cannot invoke here a universal natural law which says that only the heterogeneous seeks itself and the homogeneous flees itself. We became familiar with this law only in one single case until now—in organic nature, and so cannot yet presuppose it as a universal law of nature; but another reason leads us to assume it. The construction of the phenomenon itself compels us to assume it. For what is weight? Is *weight* thinkable within an *absolute identity*? Or does weight already presuppose *diremption*?—Every body must, indeed, have the degree of its weight in *itself*—but the cause of its weight outside of itself. If we think of a body in empty space (or all of matter in a clump), then

tendency toward one another. This mutual tendency of all parts toward one another is not explicable other than by a *common* tendency *of all parts* toward unification in a *third* (because it always remains *tendency* after all, and never achieves unification), where their mutual tendency toward *one another* is only *apparent*, almost like the magnet gives an orderly position to the iron filings toward one another. This shared tendency to unification in a third is just the binding power which holds all the parts together. Now this third would have to be something necessarily *outside* of the mass, it would have to be, e.g., in the case of the Earth, the Sun.* (According to the usual notion it is understood this way too, that is, that one and the same cause makes the parts of the Earth heavy with respect both to one another, and to the Sun.)

So, we should say that the Sun influences the Earth such that a shared tendency emerges in all the parts of the latter toward all parts of the Sun. How such a tendency is itself possible would then be a new problem whose solution can, for the time being, be indefinitely deferred. The Sun's production, by its influence, of such a shared tendency in all parts of the Earth must be explained exactly as the shared tendency of all parts of the Earth toward *one another* would be explained; namely, through the influence of a third mass on the Sun, in relation to which, consequently, the Sun [146] together with the Earth (and its remaining satellites) amount to only one mass and are only held together among themselves by the shared tendency toward unification in a third. In the same way, all of the different substances of the Earth count as only *one* mass in relation to the Sun, through which the infinite attraction becomes apparent, since it is really always only the shared tendency to unification with a higher

---

it is not *heavy*. A body, therefore, is only heavy insofar as it has a cause *outside* of itself which *makes* it heavy. Weight already presupposes an original *exteriority*. The *condition* of weight is a juxtaposition. How should this juxtaposition be explained? It cannot again be explained from the system of gravity, for it is indeed the *condition* of all gravitation. We are driven here to an *original* exteriority, which contains the ground of that derived one. This original exteriority, which is the condition of the *mechanical* exteriority of bodies, can now be of a sheerly *dynamical* kind, i.e. it must be an *original difference*. For there is a dynamical exteriority only where there is an original diremption.

The question whether we will ever be able to investigate this original difference at all, or whether we can investigate only the difference that is the condition of gravitation, e.g. between the Sun and the Earth, remains completely out of our purview. It is enough that it has been deduced from the construction of the phenomenon itself that difference is its *condition* and is, namely, the original difference through which even all mechanical exteriority is conditioned and first brought forth. In order to apply this general proposition to the individual case and to make it clear: through, e.g. the original difference between Sun and Earth, an action of the Sun upon the Earth will be conditioned by which the Earth is compelled to *fall* toward the Sun—unless a force opposed to this falling constantly hinders it.

*Thus it is shown that the cause through which an anorganic whole is held together (although it is a mere contiguity and exteriority of the parts to one another) and the cause through which one anorganic whole maintains relation to another whole, is one and the same cause.

term which holds materials together among themselves, and, whether they exist only *nearby* and *outside* of one another or not, organizes them into *one* whole.*

Here we can conveniently abstract from whatever the final cause of this tendency (proceeding to infinity) of all materials toward one another may be. We only have to make out the following. The action which sustains that shared constitution must be capable of propagation. For example, if the mass A influences B, then in order that A and C indirectly gravitate toward one another, the influence of A on C must be able to be propagated through B. Further, it is inexplicable that *all* materials of the Earth have the tendency toward all parts of the Sun down to their final parts (i.e., to infinity) unless a *shared* constitution is to be accepted in all of them, with respect to which all of their remaining specific difference disappears and which is itself only a *specific* one in opposition to the materials of other planetary bodies. In the way that the parts of the Earth in relation to the Sun are related to one another, the parts of the Earth and the Sun are again related to one another in relation to a higher third, i.e., the parts of the Earth and the Sun must once more have a *common* constitution in relation to this higher term or belong to a common *sphere of affinity*,† and so on to infinity.‡

---

*We can now state the following two principles: 1) if an anorganic whole should gravitate toward another, then all the parts of that whole can be *mutually* as different as can be in relation to *one another*—however, in relation to the *higher* term toward which they gravitate they must be *one*. The ground of their shared gravitation must lie in something common to all of them together (e.g. there are specific differences of substances on the Earth, but the gravity toward the Sun is the same for all of them). 2) If *two* anorganic wholes should gravitate *in common* toward a *third* term, then again they must have something *in common* in relation to that higher third, something common in relation to which their reciprocal difference entirely disappears. Opposed *among one another* they are *equal* in relation to the third.

†I find this expression used already by Mr. Hofrath Lichtenberg. This excellent scientist brings to our attention in his newest *Novelties from Heaven* that the effect of light on our Earth and its atmosphere is already a proof of our submersion in a sphere of affinity and shell of the Sun which has nothing to do with universal gravity.—Well, if the gravity of the Earth toward the Sun were already itself an effect of it, 1) would all parts of the Earth belong to the higher sphere of affinity of the Sun, and 2) would both the Earth and the Sun commonly belong to a still higher sphere of affinity?—(Original note.—Trans.)

‡Now, how should we designate this common element? This common element *is* just that which manifests itself as gravity, and we have no other expression for it. We could call that common element a common constitution—but then what is a *constitution*? We have not yet constructed a *constitution* of matter at all. Nor do we know at all what the ground of specific difference is. For the *dynamical atoms* through which we explained the qualities were merely *ideal* grounds of explanation. We could say it this way: the Earth and the Sun belong to a common higher "sphere of affinity"—but then what is *affinity*? We know just as little about that as we know about what a specific constitution of matter is.

We will at any rate use this expression, but not in order to explain anything by it or to anticipate an explanation, but just in order to be able to express ourselves at all.

[147] (Real *chemical* affinity should not at all be thought by this term (ultimately, chemical affinity and that higher affinity might surely have a common root); we are speaking here only of an affinity that has *contiguity* and *exteriority* as a consequence; for the problem was just how a plethora of matter, mere *coexistence* notwithstanding, could form into a unity.)

We might explain the fact that all parts of the Earth have *one* common constitution to infinity* by saying that they were all together *of a common origin*, i.e., were precipitated from one and the same original synthesis, as it were. Moreover, we would have to make clear that the materials of the Earth have again *one* constitution in common with those of the Sun in the same way; that is, the Sun with all of its satellites is a *common precipitate out of one higher compound*, and so on to infinity.

(Or, it might be thought that all planetary bodies are only the fragments of *one* infinite mass, and the various materials on them are only fragments of this one mass to which they belong.—Since I mention this image only in passing, I can also invoke the unexplained fact that the mere *contact* of two different bodies communicates to them always (or for a long time at least) a *common* constitution, as one metal communicates to another in galvanism, and still more conspicuously, as the infinitely fertile magnet communicates to the iron, where, so to speak, a contagion is in play, which the ancients notably called the *daimonic*, because it operates like a spell.)—

Generally, if the gravitation of two masses toward one another lies in a principle *common* to them both, then this common principle must extend to infinity (as far as mechanistic division goes), because otherwise the proportion of the masses and of gravitation remains unexplained. One cannot doubt on the grounds of experience that in an infinite heap of matter a common [148] constitution of all parts to infinity *is at all possible* (for that it is *necessary*, might be proven *a priori*), for the magnet demonstrates polarity to infinity, e.g., as in the newly discovered magnetic serpentine stone. It cannot be denied that the magnetism of our globe penetrates down to the smallest particle.†—(If one erects an

---

*one determination in common

†The condition or cause of gravity is an *empirical* one. However, the ground of gravity must be a *common* ground of all materials which belong to *one* whole, and this common principle must extend to infinity. Now is such an empirical constitution *common* to all matter of the Earth, present in every individual to infinity, at all thinkable?—The impenetrablility, the divisibility of matter certainly goes to infinity—but these are not empirical, they are rather *transcendental* properties—but gravity should be an empirical property. Can it be thought that such an empirical quality of all matter of the Earth to infinity is a *common* one? For example, let that empirical property have its ground in an opposition which extends in matter to infinity; can such an opposition be thought that is still the same in the smallest parts of matter? Nothing from experience speaks *against* such a possibility. The magnetism of the Earth, e.g. probably rests on an original opposition. Now, this opposition apparently extends to infinity, for the Earth is magnetic to infinity.

iron rod in our hemisphere, perpendicular to the Earth, and leaves it for a while in this position, it receives the south pole on the end turned to the Earth, the north pole on the opposite end. The reverse will occur in the southern hemisphere.*)—And yet, we would know nothing of magnetism unless two individual substances stepped out of this universal sphere of magnetism and formed a particular magnetism among themselves (why they do so is uninvestigated).†

Now, since magnetism is distinguished from the universal force of attraction in all systems of physics and is accepted as an *empirical* and empirically *determinable* constitution of matter, can there not likewise be a still higher cause, and for that reason still far removed from the *universal* force of attraction, i.e., still an *empirical* determination of all matter of our Earth, which extends to every atom—a higher cause of its gravitation toward the Sun?

It has already been remarked elsewhere (*On the World-Soul*, p. 178) that the magnetism of the Earth, excited by the influence of the Sun, is the single glimmer of hope for also understanding the gravitation of the Earth toward the Sun as a material phenomenon. It was not because I believed that the cause of magnetism is *identical* with the cause of gravity (although it is very natural to suppose a connection between them), but because I recognized something analogous therein, a proper determination *of all* matter of our Earth to infinity, but still an empirical one.

It is also quite conceivable that because that empirical constitution of matter, which is the cause of gravity, goes to infinity, no body exists [149] to which one could first *communicate* this constitution (according to the assumption). It is necessarily the case that in searching for a constitution with experiments nothing can ever be discovered about the cause of gravity of our Earth toward the Sun or the parts of the Earth toward themselves by the empirical path.

Nevertheless, it could most likely be proven that gravity has empirical conditions *generally*, e.g., in our planetary system, since we are already familiar

---

*not only that. A simple perpendicular state gives polarity instantly to an unmagnetic iron rod. What an abyss of forces we gaze into here.

†*Universal* magnetism is independent of special magnetism; for the latter is first produced by and is efficacious by means of the former.—

If, in physics, one wants to explain the phenomenon of magnetic attraction immediately by a *universal* and abstract force of attraction, then every physicist would doubtless say that such an explanation is *no* explanation. This is because one sees that this phenomenon has *empirical conditions*, that it follows, e.g. only under the presupposition of an existing opposition.—The explanation of the phenomenon of gravity from such a universal force of attraction is tolerable in physics due to the fact that the empirical conditions are seen less here, although at least traces of them can be shown in the heavens.

Magnetism generally will be viewed as a phenomenon that has its *empirical* ground in matter. Likewise, magnetism is just as universal as gravity—for the Earth is, as has been remarked above, infinitely magnetic.

with universal phenomena that indicate such empirical conditions of attraction, as for example, the fact that all orbiting planets always turn the same side to their central planet.* By gazing into the innermost construction of the heavens Herschel was brought to the thought that very *manifold* primary forces (not only *one* force) gave the universe its order.† —Even if the difference of the world regions, e.g., of the south and north, stops being merely a *mathematical* difference and one gradually comes to the idea that a *physical* cause acting universally throughout the whole solar system instituted it,‡ then why should attraction not finally also pass from a merely mathematical into a *physical* phenomenon?§

---

*Because we cannot search for this ground or this empirical condition of gravity in the matter of the Earth in an empirical way, it does not follow that we cannot *at all* prove that gravity has empirical conditions in our planetary system. Our assumption, as is well-known, is this: the force of gravity is *one,* but its conditions are diverse, and as manifold as is the universe itself. There is not just one *force* of gravity, rather there are only *forces of gravity* in the universe, e.g. our Earth can only gravitate immediately toward the Sun and not toward a higher planet, and so on. Aside from the fact that this assumption is perhaps *a priori* demonstrable, it can even be proven from *actual* phenomena that not one force of gravity, but very different forces of gravity rule in the universe, or that the one force of gravity acts under very different conditions—as stated; where, for example, the orbiting planets always turn the same side to their central planet: a proposition that is proven by almost all indications. One cannot explain this phenomenon from an *abstract* basic force inhabiting matter *as* matter; however, this phenomenon illustrates something determinate and will, if pursued further, provide great insights concerning the origin of the moons, their dignity, and the role that they play in the universe.

†Until now the phenomenon of attraction has been handled only as a mathematical problem. However, even quite a few mathematical differences have a physical ground.

‡One of Franklin's ideas,[2] to which the phenomenon of magnetism had at first probably brought him. An idea which now (according to a new notice) obtains great confirmation not only through the great differences of the two hemispheres of our Earth, but also in the moon and two other planets. (Original note.—Trans.)

§Yet another question is to be answered: if the condition of gravity is an opposition, *ex hypothesi*, then this opposition must again be raised into a higher synthesis. Then our Sun and whole planetary system would again be *one* in relation to the higher system—the common synthesis; and to that extent, the condition will again be something common to all materials of the Earth and the Sun.

Now, how should this common term be explained? How can the fact be explained that, in all substances of the Earth, the condition of gravity is the same? One could imagine that all of them together spring from one and the same original synthesis. We can say that all bodies of our planetary system were all together precipitated out of one common synthesis, out of *one* higher composite, in the same way that the condition of gravitation toward a higher system is the same for the materials of our whole planetary system.—However, all of this is mere supposition, and nothing whatsoever can be asserted about it when such an assertion cannot be proven from the *history* of the *formation of the world* itself. We find ourselves led through the phenomenon of gravity, which we otherwise cannot completely explain, to the investigation concerning the World System.

The origin of gravity ought to be for the time being investigated historically, i.e., in the history of universal world formation. Now, here one has complete freedom to accept (as does Kant) the original condition of Nature as a universal dissolution of world substance into a cloudlike shape. By this means, one can accept the universe as in a certain way preformed, for in part, an infinitely manifold diversity is presupposed in the original elements, and in part, the densest elements are placed at determinate distances from one another (e.g., in the solar distance of the present system), [150] in order (as seeds) to insert matter into the first stirrings of universal affinity and to be able to concentrate it into central bodies. Nevertheless, it turns out no better for this system of the mechanical origin of the world than for the ancient Epicurean system with the *clinamen* of the atoms; for it can satisfactorily explain neither the *beginning* of centrifugal motion nor its regularity; why, for example, all planets have taken one and the same direction. Kant's way of thinking about this is the following.[3] First, the vertical motion of the particles falling toward the central point generally gets an obliquely deflected motion from the repulsive forces of matter, which alone bring an enduring life to Nature through their conflict with the forces of attraction. Through these forces of repulsion that are expressed, e.g., in the elasticity of gases, and so forth, the sinking elements are steered obliquely from the straight-line motion mutually by one another and the vertical fall strikes out into circular motions, which circumscribe the central point of the universal depression.*—Nevertheless, it is readily noticed that that regularity which is noted in the centrifugal motion of the planets does not at all dwell in these repulsive forces, and *opposed* oblique motions should have formed by virtue of its effect rather than movements in *one* determinate direction (e.g., from evening to morning). Now it might be thought that if vortices have formed around the midpoint of the depression at greater or lesser distances, in which every particle describes for itself a curved line, these particles *could* limit themselves by their movements among one another until they all proceeded in one direction; but it is here that chance is all-too-often brought in,† in that that equality of direction, at least in our solar system (the movements of comets

---

*This is the general form of Kant's explanation of centrifugal motion, through which the motion and formation of masses is explained at the same time. Since the elements are steered to the side, they cannot fall into the central point of attraction. So for circular motion generally. However, since they are differently limited in their motions until they proceed in one direction, the motion of the elements is impressed also upon the masses that are formed from them—and thus these proceed in the same direction according to which they proceeded in the motions of their reciprocally limiting elements.—(Thus the final cause is only attractive force.)

†It can always be asked why the elements have limited themselves reciprocally precisely in this and no other direction.

excepted), presupposes a much more determinate and more powerful cause, which has been impressed upon them by this movement.

This is aside from the fact that nothing at all is to be established with mechanical explanations of the origin of the world when Nature [151] must be viewed as *product* to infinity; in which case its formation can only be of an organic kind throughout.* Since we find ourselves here in the realm of mere possibilities, we would like to present our thoughts in this regard as mere possibilities also, as long as we are joining our possibilities to actualities and are thus able to find our bearings upon this wide sea of opinions.

The question arises whether the origin of the world system ought to be thought more *organically* than mechanically, through an alternation of expansion and contraction, as happens with all organic formation. One could suppose that through one contraction the first beginning of formation happens, departing from *one* point, at once stretching through an immeasurably large part of space wherein the primal material of the world was prepared; that moreover, together with this universal *appropriation* which that *one point* exercises on the totality of matter, spread out in an infinite space, an opposite effect comes to pass; namely, that it thrusts heterogeneously constituted material from its sphere of formation, and that in such a way the universal process of formation began at many points simultaneously. Since no appropriation is at all possible without separation and both are really *one* operation in every organic formation, then one could imagine that that one point, in the relation by virtue of which it forms itself through appropriation, thrust whole masses away simultaneously with a violence that one can assume proportional with the first, still youthful and untried forces of Nature. Now, between the original and the expelled masses a *common* affinity must have taken place (because otherwise they never would have been able to contract toward a single point), but the original masses form a *narrower* sphere of affinity (in that they expell a part of their matter). If this is the case, however, then that formation of always narrower spheres of affinity has to proceed infinitely; and is not this process of organization that proceeds to infinity the origin of the whole world system?

[152] In order to follow this idea further, let us consider the first mass forming itself as the most original *product*, as a product that can splinter into new products to infinity, which is at any rate the property of every natural prod-

---

*If nature had just formed itself mechanically (and this is at bottom the case according to Kant's explanation), then it would not be so much a *product* as mere mechanical *aggregation* from already existent factors. If the world is merely mechanically *aggregated*, then, e.g. all specific difference must already be *presupposed*. If, however, the world arose not mechanically, through aggregation, but through organic development from one original *synthesis*, then, e.g. all qualitative difference in the universe itself is already a product of the *universal organism*.

uct.*—(We could also allow all matter spread throughout cosmic space to pass first through this mass (like fire, as it were), so that the parts might acquire the shared constitution which later will be the cause of the universal tendency of all materials toward one another, although this hypothesis is not required.)—As the first product of Nature (according to the laws of all synthesis) that original mass will first of all divide into opposed factors, which are necessarily *themselves products* once more. In this way, three original masses will incipiently form the first projection of the universe, still present only in germ, but it is also only three masses that are able to form among themselves a *system* of gravitation; for if we posit two original masses that are equal to one another, they will reciprocally get closer to one another and pass into one mass (supposing that no centrifugal motion is impressed upon them which is also not yet deduced); or, if we suppose the two to be unequal, then the one will draw the other into its sphere and both will once more disappear together into one mass.† If we suppose, on

---

*In general, the state of contraction and expansion is the state of productivity *passing into product*. That alternation does not only happen in organic nature, it happens also outside of organic nature—in the elementary phenomena, for example; but, as I have proven on another occasion, the elementary phenomena are not appearances of one product, but appearances of *productivity itself*, and are actually phenomena of restricted productivity. The original state of Nature was, according to the common conception, actually a state of pure productivity; it was that state where all products were still invisible and dissolute in the universal productivity. If this productivity is to pass into the product, then it must be duplicitous in itself, and here we find ourselves driven back again to our first postulate, to an original diremption as condition of all construction of matter. The deeper meaning in Kant's construction of matter out of two opposing forces is just this, that the condition of all formation is an original duplicity.

Assuming this diremption, an alternation of attraction and repulsion was conditioned by the opposition. The point from which the formation began was determined by the original opposition itself. In that alternation of attraction and repulsion, Nature really seeks only to return out of difference, which is contrary to it, into indifference. That point will then be the original point of indifference. The first product will fall within this original point of indifference. This product is, necessarily, as the first product wherein the whole of Nature contracts itself, an *absolute synthesis*—a product that can splinter into new products infinitely.—(If one asks, by what that infinite *splintering* of the product into always new products is produced, then this can certainly not be explained otherwise than by assuming that the opposition which should cancel itself in the product is infinite. If the opposition were infinite, then it would indeed cancel itself in a finite product by virtue of the unconditioned striving of Nature to return into its identity—but it will cancel itself only *in part*—the opposition will arise always anew, and thus the first product and every subsequent product will divide itself *ad infinitum* into opposed products again.

An antithesis will necessarily arise again in the first product, for it is formed as homogeneous; the absolute opposition is canceled only in part.

†Two products by themselves would form no system. It is a necessary feature of any system that both a direct and indirect reciprocal action exist simultaneously. Each individual member of the whole acts in part immediately upon every other, in part indirectly through all the rest. Therefore, the simplest system must consist of at least three products, and we can expect in advance that the total system of gravitation and every single system of gravitation in this universal one will be reduced to *three* original products.

the contrary, three original masses, A, B, and C, where the one, A, is equal to the *sum* of the remaining two masses (according to the most probable calculations, the like actually happened in our solar system), then in such a system an *equilibrium* becomes possible; but in addition, that simultaneously indirect and direct reciprocal action that belongs to every closed system will be possible in it. While, for example, the effect of A upon B is disturbed by C, once more the effect of C upon A is disturbed by B, and in the same indivisible instant the effect of B upon C is again disturbed by A, where that circulation begins anew from the beginning, without one being able to say where it began or where it ends. (*Ideas for a Philosophy of Nature*, p. 185[4])

[153] (It can certainly be said that if Nature originates not through aggregation but through evolution, and if its components everywhere first spring from the product, then throughout the whole of Nature such a universal sundering of every unity into opposing factors must take place.—In galvanism that necessary triplicity is even now proposed as a law.)

The first masses in relation had to institute an antagonism of equilibrium as they formed, i.e., they had to separate themselves into their opposed factors, and retain only the *shared* principle of both. But was it any different with these two masses? (Let them be designated B and C.) Each of these factors is itself a product, each must again divide into opposed factors. If the factors of B are designated by a and b, then a and b were reciprocally opposed in relation to the lower sphere of formation which they enclosed, but were *equal* to one another in relation to their common principle that lay in B as in a higher sphere. It is also just the same with B and C. Both are reciprocally opposed to one another, but equal to each other in relation to the higher A, their common synthesis. But where will this splintering into opposing factors finally stop?[*]—And so we would know for the time being to what extent all of the matter of *one* system had a *common* constitution. That is, any two products of the same sphere of formation are *opposed to one another*, but *equal* in relation to the *higher* sphere of formation from which they are descended.[†] The common principle[‡] is neither in the one nor in the other (for they are opposed to each other), but clearly in *both together*, i.e.,[§] contained in their common synthesis[||]—(their sun, e.g., into which they will some day return)—. Therefore, through a completely necessary syllogism (that is, because we are able to think the universe under no other con-

---

[*]Nowhere, for the opposition is an infinite one, only to be canceled in an infinite synthesis.

[†]and this is the common principle that belongs to them and which is the ground of their gravity.

[‡]of both

[§]it is

[||]and for this reason their gravitation is a common one.

dition than as organized and organically produced) we have also deduced whence the *universal duplicity* in Nature comes; namely, it *came* into [154] Nature *through* universal gravitation (but gravitation is not its cause), and this is one of those actualities to which we can conjoin our first possibilities and on the basis of which we can reason so earnestly.

We assume then, that the universe brought itself forth from one mass, conceived in formation, to a system of three original masses, and from these produced itself by an infinitely progressive organization (or formation of always narrower spheres of affinity), by means of an always advancing explosion. Now if every body thrust from the central mass again became a central body that had to divide itself into opposed products, according to its nature and necessarily, then every system in the universe must be reducible to *three* original masses. That the system numbers more bodies, in the solar system infinitely more, must be explained from the unequal force by which the explosion occurs; this proposition has a generally valid ground for itself (i.e., *analogy*), even if it only receives confirmation through the observation of our solar system.*

If one assumes that the bodies furthest from the central point were exploded by the *first* force of the Sun, then the three furthest planets of our solar system are apparently from a *common* explosion, but Mars, whose displacement from Jupiter is so relatively great, is from the second, less forceful explosion.— However, the distance between Jupiter and Mars is not made up merely by the space in between both, but by a still more conspicuous difference.† The eccentricity of the motions must obviously decrease in inverse relation to distance from the Sun, because in relation to the greater distance the centrifugal motion impressed on a body through the explosion must always become weaker. The only exceptions are Mars and Mercury. The motion of Mars is more eccentric by far than that of Jupiter. According to the assumption, however, both are also from different explosions. The same force that acted on Mars is apparently not the same force that acted on Jupiter, but rather the one which has impressed the Earth and Venus with their centrifugal motion; therefore, [155] its centrifugal motion must also be weaker than that of the much closer Earth and Venus,

---

*We suppose that the universe brought itself forth from a central point outward, first to a system of gravitation of three masses, the simplest that is possible, and from there through a disintegration of every product into new products to infinity. Then, e.g. all suns would descend from a primal sun, and the planets which course around the sun would be offspring of the suns.—Here the question arises how one would have to think that mechanism of disintegration—or the mechanism of forces that have cooperated in that disintegration; in which circumstance it is to be foreseen that the forces that have acted with that disintegration will also be the forces that have impressed the planetary bodies with their motions (and that we here come nearer to the solution of our principal problem).

†namely, through the different eccentricity of their orbits

such that among the three farthest planets the first (counted from the Sun outward) has the least eccentricity and the third the relatively largest.—Finally Mercury, which among them *all* has the greatest eccentricity, is without doubt from the final force of the Sun. (Although one must also take into account that the density of its mass and the great proximity of its centripetal force to the Sun must give it a great imbalance; for that its eccentricity is more a result of the imbalance of the latter than of the weakness of the former becomes evident from the *velocity* of its vibratory motion).—Yet another analogy contests, however, that the three planets of our solar system are from a *common* explosion; for when one compares the three most distant with the *remaining* nearer to the Sun, they are obviously superior in *mass* to them, but if one compares them *among one another*, then Jupiter is, e.g., superior to Saturn, where one cannot see a reason why this should be so, other than that all three have been exploded from one and the same force, where naturally the greater part of the mass had to underlie the centripetal force rather than the smaller. (To say with Kant that "Mars is smaller because the more powerful Jupiter withdrew too much matter from its sphere of formation," is obviously to provide a circular explanation. For "Jupiter is superior to Mars through its force of attraction," means precisely the same as "Mars is lesser in mass than Jupiter," which is just what one wanted to explain.[5]) The same striking analogy is illustrated with the three closer planets, for among them Venus, closer to the Sun, has more mass than the Earth, the Earth more than Mars; why should this be so, unless one and the same force had thrown them out of the Sun? And finally Mercury (the last explosion) has the least mass; if two planets were *visible* closer to the Sun than it is, then the first among them would again have the greatest mass.*—

---

*We said that the universe brought itself forth from one original product by means of an always advancing *explosion*. I urge the reader not to think of mechanical forces when I use this expression, which begin to operate much later in Nature. The forces which acted in this explosion are without doubt the original repulsive forces in Nature.

I cannot yet prove that which will be proven subsequently, i.e., that the cause which brought the first opposition into the universal identity of Nature—the first condition of all motion into the universal rest—is none other than the cause of *magnetism*. I assume also, then, that the first movements of that opposition were magnetic movements, and assume that even the structure of individual planetary bodies and, moreover, of our whole planetary system, leads us to this idea.

I have just recently presented the idea of Franklin (p. 84) that the differentiation of world regions is probably not merely a mathematical one, but is instituted through a *universally operative physical cause*. This physical cause can be none other than magnetism. It can be proven that magnetism was already cooperative in the first formation of our Earth from the regularity of its structure, which is still apparent enough in spite of the great catastrophies of time. Another great confirmation of the cooperation of magnetism in the formation of planetary bodies is the great differences of both hemispheres, not only on the Earth, but also on the moon and in other planets.

It is an extremely striking phenomenon that on the Earth the closer toward the north pole one comes, the more compacted the masses become, so to speak; the closer to the south pole, the more

[156] Since this view of the origin of the world seems sufficiently confirmed by the foregoing, I ought to linger momentarily in order to show how even more analogies agree with it. Such are, for example, the analogical difference in the densities of the planets; since apparently as to *time*, the less dense masses must be from the *first* explosion, and therefore are the most removed from the central body—(comets); how, further, the same materials must be (on account of the lesser effect of centripetal force on them) redirected at the last moment into the elliptical motion; and how the densities of the planetary bodies must decrease universally in inverse relation with the distance from the Sun.* Only two observations accord with our aim.

First, the origin of the centrifugal motion needs to be deduced particularly in view of its direction, neither from an immediate divine action with *Newton*, nor, with *Kant*, must it be left to chance; rather, it can be deduced from a cause dwelling in the central mass itself, which doubtlessly extends much further.

Second, it should be shown how the constant *organic metamorphosis* of the universe becomes explicable on this theory, since the universe really only endures in a continual alternation of expansion and contraction (for what is our duration compared to the periods that one solar system needs for its condensation?).†

Until now, we have brought only the formation of one system into view. We began the formation at one point of space and let it extend, admittedly, to an indeterminate size, but not infinitely far. This presupposition does not hinder us from assuming that such formations always occur from a common point outward, and that in this way the universe is conceived in infinite becoming (because a completed infinity is a contradiction). According to the

---

fragmented, as it were, since down to this pole the Earth is merely an *island*. This phenomenon is striking when one realizes that the same phenomenon is apparent in every single magnet (and the Earth is nothing other than a great magnet). In every single magnet the attractive forces of the north pole are superior by far to those of the opposite pole (approximately as in the prismatic spectrum the colors of the one pole are brighter and more forceful than those of the other).

Am I mistaken, or can this analogy even be transferred to our whole planetary system? Magnetism operated throughout our whole solar system—and determined all poles, and without doubt even the motion of the planetary bodies around their axes. The forces through which the planetary bodies were impressed with their centrifugal motion cannot be derivative or subordinate forces, but must belong to the original repulsive forces of Nature. We need not be in confusion on account of the cause which, e.g. thrust the planets from their centrality. We also cannot accept that the effect is disproportionate to the youthful, still untried forces of Nature conceived in their first development.

*Our investigation also cannot be extended to the formation of the moon and other such objects. This whole theory will obtain its full presentation elsewhere. (Original note.—Trans.)

†Yet it is also to be assumed that many primitive or independent formations must be supposed to exist in the universe, such that not all planetary bodies ultimately descended from one *primal product*, for the reason given in the immediately following passages.

laws of analogy we have to suppose that between those points strewn throughout infinite space at immeasurable distances from one another, where the first impulse to new formation happens (perhaps [157] by means of an infinitely quickening stimulus (like the electrical) through space), a reciprocal relationship will again appear to infinity, namely, a relation through gravitation. This already becomes conceivable (if a common cause of the first motion is also not to be assumed) because that central mass of new systems all form themselves through condensation out of a substance conceived in *common* solution, and simultaneously mutually exclude each other while they form themselves.—To accept a common central point of the whole universe from which *all* formation has departed would be to make the universe finite.*
If the world, however, *is* not infinite (but only *becomes*), and one assumes that *one* action (the first cause of universal motion) is propagated from one initial point outward toward all points that are capable of an independent formation, and so on to infinity, then at least that first point will be the central point of the incipient creation. Nevertheless, the original, *independent* formations will together have only an *ideal* center, because every single individual formed itself *independently*, i.e., through *its own* formation, and in the ratio that those formations progress, that center too (falling in empty space) will always be shifted to a new point.†

Meanwhile, if we turn our gaze back to‡ *one* independent system (i.e., to *one* whole of systems which have all formed themselves from one pulsating point outward), then we will be able to view the individual systems that belong to it in three different conditions at once: 1) a few in the condition of greatest expansion, where the centrifugal motion impressed on them still keeps equilibrium to the centripetal tendency without loss; 2) while others are already in a median condition of contraction; and finally, 3) others are in the condition of

---

*But this reason is *no* reason, since we can nudge back this central point infinitely (the point at which the absolutely first product of Nature exists, out of which all others have evolved). Incidentally, it is natural that our explanation can never have recourse to this first point of origin of formation, i.e. that there is no such thing for us at all. Just as our empirical consciousness is restricted to one part of the universe, all of our explanations can be related only to this part. The highest point to which our explanation can elevate itself is our *solar system*—system of planets. What holds for our system of planets holds also for the solar system, and if the latter is only an offspring of the sun, the former is also only the offspring of a central body.

†I will only make one remark, and it is that this theory of the origin of the world is at the same time a guiding thread for the whole *history* of the universe, for the history of its genesis and its gradual corruption. In addition, the existence of the universe will be a continual metamorphosis—the universe will consist only in an oscillation of expansion and contraction—it is just that *our* timespan has absolutely no relation to the period which just one solar system needs for its condensation.

‡the universe as to

highest contraction, near their collapse.*—Now, if it is asked in what relation these various conditions stand to their distance from the central point, [158] it is readily seen that the contraction must occur the fastest nearest to the midpoint; for example, between those places in the heavens where the stars seem pressed together toward one point closest to their center (perhaps, to the shared central point of all suns—for I will prove below that all worlds whose continuity with us is sustained by *light* belong to *one* system). In contrast, there are those places where the spaces between the stars are emptier at the farthest from the central point and systems of median expansion must exist in the middle between both, although the reversion of the system nearest the central point into its origin would bring the ruin of the others after it with accelerated velocity.†

If we suppose such a universal reversion of each system into its center, then according to the same law by which this system organized itself into one system at its first formation, each system will, revitalized, proceed again from its ruins; and so we have deduced at once the eternal metamorphosis running throughout the whole universe and the continuous *return of Nature into itself* (which is its genuine character).‡

From the foregoing it can be effortlessly and completely derived that and why anorganic nature *must* organize itself into systems of bodies which are constrained through the combination of opposing motions to describe regular orbits

---

*From these different conditions the various *forms* and *shapes* of the star systems can be explained, to which Herschel has principally drawn our attention. For example, the form of the Milky Way is apparently completely different than the many star clusters which have a shapely spherical look and which display an increasing densification and an increasingly intense light directed toward *one* point. We should view these clusters as systems that are already in the state of contraction and near their collapse.

†It is evident from this, that we must also think the *duration* of the universe as an organic one. The duration of a system is nothing other than an oscillation of expansion and contraction—an eternal metamorphosis.

‡I now state the results from our previous development. We began with the investigation into the essence of gravity. We assumed that gravity had empirical conditions, and that not just one force of gravity reigned throughout the whole universe. The origin of those empirical conditions was to be investigated in the history of the origin of the world. Now we found here that the organization of the universe into systems of gravitation has no other ground than the infinity of the opposition which is to cancel itself in the universe—for every original product must fragment again into products to infinity, where the higher product is necessarily the synthesis of the subordinate ones. This assumption (namely, that the universe is nothing other than the development from one original synthesis) was proven from the construction of our system of planets, for it can be proven from the mere design of this system that it formed itself outward from the Sun as the central point.

around shared central points.* But we can conveniently spare ourselves from this elaboration in order to bring more important conclusions at once into view.

Conclusions.†

A.

a) The tendency that is produced in all parts of the Earth by the influence of the Sun is a tendency toward reciprocal *intussusception*.—(The *product* of this universal tendency must be something *common* to all parts of the Earth.—Before the matter is specially investigated this can be imagined as [159] *universal magnetism*, which would already itself be a *product* of, not *cause* of that universal tendency.)—But the action of gravity brings forth the mere *tendency*, it does not go beyond the tendency.—Now, if one assumes as certain based on experience that intussusception is *actual*, the *possibility* of which we have at least postulated above (p. 24), then, at any rate, the action of gravity will be the first impulse to all intussusception—(and so the cause of gravity is the ultimate which *ensouls* the whole of Nature, as Lichtenberg already surmised)—but if actual intussusception is to be achieved, then yet another *particular action*, *different* from the influence of gravity (but standing in connection with it) must be adduced.

b) PROBLEM: to discover this action.

Solution.

α) Intussusception exists only in the chemical process. Now it is certain *a priori* that what is a *principle* of the chemical process in a *determinate* sphere, it is well to note, cannot again be a *product* of the chemical process of the same

---

*from which comes that opposition extending to infinity which, according to our presupposition, is itself the condition of gravity, and further, why this opposition is a particular one for every product, and why gravity is also of a characteristic type for every product.

I close this investigation with the reminder that all of its propositions have a merely hypothetical truth until the universal expression for the construction of a product in general is found; a problem whose solution we approach only gradually, and through whose solution everything that we have assumed until now is only then either confirmed or denied—but must be justified in any case.

†As I have shown in passing the organization of the universe into systems of gravitation is not merely a mechanical one, but simultaneously a dynamical organization. The state of an enduring activity in Nature is provided by that organization of the universe. There is an original opposition which cancels itself in every product through gravity, which proceeds in the product to infinity— and is met with in the smallest as in the largest parts.—This opposition must be thought as arising again in every moment, and becomes for that reason ground of an enduring activity in Nature. We will thus gradually deduce the whole dynamical organization of the universe from that organization of the universe which is produced through the original repulsion and force of gravitation in it—and this will now be undertaken.

sphere (although it is itself again, without any doubt, a chemical product in a higher sphere of affinity). The principle of all chemical processes that take place between substances of the Earth cannot again, therefore, be a product of the Earth. A *single* principle must come forth among the principles of affinity that is *opposed* to all others, and by which the chemical process of the Earth is *limited*. This principle must be the mediating factor of all chemical affinities. All other materials must be chemically related precisely by the fact that they strive in common toward combination with this *one*.—This principle is that which we call *oxygen*, as is evident from experience (*Ideas*[6]). Oxygen cannot be a chemical product from the sphere of affinity of the Earth.

[160] Ordinarily, one refers to oxygen as an ultimate principle, and the chemical explanation that once reaches it (is reduced to it) has the right to remain silent.—But what is this oxygen itself? No one has thought about this question at all, and in this way has simply restricted the domain of the investigation. It becomes evident that one is justified in raising this question from the preceding. Oxygen is no longer a product of the Earth. Be that as it may; but in a higher sphere it must again step into the series of products. Oxygen is *for us* irreducible, and only insofar as it is so *can* it be the mediating factor of all chemical affinites of the Earth and *limit* the chemical process of the Earth.* However, in a higher sphere it has an irreducible element to which it is itself reducible.—(Now do we see how indecomposable substances can exist in Nature without simple ones existing? See above.[7] This is not the place to explain its generality. We restrict ourselves here to consideration of this *one* principle.)— Oxygen is *by this means* opposed to all other substances of the Earth; that is, all others combust *with* oxygen, while it burns with no other substance. However, it has already been noted elsewhere that the concept of combustion is a merely relative concept, from which it follows that in a higher sphere oxygen, or an element of it, *must* itself (if it is already a *combusted* substance) descend again into the category of combustible, i.e., chemically composite substances.† It cannot

---

*just for that reason no longer a product of the Earth

†A body is combustible to us when it gives off *light* through disintegration by oxygen. But if we now think that beyond oxygen there is yet another substance that would stand in combination with light, then oxygen itself would indeed descend to the category of combustible substances.

Oxygen is the principle of combustion because no higher material stands above it, because it constitutes the boundary of our sphere of affinity—because opposed spheres of affinity contact one another in it.

Or, to be completely clear, if we think about an ideal extreme of combustion, then that material which is the *most combustible* in a given system, itself no longer *flammable*, will necessarily be the one through which all others combust. One can then view oxygen in relation to a *higher system* as the most combustible of all substances. In relation to the lower system it is necessary that precisely the most combustible substance must be the incombustible, because it has no other substance with which it could burn. Oxygen is therefore principle of combustion just because it constitutes the boundary of our sphere of affinity.

be objected that oxygen is a chemical product of the Earth because we can liberate it from a plethora of substances. We are speaking of an *original* production of oxygen itself.* The *existence* of oxygen in many substances of the Earth is rather proof of our theory that the Earth is a product of the Sun, by virtue of which a wholly novel view emerges of the specific difference of the materials of our Earth. All variety is reducible to the notion of that which is *combusted*; some are conceived in reduction—(the phenomenon of this reduction is vegetation; at the lowest stage the vegetation of metals which are maintained by the inner [161] glow of the Earth, at a higher stage the vegetation of plants)—others in permanent combustion—(the phenomenon of this permanent process of combustion is animal life).† It also follows necessarily from this that no substance on the Earth can come to light which *was* not either combusted or *would be* combusted, or was not *combustible*.

β) Supposing this, the following conclusions result.—Oxygen has the *positive* role in all chemical processes of the Earth.‡ But oxygen is a principle alien to the Earth, a product of the Sun. The positive action in every chemical process must thus proceed from the Sun, must be an influx of the Sun. Therefore, aside from the action of gravity which the Sun exercises on the Earth, another *chemical influence* of the Sun on the Earth is postulated. Some phenomenon must be demonstrated in experience, however, through which that chemical action of the Sun on the Earth is represented: this phenomenon, I assert, is *light*.

THEOREM: *The phenomenon of chemical action of the Sun on the Earth is light.*§

PROOF. We can succeed in this only through some intermediary theses.

1) For the time being, it must be asserted that if no *chance* is at all permitted to exist in Nature, then the luminous state of the Sun cannot be *acci-*

---

*Oxygen is only a simple principle in relation to the Earth.

†If oxygen is the one fixed point beyond which the chemical process cannot go, then it will be principle of all determination of quality. The division of matter into burnt and combustible or burning and such as are conceived in reduction—is a completely true division.

‡That which one can symbolically name "phlogiston" is conceivable only as the negation of oxygen.

§Light—i.e. what *we* call light—is merely a *phenomenon*, is not itself matter. I could prove this proposition to which the course of our investigation has led us on other grounds as well. I will say only this *here*: light is not at all a *becoming* matter—conceived in development—it is rather *becoming, productivity* itself that propagates itself in light, the immediate symbol of universal productivity, as it were. We have deduced productivity as the ground of all continuity in Nature. But light is the symbol of all continuity. Light is the *most continuous magnitude* that exists, and it is the most impoverished mode of thinking that treats light as a *discrete* fluid.

It follows from what has been said that I can just as little treat light as a merely mechanical phenomenon—as if the phenomenon of a vibrating medium (like Euler)—it is a wholly dynamic phenomenon.—It is remarkable to see how a few chemists who talk a lot nowadays about dynamical physics believe that one has given a dynamical explanation when one holds light rays to be *vibrations* of the ether. This mode of explanation is as little dynamic as the one that treats light as a discrete fluid.

dental to it; rather, it must be the source of light as *necessarily* as it is the central point of gravity in our system. Accordingly, all explanations will be excluded which permit the state of the Sun to depend upon something accidental or even merely hypothetical.

(For example, we cannot understand light as heat-stuff of higher intensity, and allow that the suns achieve the luminous state because (as the greatest masses of every system, and in gradual transition into a solid state) they have freed the most elastic matter by precipitation out of the common solute.—Nor can it be held that a fire rages in the suns, for on this basis it can neither be shown how the fires *must* arise on all suns, [162] nor by what means they are maintained.*—The hypothesis of light as an *atmospherical* development of the sun would only be saved from contingency by the fact that one ascribes to the sun a pure oxygen atmosphere with a high degree of elasticity and views the suns generally as the most original abode of oxygen. The latter may indeed be proven for the sun of our system, but not for suns in general.)

With the abandonment of all hypotheses I therefore propose the following proposition: *If the positive action in all chemical processes is an action of the Sun, then the Sun, in opposition to the Earth, is generally in a* POSITIVE *state.* The same thing will hold for *all* suns; namely, that they are necessarily *positive* in opposition to their subalterns.

By virtue of their positive state, the suns must exercise a positive (chemical) influence on their subalterns, *and the phenomenon of this positive influence* (not the *influence itself*), I assert, *is light*. (I could add to this that light that streams out in *straight lines* is *generally* a sign of a *positive* condition. For the time being, I can only prove this proposition through the analogy of positively charged light.—According to this, the suns would be *positive* points (for us) strewn throughout cosmic space, their light perhaps +E; so-called daylight, which one cannot even make conceivable through an accidental dissemination of sunlight in all directions, and through which dark planetary bodies are visible too—the dark bodies, *points* like the luminous bodies, appear as −E.) I assert only *on the whole* that *light* is the phenomenon of a positive state in general.

---

*The most natural explanation indeed seems to be that the suns are burning bodies. The grand image of a burning planetary body that becomes the source of life for a system of subordinate bodies while it wrestles with the destroyer can impress the imagination indeed, but not the understanding. But there is no necessity in this explanation. Indeed, Kant made an attempt, but it is not satisfying by far. Instead of the many objections that can be made to this hypothesis, I suggest only *one*: It is a very natural illusion (but a great one nevertheless) to believe that because the development of light is bound up with combustion in the chemical process of the Earth, this is also the case in the chemical process of the Sun. That the chemical process of the Earth is bound up therewith, has its ground in the luminous state of the Sun—so this cannot again be explained from a process of combustion.

Now all suns are *necessarily* in a positive luminous state, as the principles of all chemical affinity, and in opposition to their subalterns thus also necessarily in an *original* luminous state. (It is not necessarily the case that all stars' relations to their subalterns is *the same* as that of the Sun to the Earth. The universal principle of affinities in various systems must also be different. Perhaps oxygen is even a principle of affinity only for the Earth and the planets resulting from a common explosion with it. The mediating factor is thus variable, but not the positive [163] relation of the suns themselves.)—Further, by light, obviously, only *positive* light is to be understood (as, e.g., with Hunter's lightning experiment, only lightning that is positively charged is positive for the eye).

The light of the stars is *positive* only in opposition to *our negative* state. But the suns themselves are again subalterns of a *higher* system, so their light is *negative* in relation to the higher, *positive* influx which they themselves reflect in a luminous state.—It is just this that makes an infinite organization of the universe possible; that which is *negative* in relation to a higher becomes *positive* again in relation to a lower, and conversely. Light itself is *originally* a phenomenon of a *negative* condition, which presupposes a higher positive one as cause. A new world is opened up by this means, one to which only reasoning reaches, but not intuition. It is light which limits our intuition absolutely; what lies *beyond* light and the luminous world is for our senses a sealed book and buried in eternal darkness. The chemical action through which the Sun itself is again illuminated is only indirectly knowable for us.

(The assertion just presented should not be confused with an issue of Lambert's, which questioned whether the central body of our system would have to be a dark body. The chief reason that he provided for this is that a self-illuminating body of such distinctive mass would have to become visible prior to all others. I suggest, however, that not only is the central body of our system dark *for us*, but also a whole universe beyond our system, and that all self-illuminating bodies belong only to *one* system and are altogether products of a common formation.)

These concepts granted, now I can demonstrate the following proposition in experience: *light is the phenomenon of a chemical action of the Sun on the Earth.*

The proof can be established most concisely when it is shown that certain experiences are capable of being deduced from that proposition *a priori*.

a) If oxygen has the positive role in all chemical processes, then bodies which are related to *oxygen negatively* must also be related to the *luminous power* of the Sun *negatively*.

[164] (The body which relates itself *simply positively* to the luminous power of the Sun must be absolutely *canceled* for the sense of sight, as if swept away out of the series of things, because only the negative relationship to that action gives it existence at all for this sense. Nevertheless, no phlogistical body

is absolutely transparent, and conversely, every truly transparent body is related *positively* to oxygen.)

b) If light is a phenomenon of a positive action of the Sun, and active in every chemical process, then light must come forth where a transition from the absolutely negative into the absolutely positive state occurs.

(All phlogistical bodies are related negatively to oxygen. Every true process of combustion is, therefore, such a transition. *Absolute opposition* is part of the true process of combustion, i.e., the body must be *absolutely* uncombusted (e.g., nitrogen, soils, sulfur-alkali, etc. are not); further, only oxygen itself is related to phlogistical bodies *absolutely* positively, but not to an acid, where it is combined with a combustible substance.

CONCLUSION. There are only luminous phenomena where there is an absolute opposition.—Therefore, light that appears with combustion is neither a component of oxygenated air, nor of the body, but a *direct* product of the *chemical influence of the Sun* permeating everything and never at rest.—*Thus the* SUN, *or rather its* LIGHT, *comes forth everywhere only* WHERE *a positive state is produced.* The action of the Sun extends to every point of space, and the Sun is everywhere there is an illumining process.)

c) If the luminous power of the Sun acts *positively* in the chemical process, then bodies that combine with oxygen must cease to relate negatively to the luminous power of the Sun.*

(The maximum of opacity is luster, the reflecting of light off the surface in a *straight* line; a minus of opacity is its reflecting in all directions, which only happens when the body begins to iridesce. But the colors *intensify* as the positive [165] condition of the body increases. The minimum of opacity, i.e., is relative *transparency* = the (relatively) highest degree of oxydation. The most opaque body is no sooner dissolved in acids than the luminous power also begins to permeate it. The same occurs when it is dry-combusted.)

RESULT: *The action whose phenomenon is light acts positively in the chemical process*. Many effects that have been ascribed to light actually belong to the influence whose phenomenon it is.† The fact that the greatest and most noble part of our planetary body is disposed toward light-processes does not indicate

---

*Originally, all bodies of the Earth relate negatively to oxygen—so also toward the action of light. However, a body that is submitted to the chemical process stops relating negatively to that principle, thus also stops relating negatively to light, if light is that which we take it to be. Actually every body becomes transparent in the proportion to which it is permeated with oxygen.—Thus light must also be that which we take it to be—a phenomenon of the chemical action of the Sun. Only the minor premise is to be proven, as follows (see the text).

†E.g. action of light on the organic body is not *light* itself, i.e. that which we call light, rather the action whose phenomenon it is.

something accidental, but rather a universal, higher and more encompassing *law of Nature*. The action of light must stand in mysterious connection with the action of gravity which the central bodies exercise.* The former will give the things of the world the dynamic tendency, the latter the static. But this can be proven *a priori* from the possibility of a dynamic (chemical) process in general, for no chemical process is at all constructible without a cause that acts chemically but which itself is not submitted to the chemical process (we will treat of this when the time is right).

B.

a) *If all materials of the Earth relate positively or negatively to that chemical action, then they will also relate this way* RECIPROCALLY AMONG ONE-ANOTHER.

b) *Two specifically different bodies will relate positively and negatively to one another reciprocally, and their qualitative difference can be expressed through this positive or negative mutual relation.*†

(With this we have deduced that there is something like *electricity* in Nature.‡ Expressed empirically, the proposition runs: all qualitative difference of bodies may be expressed through the opposite electricities which they adopt in reciprocal conflict.§)

c) But the positive and negative relation of bodies generally is determined by their opposite relationship to oxygen. *Thus*∥ *the negative and positive relationship of bodies* AMONG ONE-ANOTHER IS DETERMINED *through their opposite relation to oxygen.*

[166] *Remark.*

The principle first proposed by the author# that the electrical relation of bodies is generally determined through their chemical relation to oxygen

---

*Namely, through the action of gravity indifference is always canceled again—the condition of gravity reestablished again. But in light we see nothing other than this reestablishment of the opposition; thus it is already clear here in what connection the chemical action might stand with the action of gravity.

†Perhaps more clearly: their reciprocal positive and negative relation will be the most original appearance of their qualitative difference, or: the qualitative difference of the bodies will be = to the difference of positive and negative state into which they reciprocally position themselves.

‡It is the single phenomenon in Nature that shows us such a positive and negative mutual relation, into which two different bodies position themselves.

§and the degree of their qualitative difference will be = to the degree of electrical opposition which they show in reciprocal conflict

∥*the difference of their electricity or*

#in the *Ideas for a Philosophy of Nature*. (AA I,7 165; and *Ideas*, 113.—Trans.)

remains true, although the conclusions drawn from it must be let go. That is, it is not because electricity itself is a *product of oxygen* (for which one can now no longer provide the electrical phenomena of light as a ground, since (p. 99) the source of light cannot at all be sought in oxygenated air), but *because oxygen is overall the determinant of quality in the chemical process of the Earth*, the electricity of bodies is determined through their relation to oxygen.

The proposition must be proposed as a principle of all theorizing about the electrical process, *that in the* ELECTRICAL *process that body which is positive* adopts *the function* that *oxygen had in the* PROCESS OF COMBUSTION.* If the body is only positive, insofar as it adopts the function of oxygen, i.e., insofar as oxygen is positive in relation to it, then conversely oxygen is *positive* in relation to it only insofar as *it* is *negative* in relation to this principle. Therefore, the positive body must relate *negatively* to oxygen (*outside* of the electrical conflict), i.e., be an uncombusted substance.—We can consider the following cases.

Either one posits two substances absolutely negative toward oxygen (i.e., absolutely uncombusted) in electrical conflict, except that they are heterogeneous and the one has more affinity for oxygen than the other; then according to the law proposed, the first must wholly necessarily become positively electrified.

(This case is really a totally pure case, because here the relationship of both bodies to oxygen is the same (namely negative), and they are opposed only *within* this relationship.† The question arises, by what means one recognizes the *absolutely uncombusted* body that relates to oxygen absolutely negatively. Electricity itself provides this characteristic feature. A body that is a perfect conductor of electricity becomes, *as soon as it is combusted,* [167] an *isolator* of electricity. It must then be concluded that all bodies which isolate electricity are combusted, as little as it may be tolerated, incidentally, by the conventional chemical divisions, although it is beyond doubt the case with many bodies (like the resins, oils, soils, etc.). We need not repeat, as it has been elsewhere, that the concepts of combustion, of oxidation and deoxidation are overall extremely relative concepts.

The only‡ exception to the law that all combusted bodies isolate (electricity)§ is water and all acids in a fluid state; but since they lose all conductivity at once in the liquid state, a still unexplained connection between conductivity and

---

*by directly engaging in it, while in the electrical process it only indirectly engages in it.

†i.e. we find a greater or lesser quantity of combustibility or of the body's negative relation to oxygen.

‡actual

§for it is no exception to that law that in chemistry many nonconductive bodies can be counted among the uncombusted bodies.

the fluid state is to be assumed. We can restrict the law proposed above, at least with respect to *solid* bodies, to bodies that are *conductors* of electricity (so that of two electrical conductors the one takes over the function of oxygen* which has most affinity with it.†)‡

[168] Or, if one posits two bodies in electrical conflict where *one* is a combusted substance which has a lesser affinity to oxygen, the other an *absolutely* uncombusted substance which then has greater affinity to oxygen, then the *latter* will take over the function of oxygen and be constantly positive. (E.g., any metal with any acid, soil, etc.)

Or finally, if two bodies are set in conflict which are *both* combusted substances, then here the law will reverse itself, the *more combusted* substance (which to that extent has a lesser affinity to oxygen) will take over the role of oxygen, i.e., be *positive* (e.g., the white band with the black; overacidic with common hydrochloric acid[8]).—In the electrical relation of isolating substances there will be a means to judge the degree of their oxydation, such that the one which is most constantly positive must also be the most oxidized.—Whether one must subsume glass under this law, so far as it is made of silica (which is perhaps the most combusted of all substances), or whether it is the same case as with, e.g., liver of sulfur, is uncertain (since sulfur is still the most persistently negative).

---

*that one that is positive

†This follows directly from the principle already ascertained, that in the electrical process that body is + which takes over the function of oxygen, and just this law (namely that the body is + which has greater affinity with oxygen) is confirmed through galvanism, where, e.g. the body which has the greatest affinity with oxygen excites the most forceful spasms. *Volta* found that by the mere contact of *two* such bodies which act in the electrical process, electricity can be produced that always has + electricity related to oxygen, the other − electricity.

‡The law that, of two bodies, the one which has the greatest affinity to oxygen becomes *negatively* electrified, was abstracted merely from the isolators of electricity. Mr. Ritter, who has followed farthest (of all who have noted it) the opposite relation of bodies in galvanism (determined by its opposite relation to oxygen), discovered for electrical conductors precisely the opposite law.—(The following laws proceed as results of Mr. Ritter's galvanic experiments. *Fluids*, which contain *oxidizable* components, e.g. alkaline salt and liver of sulfur solutions, are positive in galvanism with solid oxidizable bodies, which *are simultaneously conductors of electricity*, e.g. all metals. *Fluids*, which are already *oxidized*, like water and others, are negative with the same solid bodies. If these solid bodies are set in conflict with one another, every time the one with greater affinity with oxygen becomes positive, the one with lesser affinity becomes negatively electrified.)—Now, since the law which the conductors follow is reversed when only one body is an isolator, it is natural to conclude that the ground of this reversal must fall within the sphere of difference between conductor and isolator themselves. The illusion resolves itself when one assumes *all* isolators as such are substances which are *combusted*, not indeed absolutely, but still *relatively*, in relation to the bodies that are conductors of electricity. (Original note.—Trans.)

d) *How is the electrical process distinguished from the genuine—(chemical)—process of combustion?*

The sole difference is a consequence of the preceding, that is, that in the electrical process, the body which has most affinity to oxygen takes over the role which oxygen itself plays in the process of combustion, such that to this extent the electrical process is mediated by the chemical.

Also conversely, the process of combustion is mediated by the electrical process. *The conditions of all processes of combustion are even the same as those of the electrical process.* No body combusts *directly* or *solely* with oxygen; similarly, none becomes electrical solely or directly with oxygen. A third body is present in every combustion which takes over the function of oxygen and [169] through whose mediation oxygen is first destroyed—(This is water, in the conventional process of combustion, according to new discoveries. Incidentally, one needs only to think about the formation of alkalis through the combustion of vegetable bodies in order to be led to such a duplicity, or rather triplicity, in the process of combustion).* The electrical process is not different in principle from the process of combustion. The possibility of both is conditioned through the same ultimate principle. The simplest electrical process begins with the conflict of two bodies, A and B, which touch or rub one another, and *both* are in themselves negative (in relation to oxygen), but A, as representative of the latter,[†] becomes positive in this conflict. However, there must be a maximum of the positive state for every body. As soon as this maximum is achieved, the body must pass over into the minimum, according to the universal law of equilibrium. But the maximum is reached when the body is driven into a luminous state[‡] (see above, p. 99).[§] *Therefore,* the appearance of light is simultaneous with combustion (not just because light is a component of oxygenated air); i.e., with the transition from the maximum of the positive state into the minimum. As soon as the body is combusted (oxidized), it stops relating *negatively* to oxygen,

---

*(According to SW the last sentence is struck out in the first edition, and the following inserted.—Trans.) Every process of combustion begins with such a *merely indirect* invasion of oxygen into the process. The process of combustion is also begun through a body that comes to light first as a representative of oxygen. And so it is pretty certain that every chemical process comes to pass not through a single, but through a double elective affinity.

[†]of oxygen

[‡]i.e. combusted

[§]It is evident, to mention it in passing, that the light which appears in the process of combustion is of an electrical sort and the cause is to be ascribed to that which sustains universal electricity. The body, when it is driven into a luminous state, is, as it were, completely dissolved into positive electricity. Indeed, it is always the more combustible which becomes + —and when we consider that the same opposition which goes into the original construction of matter is shown in electricity, what then are all bodies at bottom? Nothing other than electricity.

but this negative relation is the condition of all positive function in the electrical process, so it passes immediately from the positive function over into the opposite (signaled by the isolating property and increased heat-capacity, both of which are really only *one* property). In the same way that the electrical process is the beginning of the process of combustion, the process of combustion (the ideal of all chemical process) is the end of the electrical process.

Now, however, if oxygen itself is again only the representative of a higher principle (just as the positive body in the electrical process is only a representative of oxygen), then an absolute disappearance of all dualism, i.e., a *chemical* process, will be necessary; if oxygen itself is set in direct conflict with the body, an immediate contact of the lower and higher spheres of affinity (to which that principle belongs), a transition from the one into the other [170] will occur. Oxygen will disappear as *middle-term* in the process and that higher substance* will emerge.

Further, it becomes evident that the constitution of the body by virtue of which it is capable of being heated is one and the same as that by virtue of which it is capable of electricity (for the maximum of heating up passes over immediately, like the maximum of electricity, into the process of combustion, where the heat-isolating and electricity-isolating properties enter simultaneously).

C.

Yet another question must be answered: *how is the action of gravity related to the chemical action of the Sun upon the Earth?*—We can determine two points of their mutual relationship.

The *first* is that the condition *of both* is a difference, but that the heterogeneity which is condition of the action of gravity is of a higher sort, and that the one which is the condition of chemical action is without a doubt determined through that higher heterogeneity. By virtue of the preceding we are not in a position to ascertain more precisely the relationship of these heterogeneities.

The *second* is that the action which the Sun, as cause of gravity, exercises upon the Earth, is determined by a higher action that is exercised upon the Sun, so it is not an action peculiar to the Sun; however, the action by virtue of which it is cause of the chemical process of the Earth *is* solely determined through the peculiar nature of the Sun.†

---

*that higher principle.

†(According to the SW the second paragraph is crossed out with the following remark.—Trans.) doubtful

THIRD DIVISION

The previous course of our investigation was the following[1]:

"Nature is organic in its most original products, but the functions of the organism cannot be deduced otherwise than in opposition to an anorganic world. *Excitability* must be posited as the essence of the organism, by virtue of which alone [171] the organic activity is really hindered from exhausting itself in its product that, therefore, never *is*, but always only *becomes*."

"If the essence of everything organic consists in excitability, then the *agitating causes* must be sought *outside* of it in a world opposed to the organic, i.e., an *inorganic* world. The possibility of an inorganic world in general and the conditions of this possibility must be deduced."

"Moreover, if* the organism in general is possible only under the condition of an anorganic world, then *all grounds of explanation of the organism must already lie in inorganic Nature*. This nature is opposed to the organic. So how could the grounds of the organic lie in it?—It cannot be explained except by a *preestablished harmony between both*.—In other words: inorganic Nature must presuppose for its existence and endurance a higher order of things once again, there must[†] *be a* THIRD *which binds organic and inorganic Nature together again, a medium that sustains the* CONTINUITY *between both*."

Organic and inorganic nature must reciprocally explain and determine one another. (It is evident why all explanations which have been given of this or that *individual* must generally be incomplete by their very nature—the present system's as well—and why explaining the whole of Nature in *one* swoop, as it were (as must be done), is possible only through *a reciprocal determination of the organic and inorganic*, to the mutual determination of which our meditation now advances).[2]

---

*the productive product or

†be, as it were, a common natural soul through which organic and inorganic nature are set in motion, there must

I.
On the Concept of Excitability

We have posited *excitability* as the first property of the organism without being able to explain this property itself more clearly at the time. The only thing which we could do was to dissociate it into its opposed factors, organic receptivity and organic activity. It is now time to trace that property back to actual natural causes by the deduction of that which belongs to anorganic [172] nature generally, through which the organic must be determined (which we are now in a position to do).

(It has been shown that the formation of a universal system of gravitation is the essence of the anorganic, with whose gravitation the gradations of differences in quality run parallel (for such a system signifies nothing other than a universal organization of matter in always narrower spheres of affinity). Further, it has been shown that the specific forces of attraction are conditioned through an original difference in the world-substance; finally, it has been shown that a chemical action must operate upon every heavenly body, aside from the action of gravity, which proceeds from the same source as the latter and whose phenomenon is *light*, and that this action effects the phenomenon of *electricity*, and where electricity disappears, the chemical process toward which it genuinely tends (as cancellation of all dualism)).—

1) The essence of the organism consists in *excitability*. This means that *the organism is its own object*. (Only insofar as it is at once subject and object for itself can the organism be the most original thing in Nature, for we have determined Nature precisely as a causality that has itself for object.*)

The organism constructs itself. But it constructs *itself* (as object) only under duress from an outer world.† If the external world could determine the organism‡ as *subject* then it would cease to be *excitable*. Only the organism as object is determinable through external influences, the organism as subject must be *unreachable* by them.

---

*that produces itself from itself. Organic nature differentiates itself from the dead precisely in that *it* takes *itself* as object. The dead is never object *for itself*, but for an *other*. For example, this occurs with collision, or indeed even with chemical operations, where two bodies certainly reciprocally become objects for themselves—but here we have already posited *two* bodies. The problem is that there should be duplicity in *one and the same undivided* individual, it should not be object for any other, but solely for itself.—The organism is such a whole that constructs itself (a double view of the organism—organism as subject and object).—This identity in duplicity is the one which Brown expressed by the word "excitability," but without making it clear.

†but which maintains that duplicity, and makes the sinking back into identity or indifference impossible

‡immediately

(The excitability of the organism presents itself in the external world as a constant *self-reproduction*. The organic distinguishes itself from the dead simply in that the existence of the first is not an actual *being* but rather a continual *being-reproduced* (through itself),* and that this continual being reproduced is an indirect effect of external, impinging influences, since conversely the dead (unexcitable) [173] cannot be determined to self-reproduction by impinging external influences but is destroyed by them.)

2) If (as need not be proven) organic activity really belongs to the organism only as subject, but organic activity is excitable only through external influences, then the organism as subject cannot be unreachable by external influences as we nevertheless assumed.—This contradiction cannot be resolved in any other way than this: the higher organism—(let this expression be permitted in place of the less understandable "organism as subject")—is not affected *directly* through the external influences.† In short, *the organism* (taken as a whole) *must* ITSELF *be the medium through which external influences act upon it.*‡

3) "The organism should itself be the medium, etc.," means§ (expressed more generally) nothing other than: *there must be an original duplicity in the organism itself.*

The organism is everything that it is only in opposition to its outer world. "There must be in the organism an original duplicity" means, therefore—it follows necessarily—precisely that *the organism must have a dual external world.*

4) I ask, however, how is it possible that the organism belongs to two worlds at once? It is possible in no other way, I answer, *than if every anorganic world is itself really a* DUAL *world*. But is this not so, according to what we have deduced as condition of possibility of an anorganic world? In every anorganic world a *higher order* is mirrored, a *higher world*. Where these opposing orders contact one another THERE *activity exists*.‖

---

*Viewed from the highest standpoint the existence of dead Nature too is surely a constant reproduction. However, the dead object does not exist through itself, but through the whole of Nature. Dead nature is unchangable. But the organism always perishes and always arises again. Every organic individual in every moment changes and is yet always the same.

†rather, as has already been deduced earlier, only *indirectly*

‡This will be made explicable through the galvanic phenomena. The irritable system is only the armor of the sensible, as it were—the chains in which it is bound.

§we continue to reason

‖Every anorganic world is really only the mirror which reflects a higher world to us. For this reason, as soon as the link is dissolved through which the one world is bound to the other, the higher world comes forth—like light in the process of combustion. All activity in Nature takes place only on the border of two worlds (as we have already seen). As long as this boundary remains, activity is present; if it is canceled—and this happens precisely in the *chemical* process—then the condition of all activity is canceled as well.—That boundary can never be canceled in the organism as long as it is an organism (for I have already proven that the organic product cannot perish as *organic*).

5) The resources for the answer to the question have now been found.* The answer is the following:

If the organism is to be excitable (its own object, which presents itself externally as continual self-reproduction, opposed to the externally impinging influences), then something in the organism must be unreachable by the influences of its external world, or, as we have more closely determined it, something—a part of the organism exists (if we may be permitted to [174] express ourselves in this way) which is not at all *directly* receptive to the influences of its immediate external world. The unreachable part would have to possess a cruder organism (the latter an "organism of the organism"—it would be the one that is continually reproduced through the stimulation of the higher)—and only by virtue of this lower organism must the higher be connected with its external world. In a word: *the organism in appearance must divide into opposed systems*, a higher and a lower.† Only by means of the latter would the higher remain contiguous with its outer world.

6) How could the higher be removed from the influences of this external world otherwise than precisely by the *influences of a higher world*? Now, just as the higher system‡ only connects with the (immediate) external world of the organism through the lower, the lower would have to connect with the higher order only by means of the higher organism. In short, *every organization only is an organization insofar as it is turned toward two worlds at once*. Every organization is a *dyad*.§

7) That higher influence must be more closely determined. This alone is *cause* of excitability, for only by means of it is the organism stimulated to an activity opposed to the external influences.

a) How that influence‖ acts and what its nature is we will be able to determine by the shortest route by distinguishing it from the manner of action of the external influences on the organism and to their nature.#

---

*the question was, how the organism could itself be the medium of external influences.

†In the crudest phenomenon this is shown through the so-called sensible and irritable system—but, if the organism is duplicity to infinity, then that division too will go to infinity—there would also have to be a duplicity in the nervous system.—Gall's sensitive and vegetative man. But this is no opposition, for the merely vegetative presupposes the sensitive as well.

‡the higher organism

§(According to SW the last words are deleted in the manuscript.—Trans.)

‖through which the organism is armed, so to speak, against the influence of its immediate external world.

#Even this manner of action is not *purely* knowable, precisely because the organism already stands under the influence of those higher causes. Therefore, we must ask how that influence would act upon the organism if it were a mere product, without being productive.

The external influences act *chemically* on the organism according to their nature, insofar as the organism is viewed merely as matter (as product). However, the organism is never merely *product* (mere object). Therefore, the external influences do *not* act *chemically* on the organism. The question arises, by what could their chemical effect be inhibited?

The chemical effects must be inhibited by the opposing activity of the organism which we think in the concept of "excitability." The organism itself is only stimulated to this activity by a higher cause. *This cause must exercise an activity opposed to the chemical influences.*—This is *one* provision.

[176] b) Further: *the condition of that activity operating upon the organism is the duplicity in the organism itself.* Only to the extent that there is an original duplicity in the organism itself is that cause operative upon it. *There must, therefore, be a cause which is active at all only under the condition of duplicity.* We only know *chemical action* as such a cause (which we have deduced as necessary in Nature above), and have designated it "active" only under the condition of a *positive and negative reciprocal relation*. Moreover, this chemical action must be thought to proceed from a higher order (since the action is the cause of excitability), because that which is a *cause* of the chemical process (in a determinate sphere) cannot again be a principle of *the same* sphere.* Therefore, the universal chemical influence is identical with the cause of excitability.

c) However, the cause of excitability must *work against* the chemical influences so it cannot be identical with that universal chemical influence; this cause itself must be *chemical* in only *one* respect, but *not chemical* in another respect. The question arises whether and how this can be thought.†

We have characterized the activity which is cause of excitability as one *whose necessary condition is duplicity*. An activity whose condition is necessarily duplicity cannot be thought otherwise than as an activity whose *tendency* is *chemical*, because that duplicity is necessary only for the chemical process. The activity that is cause of excitability, then, has to be a chemical activity according to its tendency.—But that activity is extinguished in its *product*. If the tendency of that activity is the chemical process then it would have to be an activity *that is extinguished in the chemical process*, which is then to that extent *not chemical*.—The chemical activity is actually *also* extinguished‡ in the chemical process (where two bodies pass into one identical subject), for a

---

*Everything in nature may be chemical except for that which is the cause of the chemical process.

†The result to which the solution to this difficulty will lead us is of the utmost importance for our entire science.

‡even itself

chemical process is possible only between bodies that *can be reciprocally subject and object*.* Chemical activity is itself an activity that is *chemical* [177] only according to its TENDENCY, but which according to its PRINCIPLE must be called *antichemical*, because it is possible only under the condition of duplicity.

The cause of excitability is *identical* with the universal cause of the chemical process, namely, to the extent that the latter is chemical only according to its tendency but not its principle.†

8) For the time being‡ the entangled strife between the systems placed in conflict above (the chemical-phlogistical and the system of vital force) is hereby resolved at least in its major points.

a) Whether life *is* a chemical process or not will be decided in the subsequent investigation. If life is§ a chemical process, how can the chemical process again be *cause* of life, or explain life? *Therefore*, the chemical system only gives us effects instead of causes (e.g., "animal-chemical elective attraction, animal crystallization," and however else the incomprehensible terms are expressed[3]). Rather, if life is itself a chemical process, then surely both must still be explained and indeed explained from a common *higher* cause, from a cause which is itself subjected to no chemical affinity and cannot enter as component part—(as individual material)—into the chemical life processes.‖—Now indeed the activity which is *cause* of the chemical process—(we are not yet speaking here about the *conditions* of the chemical process)—is in its *principle not itself chemical*. If one and the same principle is both cause of life and of the chemical process, then it still does not follow that *life* is a chemical *process*. For life certainly could be (and not only *could be* but rather *is*) only chemical in *tendency* [178] (precisely like that cause) as the advocates of vital force truthfully say—(to the extent that they persistently view life as something *sublime, beyond the chemical*, they infinitely tower over the chemical physiologists)—and this

---

*No chemical process without the existence of at least two heterogeneous bodies which themselves become objective.

†(According to SW the last passage is stricken in the manuscript and is replaced by the following.—Trans.) Thus, we have in the cause of the chemical process itself a cause which is *antichemical* according to its nature and originally, i.e., which presupposes the *opposite* of that which occurs in the chemical process.

‡the illusion which lies in the proofs of chemical physics is naturally totally dissolved,

§nothing other than

‖It remains to be seen whether now the point raised in an earlier text by the author, substantiated with proofs, is better understood—whether the whole tendency of that text is now realized at all. (So far, original note.—Trans.) Surely nothing remains to be said against this except that such a principle is unthinkable—which in any case may also be true for many who, even in physics, are not capable of thinking anything other than matter, the product.

tendency is constantly inhibited,* for which surely no vital force is required. Now, if we

b) wanted to assume a vital force (although to accept a fantasy is good neither for physics nor for philosophy), then nothing is in the least explained by this principle.† In every force we think an infinity. No force is limited in any other way than by an opposite force. Now, let there be in Nature a particular vital force that is a simple force; then by this force a determinate product would never come to light, and when one posits something already *negative* in this force‡ in order to explain the determination of its production it ceases to be a *simple* force; one has to add its factors and thus be able to submit it to construction.§

Remark.

It was easy to foresee that from these two opposed systems‖ a third, uniquely true, would have to come to light; but this third has not existed until now. The *Brownian* system, which one would at first take to be such a thing (because it is opposed to those two systems at once), is not this third system, at least if one only knows such a system as a truly physiological one that explains life from natural causes.# The following will serve to further advance this insight.

**by what it is inhibited seems to be the important question to which galvanism will give the answer.

†At first glance at least, there are the same conditions in the chemical process as in the vital processes. Why it cannot be reduced to indifference, as in the chemical process, is the major problem which already signifies that the vital process was indeed the final cause, but cannot be identical with the chemical process in its whole construction.—Wanting to explain that there is no return to indifference in the vital process by appeal to a *vital force* means absolutely nothing.

‡The organic formative drive is distinguished from every other force in Nature in that a *standing still* is possible in it, the limitation to a particular production [it is *drive* to the extent that it is directed to a determinate product]; conversely, every other force of Nature which is not proximately or distantly related to the formative drive—(for there is *one* cause which gives its form to ALL forms in Nature)—hastens into infinity, without rest and without an object in which it remains fixed. (Original note, except for bracketed addition.—Trans.)

§Further, if life is a product of an unconditioned force, then the matter in which this force acts could never cease to be alive—just as little as matter can cease to be heavy: at least there would only be an *infinite* diminution in it, such that life (to infinity) would never = 0.

‖In contrast to these two systems, the chemical-physiological and that of the vital force, the system of excitability is distinguished principally by the fact that it posits an original duplicity in the organism itself. In contrast to those systems we can understand just what this implies. According to the chemical system the *whole* organism, e.g., is subjected to the chemical process—there is nothing *inhibiting* here—no limit—one does not see why the chemical process does not lose itself in infinity and why the same organism always proceeds again from this process.

#the Brownian system is in *principle* such, but not in its *execution*.

In the concept of the organism (as has been shown in the first portion of this work) the concept of an immanent activity must necessarily be thought, an activity directed merely upon its subject, which is, however, simultaneously an activity directed to the outside. This [179] activity *toward* the outside (as an originally inner one) can be distinguished only by opposition to an *external* activity, i.e., it is necessarily at once *receptivity* FOR *external* activity. That activity, as a simultaneously immanent and outer-directed activity, can be apperceived precisely at the point at which the external resistance is met; and conversely, only at the point from which that external activity is reflected into itself is there resistance—that which does not fall within this point does not even exist for the organism.—*Brown* indicated this very well in his concept of excitability, but without being able to deduce this concept himself; that is, that the outer-directed organic activity is necessarily receptivity for an exteriority, and conversely, this receptivity for an exteriority is at once necessarily outer-directed activity.

However, since it is not enough for physiology to present this concept or to deduce it itself and instead a construction of it must be thought (i.e., reduction to natural causes, which Brown himself was not capable of explaining), one should consider how the world could not become an exterior thing (an *external world* in general)* for the organism (which is *identical* with it) except by the influence of a force that is an outer thing in relation to that world itself, i.e., a force from a higher order, where consequently the organism is only, so to speak, the medium through which opposed orders of affinity† come into contact.‡

It is thus not an activity of the organism itself but a higher activity that is cause of its excitability, acting through it as a means. Only the *excitation* can be explained (under the assumption of excitability) through the influences of its external world (which Brown calls the "stimulating potencies"), but not *excitability itself*. Those stimulating influences are only the *negative conditions* but not the *positive cause* of life (or of excitation) itself.—But after one has taken away all influences of external nature as *stimulating potencies* nothing remains as cause of excitability other than the action of a higher order, to which that nature *itself* is also an external thing; for by this means§ the dynamical organization of the universe as an infinite *involution* (as presented in the previous

---

*other

†orders

‡Nature, to which the organism belongs, becomes an external world only by virtue of the fact that the organism is snatched from Nature, so to speak, and is raised to a higher power, as it were. Dead matter has no external world, it is absolutely identical and homogeneous with the whole whose part it is. Its existence is lost in the existence of this whole. The organism alone has an external world because there is an original duplicity within it.

§(through our construction of excitability)

division), where system within system is dynamically [180] conceived, is demonstrated to be necessary in a new respect.*

II.
Deduction of Organic Functions from the Concept of Excitability

All organic activity already *presupposes* duplicity (since it is the effect of a cause that is active only under the condition of duplicity). Thus the question still remains: *how does this duplicity inherently belong to the organism?*

In order that one does not believe oneself able to get away with a mere appeal to the existence of opposed systems in the phenomenon of the organism it must be noted that this *itself* is already a product of duplicity (which is a condition of excitability) rather than a cause of it; thus, these systems are a product of excitability. In animal nature all formation proceeds from an excitable point. *Sensibility* is present before its organ has formed itself; brain and nerves, instead of being causes of sensibility, are themselves rather already its product.—The opposed systems (the irritable and the sensible) into which the organism is divided are only the *theater* of that organic force, not the force itself.—Not to speak of the fact that one cannot even demonstrate those opposed systems in one-half of organic nature unless one is able to ascribe to it the universal property of everything organic, excitability.

Therefore, excitability cannot be completely explained before the *first origin* of organic duplicity is explained.

1) We have ascertained that all organic activity exhibits itself in the organism as object. That which is the *source* of all organic activity cannot again appear in the organism as object.† Now, the original duplicity is *condition* of all organic activity and the *source* of all activity is therefore *the cause of duplicity itself.*

2) A cause that is acknowledged to be the direct source of another activity must be thought to be acting in the organism, and which is knowable

---

*Conversely, only through our efforts to connect the universal life of Nature—and even the individual life of the organism—through its final cause to the construction of Nature itself, does our theory gain inner necessity.—One has spoken for a long time about the connection of the phenomena of life with those of light, electricity, and the like, without ever being able to completely uncover this connection. The Brownians, who view this attempt of physics extremely one-sidedly, do not notice that our explanation begins its account just from the point which they leave unexplained—not the excitation, but excitability itself; but all of these hypotheses lack the inner necessity which they can only achieve through their connection to the dynamic organization of the whole universe.

†For only activity is known in the object.

only *through* activity, not knowable through and in objects like every other activity.

[181] A cause that does not again present itself directly *objectively*, but is recognized only as cause of another activity, can obviously only be a *negative* cause *returning into its subject*. But a negative cause* is only thinkable as a cause of receptivity.

The cause of all organic duplicity is thus the cause through which an *original receptivity* belongs to the organism.†

A cause by which the receptivity of the organism is antecedently determined must surely be accepted as cause of every organism. For, in terms of receptivity to external influences, it cannot be distinguished from the inorganic. In contrast, the living distinguishes itself from the dead *only in that* the latter is receptive to *every* impression, but the former is antecedently determined by its own nature to be a *special* sphere of receptivity. For the organism the sphere of its activity is also determined through the sphere of its receptivity. The sphere of its receptivity must be determined through the same cause by which its nature is determined in general.—

The cause of *sensibility* is thus the cause of every organism and sensibility itself is the source and origin of life. The spark of sensibility must have descended into everything organic, even if its existence cannot be demonstrated everywhere in Nature,‡ for only the inception of *sensibility* is the inception of life.—Although without it no organism is possible, it will become clear in the following *how* it could be present in organic nature and yet be indemonstrable.

But *how* is sensibility demonstrable in Nature *at all?* The cause of sensibility is a cause reverting into its subject, thus it cannot be known directly in the *object*. As *source* of all other organic activity it can only be known through *activity*.—

(Most readers probably do not need to be reminded that *sensibility* is for me a *completely physical phenomenon* and that it is considered here only as such.—But even viewed physically sensibility is not something exterior that one could recognize in the organism [182] as object, but something reverting into the subject of the organism, indeed, even first constituting the latter—in a word, constituting the absolutely innermost reaches of the organism itself (and, therefore, one must conclude that its cause is something that can never become

---

*a cause determining its subject.

†Our thesis, that all organic activity is mediated through receptivity, is now determined ever more closely. It has been shown that organic receptivity and organic duplicity are one and the same—it has been explained in a new respect why all organic activity is conditioned through receptivity.

‡as, e.g., in the greater part of the plant world, where it becomes indemonstrable.

objective in Nature AT ALL. But then must there be something like that in Nature if Nature is a product of itself?).*

One can only *reason* to the existence of sensibility because it is clearly nothing *outside* the subject of the organism. Then on what basis does one know it?—Perhaps from the sense organs?†—But how do you know that such organs are *conditions* of sensibility?—Only from inner experience. But here the organism is given merely as *object*. How do you recognize sensibility in the organism as *object*? This is the question. You know it only from the external *effect* which you see in the organism as *object*, you do not know it *itself*, but only its external appearance.‡

---

*Sensibility is for us, in accordance with the foregoing, nothing other than organic receptivity, insofar as it is the mediator of organic activity—in a word, the source of organic activity. It clearly follows from this that sensibility is at all knowable in organic nature not *directly* in the *object* of the organism, but only in the organic activity whose source it is.

If we also distance everything hyperphysical from the concept of sensibility (which is necessary) and think under this concept nothing other than the dynamic source of motion (which we are compelled to posit in everything organic), then it already follows from *this* concept that sensibility is something absolutely interior—reverting into the organism. (Sensibility for organic nature is just that which duality of factors is for inorganic nature, e.g., the two basic forces—the condition of all *construction*.)

†as with polyps

‡You know it only from the organic movements whose source it is.—Thus sensibility is absolutely nothing other than the inner condition of organic movement. Through this circumscription of the concept we already exclude in advance many useless investigations.

It is well-known how many hypotheses about the manner of action of sensibility have been ventured throughout history. Yet not one of these hypotheses have made remotely conceivable how a sensation produces a movement. At least *this* becomes conceivable from our assumption. The external stimulus has no other function than to produce organic duplicity; but *as soon as* this duplicity is produced, *all conditions for movement are also provided* (for the cause of excitability is active where there is duplicity); therefore, every sensation, every stimulation passes directly or indirectly into motion.

For precisely this reason sensibility is also only recognizable in motion. I wish to elucidate this through a few examples.—The state of sleep is observed to be a state of dissolute sensibility where the organism ceases to be its own object, and where it sinks back into universal Nature as mere object. But sensibility is only canceled here in *appearance*, and because it is known only in its *appearances*, it seems canceled *altogether*. But it is still not totally canceled in its appearances. The continuation of the so-called involuntary movements proves the continuance of sensibility (for these are also mediated by sensibility).

It is the same way with dreaming, and many other experiences, e.g., the resolution to awaken. Kant: dreaming is an expedient of Nature, because without it sleep would pass into a complete dissolution of life. This is true to the extent that sensibilty cannot be dissolved in any other way than by the dissolution of life itself. But sensibility can probably be minimized with respect to the degree that it, e.g., extends to the production of natural motions.

The same happens in artificial sleep, so-called magnetic sleep, as in natural sleep. The phenomena of animal magnetism relate nothing more wonderful and inconceivable than organic phenomena in general. The most conspicuous feature in magnetic sleep is the cessation of all voluntary movement while sensibility still endures. For even here it seems to happen that—as we see happen

We can most likely state what this cause is in relation to its *subject*. It is a cause by means of which duplicity comes into an originally identical thing; but duplicity is not possible in an originally identical subject (A = A) except insofar as the identity itself again becomes product of duplicity* (for where A = A, this means that A is the product of itself). Duplicity or sensibility (for both are synonymous) only exists in the organism to the extent that it becomes its own object; therefore, the *cause* of sensibility is the cause by which the organism becomes its own *object*.

With this answer we know nothing more than we knew before. For to say "there is duplicity in the organism" and to say "the organism is its own object" is to say the same thing.

The question† must have another sense, i.e., what is the cause of sensibility abstracted from its subject, *what is it objectively* or *in itself*?

If the question is posed in this way it is obvious that this *cause*, as cause of all organization, must fall outside the sphere of the organism *itself*. It can just as little fall within the sphere of mechanism, for the organism cannot be subordinated to the anorganic. Therefore, it must fall within a sphere that once more comprehends organism and mechanism [183] (both opposed principles) under itself and that is *higher* than both. That higher sphere is none other than *Nature itself* insofar as it is thought as *absolutely unconditioned* (as absolutely organic).‡ In other words, *the cause of sensibility* (or, what is the same, of organic duplicity in general) *must be found in the ultimate conditions of Nature itself*.—Sensibility as phenomenon stands on the boundary of all empirical appearances, and everything is connected to its cause as to the highest in Nature.—(One can also achieve this insight along another course.—That is, just as the organism is *duplicity in identity*, so too is *Nature*; one, equal to itself, and yet also opposed to itself. Therefore, the origin of organic duplicity

---

very often in organic nature—where one sense is extingished or becomes dim—the other comes forth the more sharply and brightly (if what happens in organic nature is not without example here according to a few traces, and which often happens even in natural sleep)—that *all senses* condense into one homogeneous sense, or that in the place of the remaining senses another, foreign to our customary condition, comes forth. Be that as it may, this much is clear: that sensibility is nothing other than the mediator of all organic movement.—Only by virtue of the fact that all organic movement is mediated by sensibility is the animal taken out of the domain of mechanics, where every force produces movement directly, and seems to become master of its movements.—Sensibility is thus = to *source of activity*—but all organic activity has *one* condition, duplicity.

*proceeds from the duplicity

†(what is cause of sensibility)

‡We have explained sensibility through duplicity, which is condition of all organic activity. But now duplicity is indeed condition of *all* activity in Nature. Thus we see organic nature connected to this highest condition to which Nature in general is connected.

must be *one* with the origin of duplicity in Nature generally, i.e., with the origin of *Nature itself.*—

Should that *duplicity in identity* really be recognizable only in organic nature?—If the origin of the organism is one with the origin of Nature itself, then it is evident *a priori* that in the anorganic, or rather in universal Nature, something analogous must become evident. But in universal Nature nothing of the kind shows itself except in the phenomena of *magnetism*—.)*

3) Sensibility is known only in another activity. Activity is its product (not an object in which it is extinguished). It should be explained once more how sensibility could pass directly into activity.

An original opposition enters into the organism through original duplicity. The organism is opposed to itself, but it has to stand in equilibrium with itself so that it *is able* to produce a product.† What we have previously called the "organism as *object*," in a word, the *product*,‡ will fall in the point of

---

*Nature is originally identity—duplicity is only a condition of activity because Nature constantly strives to revert into its identity. Organic duplicity is thus without a doubt identical with Nature in its origin—and this seems to be the common point to which we will be able to anchor the construction of organic and anorganic nature. We can say—at least in a certain sense—that if the universal activity of Nature has the same conditions as the organic, sensibility does not belong exclusively to organic nature, but is a property of the whole of Nature, and that the sensibility of plants and animals is only a modification of the universal sensibility of Nature.

The cause of sensibility is something absolutely *nonobjective*—but that which is absolutely nonobjective is just that which is the first condition of the construction of everything objective, that which recedes into the innermost part of Nature.

If Nature is originally identity—and its striving to become identical again proves it, then it is without doubt the highest problem of natural science to explain the cause that brought infinite opposition into the universal identity of Nature, and with it the condition of universal motion.

What cause this is, is at the time being not yet known, but it is likely that without this cause which perpetually sustains the original opposition in the universe Nature would sink into universal rest and inactivity.

Therefore, we can only say in advance that it is a cause which brings forth duplicity in identity. But we know of no other duplicity in identity than the duplicity in magnetic phenomena. Since, however, these phenomena are not yet deduced, it can only be noted in anticipation that magnetism most likely stands on the boundary of all phenomena of Nature—as condition of all the rest.

The organism is ultimately nothing other than a contraction of universal Nature—of the universal organism: thus, we will also probably have to accept that the sensibility of plants and animals is only a modification of *universal* sensibility.

And to that extent, the philosophy of nature is seen to be the Spinozism of physics.

†An organic product would never arise by virtue of duplicity alone; the organism would only appear at *rest* if a striving toward identity were not conditioned precisely through this duplicity, and then the unity of the organism would again proceed from the diremption.—Life for the organism is an enduring struggle for its identity.

‡the organism as subjective is duplicity itself, which cancels itself in the product

equilibrium (or point of indifference). In this way rest belongs to the organism, its condition is a condition of homogeneity, it is a world to itself, resting in itself, complete in itself.

In this equilibrium all organic activity would dissolve, the organism would cease to be its own object,* would lose itself in itself.

That equilibrium (the state of indifference) must therefore be continually disturbed, but also continually reproduced. The question arises, how.

[184] No cause lies in the organism for its becoming disturbed. The reason has to lie *outside* of the organism.—(Everything unorganized must be seen as lying *outside* the organism, thus also the fluids, e.g., that circulate in it†— which accordingly do not belong to the subject of the organism, and thus also cannot be subject of disease, for example—whose existence can only be completely deduced in the following.)—

However, disturbed equilibrium is only recognizable in Nature through the tendency toward its restoration.‡ As certainly as it§ is disturbed, a tendency to restoration must also exist in the organism. But this tendency can only proceed (like *all* activity) from the higher organism, thus the higher organism must be able to be determined to activity by the passivity of the lower. This is not possible unless a plus of activity (i.e., action) in the higher is conditioned by a minus of activity in the lower. The question arises how this activity is possible.

4) It is clear for the moment that it must be an activity that passes into the organism as *object*—(which does not *revert* into it again).—It is, in a word, an activity directed outward. Something outer for the organism (i.e., something different from it) is at all possible only through a higher influence∥ for which the external world of the organism is itself a different one, i.e., an outer world.

---

*productivity would pass into product, the organism

†—it will be shown in the following that an immanent, fitful cause of stimulation must be provided for the organism since the stimulus is not permitted to *rest*, such that the organism is not dependent upon the contingent influx of external stimulation; this happens through the fluids circulating in it—.

‡The function of the stimulus is none other than restoration of the difference. This restoration I call sensation. To be sure, we only know through our own experience that every sensation disturbs and destroys (as it were) a homogeneous condition in us, but we know it more certainly for that reason. In the cases where the sensation passes directly into motion we do not notice this at all, because precisely here the sensation is not distinguished *as* sensation, and in the same moment that the duplicity arises it is canceled again. However, where sensation does not directly perish in motion—as with the affections of the sense organs (which only are sense organs because their affections do not pass *directly* into motion), that duplicity is more conspicuous.

§the equilibrium

∥effect

Such an influence is actually acting (above p. 108) upon and through the organism. This influence is shown in experience such that it is active only *under the condition of duplicity* (above p. 109). It will thus be active in the organism only under the condition of duplicity. Duplicity will be the organic source of activity. But in the organism* the duplicity is canceled. It remains in equilibrium with itself, there is *rest* in it, but there should be activity in it, and this can only be reproduced through continual restoration of the duplicity. This continual restoration can itself happen only by means of a *third*, and therefore that cause will appear active in the organism only under the condition of *triplicity*.†

[185] (The necessary triplicity in galvanism is deduced in this way. *The third body in the galvanic chain is only necessary so that the opposition between the two others may be sustained.* Two bodies of opposite composition brought into contact establish an equilibrium between themselves entirely necessarily, and show no electricity except with the first contact and the separation following upon it. (This proceeds from *Volta's* recent experiments from which it becomes clear that in order to produce electricity at all the mere contact and separation of

---

*as object

†A few remarks should be made at once.—It is a basic law of galvanism that all galvanic activity occurs only in a chain of three different bodies. This Voltaic law has indeed been brought into doubt by Humboldt through a few experiments where only two bodies seem to be in the galvanic chain. This is the case, e.g., where only homogeneous metals close the chain. Humboldt did not consider that the final ground of galvanic phenomena lies in the heterogeneity of the organism itself, by no means to be left out of consideration. Between nerve and muscle there is an opposition. Thus, if only one homogeneous body closes the chain between both, then the effect is still to be located in three bodies. More significant proofs against the necessary triplicity in the galvanic chain were the so-called experiments *without a chain*, where the muscle began to twitch when the nerve charged with only one metal, and this is touched through a second (homogeneous or heterogeneous). Here too lurks something misleading. For it cannot be prevented that the nerve is not charged at once by two pieces—thus a chain still exists.—Now, if the homogeneous metal is touched by a heterogeneous one, then by the mere touching at least a partial dynamic alteration is brought into the chain—which can be proven through the so-called *galvanization of metals* discovered by Wells, since two homogeneous metals produce twitches as soon as one is rubbed with a heterogeneous metal or is only placed in contact with it.—If the metal is touched with a homogeneous metal, then two homogeneous metals are to be seen as two heterogeneous ones, if the one charges the nerves—(the animal organ even serves to discover heterogeneities that otherwise are presented to no sense)—and finally these attempts are all reduced to a far simpler one where, through mere contact of the nerves in one point, contractions are produced through a metal; for here too the chain is, as already said, unavoidable, because it is unavoidable to touch the nerves at two different pieces. Not only nerves and muscles, but even two different points of the nerves are already heterogeneous among themselves. Thus there is duplicity here too. Aside from this, all these attempts succeed only with a very high degree of stimulability. It remains the case here that a *dynamic triplicity* is a necessary condition of all galvanic appearances. The question is *why* it is necessary—and this question is answered by our deduction.

two heterogeneous conductors is necessary; but the electrophore is already sufficient to prove this.) But the problem is: *a connection of bodies* should be found through which an enduring action is conditioned without repeated touching and separation, thus one* IN COMPLETE REST (for the organism is just rest in activity)—and this problem can only be solved through the galvanic chain, for in this chain an *enduring* action is conditioned through its *being closed upon* ITSELF and its *remaining closed*. Because, of the three bodies A B C, no two of them can establish an equilibrium among themselves without being disturbed by the third, since between three heterogeneous bodies no equilibrium is at all possible.)

Now, since the organism is not absolute rest but only *rest in activity*, that triplicity must be assumed to be *constantly present* in the organism.† But if it is constantly present, then activity indeed exists in the organism; a homogeneous, *uniform* activity. Homogeneous, *uniform* activity appears in the object (from the outside) *generally* as rest.‡

Now an activity is postulated that passes into the organism as object (see 3. and 4.), i.e., which presents itself in the organism through an external alteration. That triplicity must be assumed *not* to be constantly present in the organism.

This contradiction can only be resolved this way: the triplicity must constantly *become* (arise and disappear, disappear and arise again), never *be*. How this continual becoming and disappearing is possible does not need to be investigated here (undoubtedly [186] because the one factor in it is an alterable and constantly altered one§).—*The condition* of that activity *is a constantly evolving triplicity* whose possibility it was our task to demonstrate.

---

*i.e., a construction

†(According to SW, in place of the last sentence the following is substituted.—Trans.) However, if triplicity is the condition of all organic activity (if the dynamic activity in the organism is raised up to a *higher* potency, perhaps through this condition and only through it—for we can already ascertain here that the organic forces are throughout probably just the higher potency of common natural forces)—thus if triplicity is condition of all organic activity, then it must be assumed to be constantly present in the organism.

‡Therefore, the organ in the galvanic chain, e.g., appears to be resting as soon as the chain is closed, and is moved only with the opening and closing of the chain, although the activity in the chain is undoubtedly enduring.

§For example, I have already proven elsewhere that the blood is deoxidized through the expressions of irritability, and returns oftener and more quickly into the organs of respiration the more organic movement there is in an animal. Now, the blood in the lungs is permeated by oxygen and this oxygen determines the electrical constitution of the body, since an oxygenated fluid is negatively, and a deoxidized one positively electrified. But the blood now appears to be a constant factor of the process of irritability, e.g., in the *heart* at rest, before the third body streams into the blood. So if a deoxidation of the blood coexists with every contraction, then the blood is surely constantly altered—the triplicity is again constantly canceled.

5) But there is yet another problem: through *which effect* (which alteration) will *that activity be presented in the organism as object?*

It is an activity whose original condition is *duplicity*. But an activity whose condition is duplicity can only be such as proceeds toward *intussuception* (because the condition of *intussusception* is duality). That activity will appear externally as a tendency toward intussusception. But no intussusception is possible without a transition toward a common occupation of space, and this transition does not happen without *density* or *shrinking of volume*. That activity will appear externally as an activity of shrinking in volume, and the effect itself as *contraction*.*

(Much has been devised to explain the mechanism of contraction but which upon closer inspection dissolves into nothing. The opinion that with each contraction a transition from a vaporous into a liquid or from a fluid into a solid state (and therefore a solidification) is exhibited has a few things going for it, namely, that Nature even in such transitions is bound to show great force[†]—that the animal and the plant, seen objectively, are really nothing other than a continual leap from the fluid into the solid form (just as all organisms are like amphibians, placed between the solid and the fluid)—that with age the fixity of the organs of movement increases, and so forth.[‡]—But all of these mechanical modes of representation remain far from reality; in particular, a plethora of phenomena which galvanism approaches cannot even be conceived by means of them.—Undoubtedly, the ingenious mode of representation of Erasmus Darwin (in his *Zoonomia*[§]) is closer to the truth—at least *to the extent that* an alternation of attraction and repulsion is observed in electrical phenomena just as an alternation of contraction and expansion takes place in the phenomena of irritability, and *here* too the restoration [187] of a *homogeneous* state is the condition of reexpansion.[||]—Although it is certain that both can

---

*Here I come upon the most enigmatic phenomenon of organic nature—the organic power of contraction—which seems to be totally and exclusively proper to organic nature, and to which nothing similar in the rest of nature can be compared.

[†]that it makes conceivable to some degree the intensity of muscular force.

[‡]Nothing at all exists similar to this phenomenon, as was said—except maybe the chemical appearances, e.g., as an oxidized metal loses volume through deoxidation. I have ventured the supposition in the text on the World-Soul (see AA I,6 553.—Trans.) that for every contraction a deoxidation of the organism exists, the "agent electricity" (which now I also still have reason to accept), but I doubt that the contraction is itself *explicable* by the deoxidation.

[§]He explains contraction by analogy with electrical phenomena, and in fact these appearances are the only ones with which, as will be shown shortly, matter seems to stand at the same level on which it undoubtedly stands in the expressions of irritability.

[||]It undoubtedly comes to pass that the organism first contracts and then again expands through the same mechanism according to which two electricities attract and again repel one another.

only be analogically compared with one another (like the phenomena of electricity and of irritability in general) in the way the higher can be compared with the lower.)*

6) But the *tendency* of that activity is intussusception, and *precisely because* every activity is extinguished in its product it would be extinguished in intussusception. Thus intussusception can*not* be reached.—The question arises how this is possible.

Only in the following way. The condition of intussusception must again be negated by the tendency to intussusception itself. (In what way this happens is, again, not to be investigated here.† It could happen, e.g., that the third body in that conflict is always and necessarily a *fluid* one *through* which the contraction itself would be *propagated*. For then its condition would be recanceled by every contraction—mere duplicity would exist once more, and no longer triplicity.)

However, if the condition is canceled, then the conditioned (activity) also ceases. This mere *cessation* of activity cannot be the *cause* of the restoration of the former state of the organ. Rather, an *opposite* action has to step in with the cessation of the action that is the cause of contraction, which becomes the cause of the opposite state of the organ.—This action is not admissible so long as an action opposed to it maintains the equilibrium, but it must come forth just as its opposite disappears, i.e., it must be an *always present* action and must be grounded in the subject of the organism itself.

Its effect is the opposite of contraction, i.e., restoration of the volume or *expansion*.

That activity‡ would be exhibited in the organism as object by an *alternation* of *contraction and expansion*.

Remark.

Irritability (in the narrow sense of the word) has not only been deduced *in general* by the preceding, its conditions of possibility have also been provided.

a) Its *ultimate* condition is organic *duplicity*. It is thereby explained why irritability *appears* connected to the existence of opposed systems (the nerve and muscle systems) in the phenomenon of the organism. I say *appears*, for no

---

*The second stage of the transition of productivity into product is exhibited for us by the phenomenon of irritability. It is to be expected that there is still a deeper, third level. Irritability is still something inner, is an activity that is not yet completely transferred into the product. If we suppose that an activity that expresses itself in that alternation is fixed, and *completely* passes into the product (how this transition happens is not yet explained), then it will immediately appear as *productive activity* or as process of formation.

†and will be investigated in what follows

‡mediated through sensibility

experience reaches to the *first* origin of duplicity *itself*.—Just as everything visible is only the manifestation of an invisible, that higher system only represents that which will never itself become an object in the organism. In that system (the nervous system) the organic force can only present itself to its *object externally* because it is itself simply the bridge over which that force reaches into the world of sense. (The organism is the mediator of two worlds.) Just as the Sun, through rays thrown out in all directions (the image of itself), only indicates the direction of its higher influence, so the nerves are only the rays of that organic force, as it were, through which it indicates its transition into the external world. Since the nerves are also its first product, that force is as if chained to the nerves and not to be separated from them. Because the cause of life has also identified itself with them, it is *impossible* that they present themselves *externally* to themselves—(as if this is what happens in contraction, a shallow representation that now is beginning to become universal).

Now, according to the preceding, what is *sensibility* as such? All connotations that are attached to this word must now be excluded, and nothing is to be thought under it except the *dynamic source of activity* which we must posit in the organism as necessarily as in universal Nature generally. But it also results from our deduction of irritability that *sensibility* actually *disappears* in irritability as its object, and that it is therefore impossible to say what it is *in itself*, since it is itself nothing in appearance. Only the positive is known, the [189] negative is reached by reasoning. Sensibility is not itself activity but is *source of activity*, i.e., *sensibility* is only *condition* of all *irritability*. Sensibility is not *in itself* knowable, it is knowable only in its object (of irritability), and therefore surely where the latter exists so must the former, although where it immediately passes into irritability only it is knowable.—Incidentally, how sensibility passes into irritability is explained precisely by the fact that it is nothing other than organic duplicity itself. The external stimulus has no other function than to restore this duplicity. But *as soon as duplicity is restored, so too all conditions of motion are restored.*

Just as sensibility is the condition of irritability, conversely, irritability is the condition of sensibility, for without activity directed outward there is also no activity reverting into the subject. It was ascertained above that the organism as object would fall into the point of indifference without excitation from outside. Thus all excitation from the outside occurs only by the disturbance of that state of indifference. But this state of indifference is itself only a product of irritability. The activity whose tendency is homogeneity is just that which manifests itself in irritability as an activity of intussusception. Thus irritability, or rather the activity which is active in it, is conversely not at all the positive, but the negative condition of sensibility. Every sensation is only thinkable as the disturbance of a homogeneous state.

(Therefore, because a *homogeneous* activity is disturbed by every excitation from outside and, as it were, dissociated into opposites, in every SENSE there is a necessary *duality*, for to me *sensation* means precisely nothing other than the disturbance of a homogeneous state of the organism. Therefore, for the sense of sight the polarity of colors (the opposition between warm and cold colors) is the duality that becomes objective in the prismatic spectrum*—(just as it is quite certain that in Hunter's experiment the negative lightning is not a mere privation but a real opposition to the other; although in every duality aside from the actual opposition [190] there is still a *more* and *less*, as, e.g., the prismatic colors of one pole are also the darker colors, one pole of the magnet is also the weaker). For the sense of hearing the major and minor tones are the duality, for the sense of taste the acidic and alkaline tastes (for all other kinds of taste are only mixtures of both of these in various proportions). For the sense of smell there undoubtedly exists a similar opposition which is not clearer only because this sense is generally the darkest (thus most fitted for associations of ideas) and (on account of its thanklessness) the least cultivated.—One can employ this necessary duality in every sense as a principle of distinction for the senses generally. Therefore, the feeling of heat, e.g., does not serve as the name of a *sense*, because there is no opposition possible in it, but only a mere more or less.—(Opposition only exists where factors in the connection *neutralize themselves*, like the opposed colors of the prism, the acidic and alkaline taste, and so forth.)—For the sexual sense, its opposition does not fall within it but outside of it.)†

If irritability (or rather its product) is a homogeneous state (a negative condition of sensibility) and the former is proper only to the lower organism, then we have explained how the organism itself becomes the medium of external influences (above p. 107). Galvanism finally makes it obvious, for in it the irritable system appears only as the armor of the sensible, solely as the middle term by means of which the latter is in connection with its outer world.

7) *Irritability* (by which the organic appears to be moved inwardly) is still something *inner*, but the activity must totally become an *external* one, must present itself completely in the *external product*, and, when it is presented in it, *dissolve* in it. But this activity (in which it passes over completely into the product as an external one) is nothing other than the *productive activity itself* (the activity of the *formative drive*). *Irritability* must pass directly into *formative drive* or *force of production*.

---

*On other occasions I have asserted that electricity, or that which corresponds to electricity in organic nature, is doubtless the single immediate sensation—for which the galvanic phenomena are proof, if their basis is identical with that of eletricity.

†A genuine sexual sense must at any rate be assumed in those animals which unify both sexes in themselves.

With what, then, does all formation in organic nature begin other than irritability, i.e., with an alternation of expansion [191] and contraction? By what means does the metamorphosis of plants occur if not by such an alternation of expansion and contraction (Goethe on the metamorphosis of plants[4])? And is not this alternation of expansion and contraction almost more evident in the metamorphosis of insects than in plants?

If irritability appears only at its extreme in the force of production—in direct transition into its object—then irritability must totally dissolve as soon as the production is completed. But the production *must* be completed because it is a *finite* production. If it is to endure after the completion of the product, then it must be finite in one respect, in another infinite. It must have an *infinite production at least within* its *determinate sphere*—the existence of organization has to be a constant being-reproduced, in a word, the force of production must be *force of reproduction.*

8) The question arises, how does the force of production pass into the *reproductive force?*

For the moment it is not thinkable otherwise than as a constant rekindling of irritability and (through irritability) force of production.* This rekindling (because the condition of all irritability is *heterogeneity*†) is not possible otherwise than through an ever-renewed heterogeneity sustained in the organism, and the means to always renew and to sustain this heterogeneity—is *nutrition.*

Thus, the aim of nutrition can be neither the universally accepted one, replacement of the parts abraded and used-up through friction, nor even the maintenance of the chemical life-processes (like the flame) by an ever-renewed influx of material.

Others have already shown how highly inconceivable the loss of solid parts through friction is.‡ Then where is the friction in, e.g., plants, who need nutrition too? And what an unfitting means to this end! If one further ventures that with *stimulation* the requirement for nourishment is actually increased in every living being, that in the same [192] proportion in which nutrition is increased the respiration becomes faster and heavier, and that every animal most corrupts pure air in its state of digestion, and so forth—if one ventures this, then one is led far sooner to the thought that the aim of nutrition is the constant rekindling of the process of life.

---

*It should also be noted that formative drive is only formative *drive* because it emanates from *irritability*, or expressed otherwise, because it occurs through the mediation of excitability. In the dead realm of Nature formation happens through blind formative force—unmediated through the higher, which appears as excitability in the organic realm of Nature.

†a never canceled difference

‡I refer particularly to the text of Brandis on vital force.

It is by no means proven* that the process of life is *actually chemical* (for that it is chemical in *tendency* we ourselves assert, and thereby we explain the superficial appearance of truth which the arguments of the chemical physiologists have); it could perhaps be said that that process which appears in irritability as a process of a yet higher kind finally *becomes* chemical in the processes of nutrition and assimilation (according to their tendency). At most *apparent* grounds for this assertion can be adduced, but then they are refuted upon first inspection. It is not as if the products of nutrition and assimilation were not chemical *products* (for what natural product is not chemical? only that which is not even a PRODUCT of Nature any longer is *nonchemical*, that which is first *cause*), but that the *emergence* of these products in the organism is not explicable through a chemical process.—That chemical products are produced, i.e., products susceptible of chemical analysis, surely every physiologist has recognized; but they have not known the *cause* by which they are produced.

If life is not a chemical process, then no function (including nutrition) can have the chemical process as its aim.

The aim of nutrition must be something totally different, namely, the following. What comes into the organism by its means acts as a stimulating *potency*, thus acts only *indirectly* in a chemical manner.† Its stimulating force is at any rate determined by its chemical quality, but for this reason this force is not itself of a *chemical kind*, just as little as the electrical force of a body (because it is determined through its chemical constitution) would for that reason *itself* be of a chemical kind.—And even the mode in which it acts as stimulating force is physically explicable since the discovery that the activity of the members in the galvanic chain is determined by their chemical quality.‡

[193] The aim of nutrition is the ever-renewed stimulation of the organism, i.e., determination of the organism to constant self-reproduction (above p. 106); but the organism is itself again a *whole* of systems, every system in this whole has its *own*, *proper* function, so each must also be stimulated

---

*The preceding is proof that causes far higher than chemical ones act in sensibility and irritability.

†I deny just as little that that which comes into the organism through nutrition acts chemically—it is not as if I assert that its chemical nature and force is canceled (which is nonsense), rather: it comes not directly, but indirectly in a chemical way—as a *stimulating* potency.

‡Moreover, one hardly needs to remain with the mere assertion that the nutriments act as stimulating potencies. This is physically explicable too, since we see that even the function of a body in the galvanic process is determined through its chemical quality, i.e., even in the process of irritability. The galvanic process is just for that reason the connecting link that lets chemistry and physics communicate with the principles of physiology. It is a very natural illusion that deceives the chemical physiologists when they are able to explain the effects of so many materials on the organism by their chemical effect, and now believe themselves permitted to conclude that organic life is itself a chemical process.

in its own way.* As many different products (causes of excitation) as there are different systems in the organism (secretion) would have to emerge from the homogeneous material,† but also conversely, the emergence of these different products is conditioned by the existence of the different systems and their special activity. This process thus flows back into itself. One need not ask any further about its aim. It is itself the end, and sustains and reproduces itself.‡ There are really two propositions contained in this assertion which require special consideration.

a) There are individual systems of *specialized excitability* in the system of the organism. We thus deny the *absolute identity* of excitability throughout the whole organism; but not because we deny that that which acts as stimulant on the organ would also act as stimulant on the whole organism.§ It does not happen that every excitation of the part propagates itself to the whole organism‖ on account of the absolute *intensity* of excitability,# but is due to the *synthetic* relation of individual systems of the organism to one another, in which they must all be thought in reciprocal relations of causality. We do not think any occult quality by the concept "specialized excitability." The excitability of any organic system is determined through the (chemical, better *dynamical*) quality of its factors, which provides that it can only be excited through such and no other cause** (just as the power of excitation of a metal in a determinate

---

*We can see all the individual organs of an *animal*, e.g., as individual animals that all mutually nourish one another *parasitically*, as it were. This is not merely a figurative expression. Other very notable phenomena of organic nature too—not only the phenomena of secretion—point to the fact that every such organ has its proper force of reproduction, indeed even its own productive force. The origin of various animal species, e.g., which are found in various organs—in the digestive tract, in the heart, in the brain—found in many, perhaps all animals, cannot be explained by previous hypotheses. One should probably not venture to assume an actual power of production of these organs that would belong to them independently of the whole of the organism as the reason for this.

†of nutrition

‡In short, the phenomena of secretion can be explained only from a specific power of reproduction of diverse organs in which the power of reproduction is determined generally through excitability—finally, only as effects of a specific irritability.

§or, that the *degree* of excitation which is produced by any stimulus whatever in individual organs is proportional to the excitability of the whole organism.

‖and that the *intensity* of the effect of a stimulus on an individual organ is determined through the temperature of excitability in the whole body

#through the whole organism—and Brown did not even think about such a thing

**Under the *specificity* of the excitability of an organ I think nothing more than that the receptivity of this organ for a stimulus is determined by the dynamical quality of factors out of which the organ is constructed.

galvanic chain is determined through the chemical quality of the remaining factors of the chain);* e.g., the power of excitation of bile for the system of the liver is also determined in this way through the quality of the remaining factors of this system. There is thus nothing inexplicable or physically indeterminable here.

[194] b) Now, however, the assertion that this *specialized excitability* is again *cause of a specialized power of secretion* particularly requires to be proven.†— The proof lies in the preceding. What is the power of secretion other than a specific power of reproduction? But the power of reproduction is originally not at all different from irritability, thus specific irritability = specific power of reproduction.—And is this transition without example in organic nature? All infectious diseases act‡ only on irritability;§ moreover, beyond their general agitating force, they act *specifically*, irritability is specifically affected by it—and the product of this specifically affected irritability is homogeneous with the affecting cause—is again the same poison.—Thus for the liver, e.g., the bile is a kind of *contagion*,‖ it is a stimulating potency for the organ, and through this is the cause of its reproduction.

Thus, here there is a galvanism that reproduces itself. How that transition of specific excitation into specific force of reproduction occurs (for that it happens is understood) has been unexplained until now, merely because there is still no concept# of that *higher* chemical process (for the product, but

---

*Where the mode of action of the exciting body is thus never an absolute, but is always merely *relative*—or just as, e.g., the right ventricle of the heart calls for deoxidized, the left for oxidized blood as third member in the chain, in order to be determined to contraction. In this sense specialized excitability must be proposed—and is also proposed by Brownians in that they admit that an organ will be affected at any rate more easily by one stimulus than by another.

†which is harder to make conceivable, although it is a necessary result of the assertion that irritability passes directly into power of production.

‡firstly

§no alteration enters into the fluids, but with smallpox, e.g., probably an alteration of irritability.

‖infectious disease

#Since it is still undeniable that there is chemical production in the animal body—how does this arise, if nothing of a chemical sort happens in the organism?—. I assert that this production too comes to pass through a process higher than the chemical one—through the process of the formative drive: so I assert that just as irritability is perhaps a higher potency of the inorganic, formative drive is a higher potency of the chemical process—that there is in the organism a *higher* chemical process (for the product, but not in *mode* of production), but concede that we cannot characterize this process more precisely, which is undoubtedly an effect of galvanism determined by irritability, because we have learned to affect the two higher organic functions (sensibility and irritability) through galvanism, but not yet the power of reproduction, since admittedly the process of the formative drive is just as much a galvanic process as, for example, the process of irritability.

not the production). This process is an effect of galvanism, and for the time being can only be argued *analogically* from the action of galvanism on dead chemical substances (about which, moreover, still little is known) to the higher action.*

Since the excitation in the object presents itself as a constant self-reproduction the excitation through the stimulating potencies of nutrition surely passes unavoidably into an annexation of mass through *assimilation*. Since the excitation becomes self-reproduction, the annexation of mass can only happen through assimilation, and the original form is not altered, but only the volume.—(Necessity of growth, the second stage of the organic power of reproduction.)

[195] *Remark.*

The following elucidations are necessary.

a) I say that the annexation of mass is an unavoidable consequence of excitation. Thus neither assimilation nor growth are Nature's aim in nutrition. The *aim* is just the excitation itself, the constantly renewed kindling of the higher process of life, and this process of life is not a means to something else, it is life itself. Annexation of mass and growth is simply an unavoidable result of that process, and to that extent is something contingent with respect to the process itself; so although the result itself is not to be denied, it is still not to be seen as the aim of nutrition.

b) It should be noted that it is only denied that assimilation *occurs* in a chemical fashion, not that its product is chemical and is open to chemical analysis. Thus all the discoveries of chemistry retain their value, e.g., that the mechanism of "animalization" consists in the separation of nitrogen from the remaining substances, and so on.[5]

---

*However, altered secretions are known which would be submitted to galvanism, e.g., the lymphatic serous moisture in wounds. (To this point original note.—Trans.)—But as long as more deeply penetrating experiments do not exist concerning this point, one can at any rate indeed *deduce* that the process of secretion (e.g.,) ultimately comes back to the process of excitation, but *how* it emerges from the latter cannot be observed. One could for the time being perhaps invoke the chemical action of galvanism on *dead* substances, but hitherto little is known about this either, as stated above. This is related to the experiment performed by Humboldt, since, e.g., the water between two homogeneous silver plates remains undecomposed, but, e.g., while enclosed between silver and zinc—just like the animal organ—is decomposed, which decomposition happens undoubtedly through galvansim, as I already supposed in my text on the World-Soul (AA I,6 244f.), where the process is chemical for the *product* but not the production.

One must assume *a priori* that galvanism also affects the power of reproduction just as it affects sensibility and irritability; that all secretions, the process of assimilation—even the formation of the embryo—happen through a law of galvanism.

c) Finally, a new view is established regarding the function of all fluids in the organism, namely, that they are stimulating causes both of the organism and of matter, through which it produces and reproduces itself.—The fluid oozing around the embryonic heart in the chicken egg is at once matter and (as stimulating potency) cause of formation; therefore, with the stasis of formation the matter is also exhausted at once.—Thus in *blood* (this powerful cause of excitation) the triplicity of all organs of life is simultaneously recognizable; for if the threadlike part contains the substance of the muscle, then to argue analogically, the serous part contains the substance of the nerve fibers, and finally the globulus part contains the substance of the brain (by which the contingency of these organs becomes perfectly clear, and that they are a *product* of force, not the force itself).

9) The force which appears as active in reproduction is a force infinite in its nature, for it is joined to the eternal order of the universe itself, and is active anywhere its [196] conditions are given. But its conditions are always given in the organism. It always has to produce more. This progressive production would

either be *limited* to the product, would *not endeavor beyond* it, i.e., an *unlimited growth* would have to take place since the organic form cannot be overstepped.

And such an unlimited growth is also actual in Nature, in animals and plants, to the extent that they are merely bud-bearing (gemmiferous); for all polyps in the world are only buds of an original stem (and under this category are arrayed a great many of the examples set out above (p. 36) concerning sexlessness in organic nature).—

Or, the production would endeavor beyond its product. But the condition of that force is duplicity. If it does go further, then there must be a duplicity in the product whose one factor would fall *outside* of the product.*

If there were *no* such duplicity in the product (one of whose factors lies outside of it), then the productive force could indeed go futher, but it could present itself only in products that (because the condition of everything organic is duplicity) *by all accounts* would be *inorganic products*—and these would be the products of the so-called *technical drive*.

Note.

Since we find ourselves led to this issue by our investigation it is doubly necessary to linger with it, because this phenomenon of organic nature seems the least directly explicable from our point of view.

---

*Or rather, as will be shown shortly, the *product itself* must be the factor of an opposition whose other factor lies outside the product.

This whole theory *everywhere* presupposes the principle that we observe in organic nature nothing other than the play of a *higher* mechanism, but indeed a mechanism still always explicable [197] from natural causes and natural forces, as wonderful (i.e., unexplained up to now) as its phenomena may be.—How would it look for this whole theory if we could not make comprehensible these particular productions of organic nature from our principles, which so many philosophers have assumed to be a *degree* of, or at least to be *analogous* to, reason?

Probably everyone will concede that the phenomena of irritability or force of reproduction, and even of sensibility, are still grounded on natural causes; for even those who attribute ideas to animals (and for that purpose a soul after whose *locus* they inquire) still believe that certain organic movements correspond to these ideas and even undertake to determine these motions. But the technical drive of animals seems to them to be something extending beyond all those merely organic forces. Now, how could I assert that sensibility too has its cause *merely* and *solely* in Nature if I cannot trace back to natural causes that which seems to be its most immediate product (the technical drive)?

The path to it is laid out by what has already been said. I have shown how, from sensibility on, *one and the same force* fades first into irritability, from there into power of reproduction, and from this (under a certain condition) into *technical drive*. The technical drive ceases to be a special drive and different in kind from the others; it is merely a modification of the *universal* formative drive, and finally, like the latter, is itself a modification of the *universal* cause of all organization, of sensibility.

It is not enough that the products of this drive themselves confirm this view far more evidently than that analogy with reason. All products of the technical drive have the peculiarity that they are perfect in their kind and are genuine masterpieces. Every animal that has such a drive steps onto the stage with its art and is born cultivated. Nothing is halfway here, incomplete, or demands improvement. Just as the incomplete is also the perfectible, the complete is necessarily the flawless.—*Flawlessness* is the principal characteristic of all the technical products of animals.

[198] This *single* characteristic is also clearly sufficient to reject all share of analogy, of degree, or of a kind of reason as belonging to these products.

a) That something is *analogous* to *rationality* in these *products* is not at all denied, for every eye perceives that. But to reason from this to an analogue of *reason* in the *animal itself* is to reason too hastily. We see the same analogy in the regular motions of the planets and in *all* organic production, and would have to ascribe to the planets a reasonable soul on the same grounds, which leads them around the sun, or believe that every animal and plant soul also builds its own organs itself.

b) To accept a *degree* of reason as basis of explanation would itself be unreasonable. It is not as if we do not see the animal accomplish by instinct in its narrower sphere even more than that which we accomplish through reason in our broader sphere—but this is because reason is absolutely *one*, because it does not admit of degrees, and because it is the *absolute itself*.

c) "But if there are no degrees—then it is still a kind of reason!—That is, just as the human reason represents the world only according to a certain form, whose visible expression is the human organization, so every organism is the expression of a certain schematism of the intuition of the world. Just as we surely see that our intuition of the world is determined through our original limitation, without our being able to explain why we are precisely limited *in this way*, and why our intuition of the world is precisely this and no other, so too the life and the intelligence of animals can be just a peculiar (although inconceivable) kind of original limitation, and only their *mode* of limitation would distinguish them from us."

It was certainly a powerful dream that dead matter is a *sleep* of the intelligent forces, that animal life is a *dream* of the monads, that the life of reason is finally a state of general wakefulness. And what is matter other than *extinguished* [199] *spirit* ? All duplicity is canceled in it, its state is a state of absolute identity and of rest. In the transition from homogeneity to duplicity a world already dawns, and with the restoration of duplicity the world itself opens up. And if Nature is only *visible* spirit, then the spirit must become visible in it generally (as the beauty in it comes forth as soon as it admits the mechanism of natural laws), as soon as the identity of matter, by which it is suppressed in itself, is canceled.

But what good is this dream of physics?—For it the animals remain, now as before, *selfless objects*, whether their life is now a dream-state of the monads or a mere play of natural mechanism, for only what intuits itself steps out of the sphere of the merely *intuited*. That which does not set *itself* outside this sphere remains the captive of an alien intuition, something to be dealt with and explained according to laws of matter.

Therefore, all ways of thinking a rationality in animal activities fail us, and with them all those explanations of the technical drive which presuppose a deliberation, a possibility of experience, of a tradition, and so forth, among animals.

We must assert that they are driven to all of their acts, as to their productions, by a blind exigency; all that is left is to determine the mode of this constraint.

a) Philosophers who deny all rationality to animals have allowed them to be driven not only to their actions, but also to their productions, by the feeling of pleasure. They did not know that instinct and impulse do not exist together in the feeling of pleasure, and at bottom they cancel all instinct, while they carry human baseness into Nature.—It is no better to say that the bees, for example, are driven by *pain* to build their cells. What occurs through impulsion from pain or from need also only occurs carefully and slowly, but conversely "the swiftness of a force comes from self-impulse."[6] Is there anything laborious or clumsy and born of compulsion recognizable in those [200] productions?

b) Therefore, we will assume that the technical drive of animals results from the determination of their *physical* forces with respect to the mode of their actuality—(with the exclusion of the soul-forces which Reimarus has mixed in here and whose existence is refuted by the preceding); or more clearly, we will assert: it is *physically impossible*, by the very nature of the animal, that it produce anything other than the *regular*. We appeal to the fact that even in those animal taxa which possess technical drive above all others, all instruments of motion are so limited as to their use that the instrument and its use are *one and the same*; that, therefore, in organic nature generally, because everything in it is infinitely interconnected and everything else is altered with the alteration of one thing, nothing disharmonious or contradictory in itself *can* arise in it and through it. Further, we appeal to the fact that the animals which possess the technical drive are determined through the sphere of their irritability, as of their sensibility, which means that such an animal cannot be stimulated to movement through a sensation that is *irregular* or not completely measured to its inner nature—(which is already possible with animals of a higher kind (where the technical drive also disappears) on account of the disproportion of sensibility over the other organic forces)—. Finally, we appeal to the fact that the sensibility of these animals has an infinitely narrower circle, that the various rays into which that force splays itself out in the higher organisms only run together into one point in them, and so one sense seems to replace the other, the one seems to govern the other, through which an error of sense (if it be permitted so to express it), or rather a blunder in animal actions generally is impossible, and so forth.

It is presupposed by this explanation that in the animal a *productive* force acts overall; the *problem* is only to explain why this force acts necessarily in a determinate *form*, and [201] reveals itself only through regular actions. It is now quite evident, for the reasons just stated, that regularity *overall* has to exist in the organic motions of such an animal, but not why these movements also produce *externally* regular products, analogous to artworks; and the objection which Mendelssohn brought against Reimarus comes to mind, namely: even if one presupposes a certain determination and direction in the organic forces of an animal, still one can construct no concept of the direction to design a hexagon (like the ones bees design in their wax cells) or any other regular figure.

I respond: granted that there is a force in the animal which endeavors to go beyond its product, this force must be consumed, like every force in Nature; it must (since it is an original productive force) extend to a *product* (i.e., to a determinate thing) in which it is extinguished. But with its necessarily determinate MODE *of effectiveness* its *product* is also determined; this determinate mode of effectiveness and this determinate product are *one and the same* thing, are in no way different. The product already lies in that determination of the organic forces, and the product which you see is only the visible expression of the determination of those forces.

"But granted that given the organic forces its product is already determined as well, how does this regular determination precisely belong to that force—this directive toward the production of a hexagon, for example?"—I answer: this hexagon is *not* a hexagon for Nature. It is a hexagon only for you, which you question, and which you read into Nature. The mistake is that you only pronounce what it is, for only while it runs through your head does it take on the appearance of rationality. For Nature it has nothing at all to do with "producing" a hexagon, just as little as Nature "produces" a snowflake.—

"But granted that this regularity exists only for *me*, why does Nature produce precisely what is regular for *me*?"—This question is more penetrating, so the answer must take a [202] higher standpoint as well.— —What you see here in the products of the technical drive is only the final work of the same force that has produced the organism itself, and that still uses it only as an instrument of its formative tendency after this first product is finished. (In the majority of insects the proof is clear; you see that this insect in which that drive is active will soon *cease to exist* (at least to *be* what it was), if it endures it must be transformed).

Now, in organic forms we only observe products in which everything is reciprocally means and end. We have no other name than the *organic* for this kind of inner perfection, because organic nature is *a unity* with respect to it.— *Where the organic form stands at its limit and the organic force extends out beyond this limit** *it no longer produces that inner perfection, but only an outer perfection.—* This external perfection is the *geometrical* type, and you observe this everywhere in Nature either where the organism stands at its limit (as, e.g., in the casing of

---

*but this is just the case with the technical drive, which extends out beyond the organic product (which is also the case with the reproductive drive, but which finds the factor of duplicity that is its condition outside itself).

shelled animals), or where mechanism begins, e.g., in the motions of the planets, generally in the laws of all motions, with respect to which Nature is the most perfect geometer.

The question really extends to the whole of Nature, for Nature produces this external, geometrical perfection for no other reason than that for which it produces inner, organic perfection. But this reason is none other than *blind necessity*, with which Nature acts generally. If there were chance in Nature—just *one* accident—then you would catch sight of Nature in universal lawlessness. Because everything that happens in Nature happens with blind necessity, everything that happens or that arises is an expression of an eternal law and of an unimpugnable form.—Therefore, you see your own understanding in Nature, so it seems to you to produce *for you*. And so you are only right to see in its lawful productions an analogue of freedom, *because even unconditioned necessity becomes freedom once more.*

[203] This explanation still remains too much in the *universal*, and even if everything were demonstrated by the fact that the technical drive of the animal (and with it all behaviors) is effected through merely natural forces, then the question still arises *how* they are effected and *through which* natural forces.

However, we also do not need to remain at the level of this universal explanation. Since the technical drive (to limit ourselves to this) comes forth from the continuity of *all other* natural forces, since only the *universal* force of production dissipates in it—(it is clear from this that it first appears in the series of organisms where this* force begins to achieve a preponderance over the higher; for why are the most sensible animals robbed of the technical drive, and conversely the most richly artistic animal outside the sphere of this drive the most restricted with respect to its sensibility?)—since further, this drive, just where it expresses itself most conspicuously, only constitutes the transition to *metamorphosis*, then its cause will be no more enigmatic to us in the future than is the cause of the higher organic functions and of the force of reproduction and all of its quite manifold appearances.† Are not buds and blossoms,‡ is

---

*blind productive

†It seems rather that precisely through this drive the organic technical drive reverts into a mere drive of crystallization: and thus *one* chain stretches from the perfect organic crystallization over to dead corporeality. The lesser the sensibility, the more the technical drive—is a universal law. Therefore, one could say that the crystal, which crystallizes far more completely and quickly than the hexagon of the bee, has far more technical drive than the insect.

‡(According to SW, in place of the last words it reads the following.—Trans.) are not many crystallizations.

not the shell of the mollusk a more perfect artwork than even the cells of the bees,* and don't all of these appearances have their common cause in Nature?

If it is now *demonstrated* by the foregoing that the technical drive of the animal (and to reason analogically, all of its instincts) are blind effects of Nature, do we need to worry about any more objections, be they taken from experience or from the prejudices of the common standpoint? Only a few of them will be answered briefly, because they provide the opportunity for further elucidations.

The principal objection that we must expect, and to which all others are reduced, is that we degrade the animals to mere Cartesian machines, that we run into all of the triumphant arguments which one has brought before these philosophers in ancient and modern times. Whether the animals really become machines in our theory will become clear through the analysis of this objection. For the moment, the theory of the existence of ideas in animals (and all that goes with it) at any rate collapses on our theory. Then also

[204] a) the view of the so-called *sense organs* as such, through which ideas are awakened, simultaneously collapses.—From this we have nothing to fear, at least until someone has made *at all* comprehensible the origin of ideas by an external stimulus of these organs, since we deny, even where the existence of ideas is certain, that these ideas arise from external impressions. We assert rather that an activity of the organ excited by an external stimulus is only the *necessary correlate* of the idea, because this coexistence is the sole means by which our original idealism is transformed into realism, for without it we would believe that we intuited everything only in ourselves. Therefore, the self must already become material for us in our *originary* productive intuition, i.e., must become an object for us that is affected by external nature. Now it is certain that what corresponds to a representation in its organ is an altered receptivity of this organ. Why is light only light for the *eye*, and not also, e.g., for the dead body, and why does the eye itself produce (in the galvanic chain, e.g., where one does not have to think about a material development of light) an illuminated state when the normally present external condition of this state is lacking?

The alteration that is produced in the organ through external stimulus (which for brevity's sake I call sensation, with exclusion of all the connotations that may otherwise be attached to this word) is an *inner* one, an absolutely unrecognizable alteration seen from the outside, or as we have expressed it above: sensibility is an activity reverting into its subject. It is cog-

---

*where it becomes manifest that precisely in those products where sensibility no longer has a share, the products become more perfect.

nizable in the object only *indirectly* in the expressions of irritability whose source it is, and in many animals, indeed even in individual organs of an animal (the so-called involuntary organs) it disappears so directly into external movements that it can hardly be distinguished from the latter, and so is also no longer recognizable.

Now, we would admittedly degrade animals to the status of machines if we asserted that they were set in motion *directly* by an external impulse (under which one can conceive everything that acts in a straight line, including attraction), for every merely [205] mechanical impulse passes *directly* into motion. However, I assume that even where sensibility *disappears directly* into external movements (i.e., where the movements appear as completely involuntary) they are still not directly *produced* through the external impulse, but are *mediated* by sensibility (as the universal, dynamic source of motion). Every external force first passes *by way of* sensibility before it acts upon irritability, and sensibility is the source of life itself, *precisely because* through it alone the organism is torn away from *universal* mechanism (where one wave pushes the other forward and in which there is no standstill of force) and by this means becomes its own source of motion.

The animals would become machines if we concurred with the absurd opinion of the Cartesians that allows all external causes of excitation to act by impulse or attraction upon animals (in mass), for then these causes act only mechanically, i.e., in *straight lines*.—Now, for us sensibility is still something no less grounded in natural causes (though we allow all external causes to reach the organism only through it), although we confess that because we know sensibility only as SOURCE of all organic activity, and because all forces act through it as their *common medium*, it disappears for us into the ultimate conditions of Nature; from this it is understood that *sensibility* is probably the UNIVERSAL source of activity in Nature, and therefore is not a property of the individual organism but of the *whole of Nature*.

b) What the so-called "voluntary" movements of the animals are according to this view (with respect to which a second objection will be made against us) is clear from the preceding, and will become still clearer from the following.

c) "But this opinion robs the greater part of Nature of life and deposits it into the realm of the dead."— [206] Supposing it were so, then this consequence could prove nothing against the principles demonstrated.—But is it so?—In order to present the matter from *one* side we have placed the technical drive in continuity with the universal force of production. But this force is, above all other organic forces, submitted to the universal organism. (How is it otherwise explicable that, although in the animal kingdom—one can say *universally*—*separate* sexes are produced, nevertheless a proportion of both sexes of each species is maintained—that generally, with a view to the reproduction of the species—(this is certain at least in the human species)—such a conspicuous lawfulness is noted that reproduction in the organic realm of nature is solidly connected to certain seasons accompanied by universal alterations in Nature?) If it is certain that the force of production is intertwined in the most intimate way with the *universal* organism, then this will hold as well for *all* drives of the animal—(should we believe that a universal alteration of Nature, e.g., correlates with the drive of the migratory bird, which, in the very season when the magnetic needle reverses in order to point in the opposite direction, initiates the flight to another climate?)—It has to hold for *all* drives, for they are all only modifications of the universal formative drive, because the latter alone has a *direction toward an external object*. But this is even more the case for the *technical drive*, and—we will see the products of this drive as products of that UNIVERSAL formative cause that acts on Nature through the organism as through a middle-term, and joins the whole of Nature in a *universal* organism—in short, we will be able to see them as products of that cause which is the *universal* soul of Nature, as it were, whence *everything* is set into motion.* Our opinion, then, is just [207] that no *individual*, *unique* and *disconnected* life belongs to the animal, and we simply sacrifice its *individual* life to the *universal life of Nature*.

10) It is presupposed that the technical drive extends out beyond the product, without the existence of a duplicity where one factor falls outside the product. But if there were a duplicity in the product whose one factor *actually* falls outside of the product, then it could only lie again in an organic product, for the duplicity must be of an *organic kind*. This product must be opposed to the first with respect to this factor, but just *for that reason* has to be *equal* to it generally† with respect to the higher factors of the organism. With a view to *this* duplicity, of which in each product there is only one factor, both *individually* must express the universal character of their stage of development incompletely, but *both* together completely express it.

Individuals that are related to one another this way are individuals of *opposite sexes* of one and the same species (above p. 42f.).

(What could only be postulated above (p. 42) has now been deduced; namely, the universal sexuality in organic nature, which is, as it were, the extreme boundary of universal organic opposition.)—That force, whose sole condition is duplicity, is effective only where its conditions are given. But its conditions are given. It will then proceed to act. What was its object becomes condition of its possibility, or its instrument; these are the opposite sexes. The question arises as to what their product will be.

Esse apibus partem divinae mentis haustus/Aetherios dixere. Virg.[7] (Original note.—Trans.)

†in relation to a higher concept in the stage of development.

Their product is a *new duplicity*, i.e., it reproduces its *condition* to infinity. How *sensibility* belongs to the *individual* organism is certainly conceivable. The *individual* serves only as *conduit* in which that *one* incendiary spark of sensibility propagates itself to infinity. But what is the *ultimate* source of that force?— Through the act of fertilization the force of production is in no way awakened.* It is sensibility that is first awakened, and that next passes into [208] irritability, and finally into formative drive. The fluid† matter is only the *stimulating cause*;‡ it seems that the *merest* contact acts as a kind of contagion§ in fertilization, through which sensibility is awakened,|| just as *polarity* can be produced by the mere *contact* with the magnet.

In this way the circle of organic nature closes. The force of production is the *furthest* force of the organic forces. Sensibility can be lost in irritability, irritability in force of production, but in what can the latter be dissipated? It must be absolutely extinguished if it cannot revert back into its origin (sensibility).# It is only possible that it revert to its origin if one factor falls outside of its product. But it only happens *that* its one factor again falls outside its product if it does not lose itself in another force but directly in the product itself.

Now the *product itself* must separate into opposites.** If it is only *one* product that separates itself into the opposite sexes, then the production is only *one* as well. The production is distributed in distinct individuals. These individuals must themselves be submitted to a higher order, by virtue of which it is impossible that one sex is born without the other being born simultaneously (or more generally expressed: by virtue of which a proportion of opposite sexes is maintained).†† The cause of this order cannot again fall within organic nature itself, it must fall outside its sphere, but can just as little fall within anorganic nature, so it must fall within the higher order which unifies both, or in a *universal* organism. Thus organic nature is intertwined in a *universal* Nature with

---

*for it is a subordinate force

†reproductive

‡the process of fertilization is not a chemical but a dynamical process

§Grounds for this assumption are already found in Harvey's famous work. (Original note.—Trans.)

||through which a duplicity is first rekindled and through this a new process of excitation.

#through which the eternal circulation is conditioned

**And since we had a simple duplicity before, we now have a duplicity of products, a duplicity of the second power. The extinction of the force of production is inhibited by it alone. For through it, it becomes possible that it return into its origin (sensibility).

††Since universally almost always *four* individuals (at least where there are separate sexes) are required in order to reproduce the *species*, then perhaps it would not be mere caprice to point out how the original duplicity progresses first to triplicity (in irritability) and finally to quadruplicity (in force of reproduction). (Original note.—Trans.)

both of its extreme ends (sensibility and force of production), a universal which we at the moment can only postulate.

[209] 11) For the organic activity now deduced, one factor certainly lies outside the* product, and this one factor is transferred into a new product. The *activity* thus endures (for it reproduces its condition to infinity), but not the *product*. The latter as individual is only a means, the species end.

The remaining organic activity of the individual perishes in the reproduction of the species, for all higher forces are dissipated in the latter as the most extreme point.—But the tendency toward this extreme already manifests itself in the earlier modifications of the force of production; for, throughout the whole of Nature, is not the technical drive (which in a few species is the equivalent of the formative drive; above p. 36) only *harbinger* of the awakening *formative drive*, from insect up to human being? The insects possess technical drive only before their sexuality is developed, just as the worker bee always possesses this drive because it will never arrive at sexual development. As soon as the insects have passed through their metamorphoses—and these are only phenomena of sexual development—all technical drive is extinguished in them.—But the bird too builds its nest, the beaver its den before the mating season—could it be the result of a special foresight? Not hardly. One and the same blind drive guides all actions of the animal. The technical drive is a modification of the productive drive generally, and that which passes directly into the mating drive.†

With mating completed the final heterogeneity‡ has passed into activity, and the cause whose tendency is *cancellation of all duality* (and which appears *for this reason* as active only under this condition) is no longer inhibited by anything—disappearance of all duality is therefore necessary.—But a disappearance of all duality exists only in the *chemical process*, i.e., in that which, in the anorganic world, corresponds to the organic formative drive.§

---

*individual

†The technical drive is a drive just as blind as the mating drive. Therefore, all products of the technical drive are invariable and cannot be improved upon.

‡duplicity

§The product reverts into universal indifference. But indifference is only produced in the chemical process through that which corresponds to the organic formative drive in the anorganic world.—The product does not perish *as* organic, i.e., as product of the first potency, it must first decline into a product of a lower potency if it is to pass into indifference. And with that the stages through which the productivity in the organic kingdom of Nature gradually transits into the product have been deduced.—With every organic product Nature passes through all of those stages. This does not exclude the fact that those various stages can be *distinguished* in any one product more, and less in another. This would provide not merely a graduated series of production but of products as well.

The three stages of organic production deduced by us: sensibility, irritability, and formative drive, are conditions of the construction of an organic product generally and to that extent functions of the organism itself.

[210]   III.
The Graduated Series of Stages in Nature

And so at least a *part* of the general problem set forth above (p. 53), TO DEDUCE A DYNAMICALLY GRADED SERIES OF STAGES IN NATURE, is solved. At least the first stages through which Nature gradually descends from the organic to the inorganic are known to us, and presently we have no other business than to demonstrate that graduated series in Nature itself.*

The functions of the organism must be *opposed*; therefore, they exclude one other reciprocally in one and the same individual, for they are either distributed into different organs, or are totally supplanted by one another. This was proven right at the start (p. 51f.).

It is only now explicable *how* those functions are opposed. Since, according to our preceding investigations, sensibility, irritability, and force of production are, with all of their modifications, really just *one* force (at the least because every lower force has *one* factor in common with the higher), it follows that they can be opposed only with respect to their *emanation*, or their appearing in the individual or in the whole of organic nature.† Force of reproduction is also irritability and sensibility and supplants both of them only in *appearance*, for the final term in which these two are lost is just the force of reproduction.‡

---

*If the higher function is gradually displaced by the lower *in the appearance*, then indeed there is only *one* organic product, but there are as many stages of the *appearance* of that product as there are stages in the transition of productivity into product. This leads to the idea of a *comparative* physiology, which seeks the continuity of organic nature not in the transitions of shape and of organic structure, but in the transitions of the functions into one another. (The previous remark was added to the original note, which follows, until the next parenthesis.—Trans.) The idea of a comparative physiology is already found in Blumenbach's *Specimen physiologiae comparatae inter animalia calidi et frigidi sanguinis*, and further explicated in the discourse on the relations of the organic forces by Mr. Kielmeyer, whose major idea is taken from Herder's *Ideas for the Philosophy of the History of Humanity*, first part, pp. 117–126; namely, that in the series of organisms, sensibility is displaced by irritability, and as Blumenbach and Sömmering have proven, by the force of reproduction. But HOW sensibility is supplanted by irritability, and both finally by the preponderance of reproductive force, has not been explained at all by any of these investigations. (Note appended to the original note now continues.—Trans.) Neither the mechanism nor the ground of this graduated series has been discovered up to now. This has in part already come to pass in our deduction and will proceed from this point.

†Since, that is, sensibility, irritability and formative drive are only various *stages* of a limited productivity—or productivity passed into the product—then it follows that they can be opposed only in *appearance*, that the higher stage can only be suppressed in appearance by the lower, only because it is conditioned by the latter.⁸

‡The graduated series of functions has been deduced *a priori* from the bare concept of productive, organic productivity. Thus nothing remains but to confirm the series in experience.

[211] However, since those functions of the organism exclude one another at least in appearance, the proof of the actuality of such a dynamically graduated series can be executed only

a) in part through an examination of the various *organs*,

b) in part through the various *states* of the same individual (i.e., to the extent that in both the organs and the states the dominance of one function excludes the others),

c) in part, finally, from the diversity of organisms themselves and the corresponding diversity in the proportion of organic functions; and we will also actually employ this threefold mode of proof.

The functions of the organism appear to be exclusive of one another and opposed to one another. Therefore, all possible relations will be exhausted by means of a reciprocal determination of these functions by one another.

A. *Reciprocal determination of sensibility and irritability.** Sensibility and irritability determine each other reciprocally to the extent that sensibility is expressed in irritability as its most direct phenomenon. However,

1) both sensibility and irritability must have at least *one* factor in common precisely because the one passes into the other and is presented in it only as its *object*.

2) If irritability is = to the product in which sensibility presents itself most directly,† and if every activity is immediately extinguished in its product, then as *irritability increases in the phenomenon, sensibility must decrease*, and inversely *in the proportion that sensibility increases, irritability must decrease in the phenomenon*.‡ (The latter qualification must always be added, because *originally* irritability is possible without sensibility just as little as sensibility is possible without irritability.)§

[212]   *Proof.*

This can be carried out

a) from the various *organs* of the same individual.

---

*Until now "irritability" has been used for the appearance of contraction and expansion. According to the original use of the word, irritability is the mere capacity to be stimulated. But since the conventional usage of the word has already for a long time denoted this phenomenon of stimulability with it, and all the terms used in place of it until now—like, e.g., "capacity to be affected," among others—just as little signify the matter at hand, I will nevertheless still retain this expression, perhaps until a more correct and more fitting one is found.

†If irritability is = to that activity into which sensibility immediately passes.

‡(for it)

§(According to SW, the last sentence is replaced in the manuscript by the following.—Trans.) This is a universal law which can be derived *a priori* from the relation of the two functions that we have deduced.

aa) Since sensibility is an activity that recoils back into its subject it can only be distinguished in *opposition* to an activity *directed outward* (irritability). So where sensibility achieves a preponderance in organic nature an organism must originate that is *only* sensibility, i.e., whose function is not exhibited as irritability (by activity directed outwardly).* This explains why sensibility is only conceivable as the *negative* of *irritability*, as has been said elsewhere. Sensibility as such becomes unrecognizable through the fact that it is lost directly in irritability, so it is *recognizable* only when it does not pass directly into outer movements (or when by means of it the excitation from outside does not pass directly into outer movements). Now, if sensibility is thinkable only as the negative of irritability, then where there is a preponderance of sensibility there must be an organization which is an absolute negation of irritability (which is not at all subjected to irritability)—such an organization is the *brain* and *nervous system*.† (If there is a gradation of organic forces, as we have proven in the preceding, then there must also be a gradation of organs. And if the organism is only the contracted, miniaturized image of the universal organism, then such a gradation of forces must also be found in the world-organism, as we will see later.)

The brain and its extension, the nerves, have devoted themselves entirely to *sensibility* alone, irritability is totally squeezed from them by the preponderance of sensibility, for not a single man has yet proven the opinion‡ that all functions of the nerves§ are contractions.

bb) Conversely, since sensibility is thinkable only as the negative of irritability, then it must absolutely disappear where it passes directly into [213] irritability.‖ With the organization (organ) that is exclusively sensibility another organization (organ) must coexist that is only *irritability* in order to maintain equilibrium with it; this organization (organ) is that of the heart and its

---

*Here too the qualification must be added that through the preponderance of sensibility irritability is canceled only in *appearance*. Those three functions belong to the construction of all organisms, and so does irritability.—Where sensibility has a preponderance, irritability is naturally canceled for the appearance. In the *appearance* the higher function is always suppressed by the subordinate one.

†(According to SW, the last sentence in the manuscript is corrected.—Trans.) So where there is a preponderance of sensibility there must be an organization in which irritability is totally canceled for the appearance, i.e., an organism whose excitation does not pass directly into movement. The brain and nervous system is such an organization. In the latter, productivity still seems to stand on the first stage of that transition.

‡and it is nonsense

§(in the narrower sense of the word) i.e., the genuine function of sensibility

‖(SW notes a correction in the manuscript.—Trans.) According to the same law, conversely, sensibility will absolutely dissappear for appearance where it immediately passes into movement.

extensions, the *arteries*.* Since this organization has dedicated itself wholly to irritability, all *sensibility* must be squeezed from it by the preponderance of the latter.† That is, here all sensibility immediately dies away in movement. No *reflex* whatsoever occurs any more and all organic activity is only an *activity directed outward*. But this activity directed outward is itself possible only under the condition of sensibility; thus, sensibility exists only in that it is extinguished in irritability, and only to that extent can the heart, e.g., be called an "involuntary" organ with any meaning.‡

b) The proof can be carried out by consideration of the various states of the same individual, e.g., in disease, where all power of movement perishes in increased sensibility, or conversely, where sensibility sinks with increasing irritability. Even the state of *sleep* belongs here, where with the sinking of sensibility the irritability of the heart and the arteries is increased.§

c) The proof can be carried out by consideration of *different organisms*. If it is certain from the preceding that sensibility (as the negative of irritability) is bound to the existence of an *organism* which is not at all subject to irritability, then we see the *brain*, as the seed, so to speak, from which that organism evolves, at its greatest and most perfect form in man, and dissipating backward from him in an increasingly smaller volume and more imperfect organization. In the whales it is almost = 0 in comparison with their remaining mass, surrounded by a fatty, oily fluid, so the dullness of their expressions of sensibility follows. In the race of birds one observes little manifoldness of structure anymore, few protuberances, concavities, and sinuosities.—In the reptiles (where the nerves first cease showing nodes (secondary brains)), it becomes very small, and likewise in the fishes, which with respect to [214] sensibility stand still *below* the former, because their brain too becomes more inaccessible through their environment. In the insects it begins to be quite problematic; with certainty one knows only the extended medullary substance furnished with many nodes. In the greater part of the worms it becomes completely

---

*The so-called sensible and the irritable systems thus exhibit *writ large* the opposition that takes place in every single organ on a small scale, e.g., in every nerve.—Insofar as every nerve is an organization, there will be those three stages in it, and to that extent there will again be a triplicity in every organ; but for the organism as *total product* the nervous system is *merely* reproduction of sensibility, just as the muscular system is *merely* reproduction of irritability—although every individual organ, e.g., every nerve, *again* has, if I may say, its nerves, and that threefold graduated series must be thought as present in every organ generally, e.g., there is a sensible and irritable system in every nerve.

†for the appearance.

‡The movement of the heart is mediated just as much by sensibility as are the voluntary organs, only here a direct transition takes place.

§This second proof is especially to be carried out through the theory of disease.

indemonstrable, and in the zoophytes all external signs of sensibility disappear simultaneously with it.

Now, just as the brain gradually dwindles away throughout the whole organic world and finally disappears, it is the same with the external organs of sensibility. The eye, e.g., is preserved down to the insects, and comes forth in a few races, e.g., in the birds, more perfectly. In the insects the structure of the eye begins to abandon its regularity, for here it appears sometimes very large and sometimes very small, sometimes it is only an eyelike organ, sometimes hundreds of them at once, in which that sense spreads itself out. In most of the worms, if they have eyes, they are at least covered. In the polyps no such organ is demonstrable, although they appear to seek the light.

It is uncertain through what medium that *one* force which is the cause of sensibility refracts itself into various rays; nevertheless, the diminishing manifoldness in the structure of the brain teaches us the mounting preponderance of one sense above all others, and the final contraction of all senses into one homogeneous sense (as in the polyps) shows that that force begins to become always more uniform backward in the series from man, and finally disappears in completely involuntary movements.

If, in this way, sensibility gradually fades throughout the whole of organic nature, then according to the law proposed, irritability must rise in the same proportion. But where sensibility absolutely disappears, it is only because it loses itself immediately in movements, in which case the movements can be called *involuntary*, although for the genuine physiologist the concept of a voluntary movement is a meaningless concept.* The movement of the heart indeed appears as involuntary, not as if this movement were not mediated through sensibility just as are all organic movements, but [215] because here sensibility loses itself directly in its effect, and instead of the cause we see only the effect. Conversely, other movements appear as *voluntary* because they are not produced by any *determinate* stimulus (e.g., that of the blood, through which the heart is moved), but only by *the sum total of incessantly acting stimuli* (of light and other universal causes). Since these stimuli continually propagate effects without every single one passing into movements—(in which alone one recognizes sensibility, for sensibility *is* nothing other than the negative of irritability), then by this means a *sum total* of force of motion must arise which the organism seems to be able to dispose of, since for it its utilization is just as necessary as in the so-called involuntary movements. Therefore, simultaneously with the exhaustion of that sum of stimulation which follows upon strenuous efforts (and which is called fatigue)—likewise, fully in accordance with the proposed law, in the mounting irritability of the involuntary organs (which is produced through intoxication)—sensibility seems to be extinguished too (in sleep), although it is certain that sensibility is not

---

*Since there are, in the strict sense, as little voluntary as involuntary movements.

extinguished, judging by the (uninterrupted) dreams during sleep (which one must conclude exist also in animals from the many movements they make while in this state), and that sensibility (as source of life) cannot be extinguished *otherwise* than with the extinction of life itself.*

Assuming this correction of the concepts of voluntary and involuntary movement, then instead of sensibility, irritability must alone come forth where sensibility fades in organic nature; i.e., sensibility must be completely lost in irritability. According to the conventional terminology, the movements must become increasingly *more involuntary*.

And so it is. In the plants, their saps are indeed circulated by stimulation of their vessels, but only in a few vessels, and only in a few plants (e.g., in the *hedysarum gyrans*); in others, there are traces of something similar to voluntary movements only in certain conditions (e.g., at the moment of complete sexual development). The movements of the *mimosa pudica* and of *dionaea muscipula*, among others, because they follow upon a *determinate* external stimulus (usually *contact*), are only to be seen as involuntary movements (and therefore the debate over the sensibility of plants is settled. Sensibility (as *universal* cause of life) must also belong to plants. But it *must* be *indemonstrable* in organic nature [216] to the degree that the preponderance of the subordinate forces increases, because it is only assumed to be there when it does not perish immediately in movements).

It is precisely the same in the lowest classes of the animal kingdom, for here too all movements contract themselves into such a narrow sphere and in such regularity that the last trace of choice disappears too.—Where sensibility gradually comes forth more visibly, for example, in the class of insects and the amphibians, the movements become less uniformly regular† and more manifold (one should remember that many insects unify in themselves every sort of movement), but irritability still asserts its independence from sensibility, since even after the disturbance of the whole organism its effects endure in individual organs and the lesser *vulnerability* of these animals proves the restricted reign of sensibility.‡ Finally, with mounting vulnerability the sub-

---

*Cf. pp. 115–116, note ‡.

†seemingly freer

‡It is generally well-known how far this vibration of individual parts goes after the disturbance of the organic constitution, particularly in the classes of amphibians and insects. This independence of the subordinate organic functions from the higher here goes so far that insects, even after the major organs (head and heart) are taken from them, still exercise technical drive and reproduce. It is generally well-known that worms, maggots, butterflies, and snakes, even after the separation of the hind-part from the head, still undertake all sorts of motions.—Ridley tells of a turtle which after having its head cut off lived for six months and ran around as if a burdensome ballast was taken from it. Afterward, its heart and guts (excluding the lungs) were torn from its body and it still lived for another six hours, and still displayed many of the motions which it undertook in natural conditions.—Here, then, the whole organism is almost nothing but irritability.

ordination of irritability under sensibility also increases, but such that at the same time the velocity, manifoldness, and force of the movement increases (as in the most agile animals, the birds and most of the warm-blooded animals, whose irritability declines at the same time as sensibility). Only gradually does the *mobility* decline, but sensibility only comes forth at the apex of all organization as master of the whole organism in absolute independence from the subordinate forces.

It is thus proven through general induction *that throughout the whole of organic nature, as* IRRITABILITY *increases* SENSIBILITY *decreases, and as* SENSIBILITY *increases* IRRITABILITY *sinks*.

Sensibility is lost indirectly through irritability, but irritability is lost immediately at the most extreme boundary of organic force, where the organic and the anorganic world are divided—the *force of reproduction*.

B. *Reciprocal determination of sensibility and force of reproduction*. If sensibility loses itself in the force of reproduction only through irritability, then the force of production must increase in the same proportion in which irritability achieves preponderance over sensibility; and so it happens, for from man backward we see the force of production on the increase throughout the [217] race of four-footed animals, the birds, and so forth, down to amphibians and fishes; for slowing nutrition, the decline in irritability, the manifoldness of peculiar secretions (the animal venoms, e.g., among others), an altered power of assimilation, finally sometimes the size of the produced individuals, sometimes their more complete formation, sometimes their ever-increasing numbers (becoming incalculable at the lower levels), already announce the preponderance of the force of production in this part of Nature. Where the force of reproduction again decreases with respect to its intensity (in the insects) the drama of metamorphosis steps in, and with it the technical drive; and where the latter is extinguished, an unlimited regenerative drive steps into its place.*—But in the same proportion sensibility sinks.

C. *Reciprocal determination of irritability and force of production*. Even where irritability scarcely remains (in completely involuntary movements), the force of reproduction, the furthest of all organic forces, must be present in the phenomenon. There must be a third system in every organism, therefore, which one can call the *reproductive* system, and to which belong all organs of nutrition, secretion, and assimilation.—Why is the excitable heart not an organ of secretion, but instead the sluggish liver? Further, Blumenbach and Sömmering have proven that only those parts that are independent of the brain, and *all* parts only of such animals as do not even have a brain at all, or a very imperfect one, *regenerate themselves*. This means, expressed more generally: the force of

---

*Polyps cut up, quartered, turned inside-out like gloves.

reproduction in all of its perfection first comes forth where irritability and sensibility are either already extinct or are at least close to extinction.* This level of organic nature is indicated by the race of the zoophytes and the plants (of which each individual part is *identical* with all others, and almost all heterogeneity has disappeared).†

[218] *Conclusions.*

In sum, all of the foregoing furnishes the following *result*: "The organism, in order to be stimulable, has to be in equilibrium with itself, and the organism as object falls in this point of equilibrium. If the organism was not in equilibrium with itself, then this equilibrium could not be disturbed, and there would be no dynamical source of activity in the organism, there would be no *sensibility*. But precisely because sensibility is only the *perturbation* of the organic equilibrium, it is only recognizable in the continual restoration of the equilibrium. This restoration is displayed in the phenomena of irritability; thus, the most original factors of excitability are sensibility and irritability, which necessarily coexist. But because the product of every restoration is always again the organism itself, it appears at the *lowest* level as the constant *self-production* of the organism, and its cause appears as *force of reproduction*; but *that* it appears as such is, finally, only conceivable through the influence of a higher order, by which

---

*Surely the force of reproduction is not conditioned by the absence of *nerves* (for otherwise the naiades, e.g., could not exhibit regeneration), but by the sinking of sensibility down to a determinate degree which one must investigate in experience, and which itself exists with the existence of the nerves. (Original note.—Trans.)

†There still remains the proof touched upon under b), p. 144, which can be carried out with reference to the *various states of one and the same individual*. As Nature, along with the whole organic world, runs through those three stages admitted by us (Nature repeats itself constantly—only it begins in the one where it leaves off in the other)—so also with every individual. The same graduated series in the whole is again in the individual. The individual is only a visible expression of a determinate proportion between sensibility, irritability, and force of production.—*Shape* is only the expression of a dynamic relation—e.g., with sinking irritability the whole system of repiration is limited, with sinking sensibility the organ of the brain is limited.—Now, if every organization is only an expression of this proportion, then it too exists only within these limits—neither short of them, nor beyond. If the proportion were not determinate then no deviation from it would be possible either. If the existence of the individual were not *restricted* to this determinate proportion, then a deviation from it could exist with the existence of the product. Conversely, to the extent that the proportion is a determinate one, from which no deviation is permitted to occur, the product is capable of disease. Those states are thus the opposed states of *health* and *sickness*, and so we clearly find ourselves here—led to the concept of disease by our theory of the dynamic graduated series in organic nature. (The SW editor notes: "The derivation of this concept from the graduated series is presented later in the *Outline*. In the lectures, however, Schelling inserted the chapter on disease here, after the remark in the manuscript."—Trans.)

the organism protects itself against the influences of its immediate external world, and is, so to speak, armed against it (i.e., by *excitability*)."

The following principles flow directly from this:

If there is a gradation of forces in the organism, if sensibility presents itself in irritability, if irritability presents itself in force of reproduction, and if the lower force is only the phenomenon of the higher, *then there will be as many stages of organization in Nature overall as there are various stages of the appearance of that single force.*—The plant is what the animal is, and the lower animal is what the higher is. In the plant the same force acts that acts in the animal, only the stage of its *appearance* lies lower. In the plant it has already wholly dispersed into force of reproduction, which is still distinguishable as irritability in the amphibians, and in the higher animals as sensibility, and conversely.— —

THEREFORE, THERE IS ONE ORGANISM THAT IS GRADUALLY ATTENUATED THROUGH ALL OF THESE STAGES DOWN TO THE PLANTS, AND ONE CAUSE ACTING UNINTERRUPTEDLY WHICH FADES FROM THE SENSIBILITY OF THE FIRST [219] ANIMAL DOWN TO THE REPRODUCTIVE FORCE OF THE LAST PLANT.

If every point in this evolution were not a point where the force becomes productive force and *necessarily the point where the force refracts itself* (above p. 139), *then there would be nothing but plants and reproductive force*; for only in that that force, as productive force, has to divide itself into opposed individuals does it become possible that it reproduce its condition to infinity, and thereby reproduces its product.

Instead of the *unity* of the PRODUCT which we above sought in Nature and which we could not assume precisely because of the separation into opposite sexes (above p. 46f.), now we have *a unity of* FORCE of production throughout the whole of organic nature. It is indeed not one product, but still ONE force, that we observe to be inhibited at various stages of appearance. This force originally tends toward only ONE product; that the force is inhibited at various levels means just that that *one* product is inhibited at various stages—and it follows necessarily that all of these products inhibited at various stages amount to only *one* product.

Thus it is high time that we demonstrated and justified the thought that the graduated sequence of stages in organic nature, the organic *forces of sensibility, irritability* and *formative force,* are all only branches *of* ONE *force*, just as, without doubt, *only one force is expressed in light, in electricity, and so on, as in its various appearances.**

**On the World-Soul*, p. 252f. (Original note.—Trans.)

If the *universal* organism contracts itself into organic nature, so to speak, then at least the *analogues* of all of those organic forces must be manifest in universal Nature. And thus [220] 1) LIGHT is that which corresponds in *universal* Nature to the cause of the FORMATIVE DRIVE in organic nature. If light is the final cause of all chemical process (above p. 96), then the formative drive (like the organic for the anorganic generally) is itself only the *higher potency* of the chemical process, and so, since all inorganic formation only occurs chemically, one action gives *all* natural formations their lawfulness.*

Now, under no circumstance is anything at all *material* to be thought by this action, as little as by *light* itself. It is simply not itself material, only its immediate products are. If light were its *product*, then it would be *matter*, in the general sense that something is matter. Since all matter is occupation of space, i.e., *is* action of a determinate degree, then to that extent all *matter* is *immaterial*. Light is not its *product*, but only its phenomenon. Light, i.e., that which we *call* LIGHT, is not matter at all, not even a "materialization" (matter conceived in becoming), it is rather *becoming itself*; illumination is the most immediate symbol of the never-ceasing creation.—Since light requires no higher light, and since it is really that which signifies the extreme boundary of our sensibility, it can no longer itself be an object, i.e., matter. However, it is self-evident that some sort of substrate, some substance must lie beneath that *becoming* which we call light. What we call light is not that substrate, but the *becoming itself*.

(Naturally, the question arises, how does this perspective on light harmonize with its chemical effects, and even with the optical phenomena which are supposed to prove the materiality of light?)

[221] What count as the a) chemical *effects* of light are all reducible to the *deoxidizing* property of light. The reason for this property must be sought in the relation of light to oxygen. Now what is this relation?

Since light comes to the fore in the chemical process, just as oxygen, the middle-term of this process, disappears, then oxygen must be the mediator of the opposed spheres of affinity (of the Earth and of the Sun).—As long as both are separated and only *indirectly* come into contact, i.e., as long as that mediator (which separates the two) is still present, there is *duality* too, and with it electricity. As soon as the mediator is canceled and the opposed spheres of affinity pass over into one another—the phenomenon of that transition is the Sun exposing itself, so to speak, in light—all duality is canceled and the chemical process begins.

---

*Influence of light upon crystal formation. Prévost's new light experiments?—Universally, with a more robust in-pouring of light, movement increases in organic nature, and so on. (Original note.—Trans.)

Now, since light is only the phenomenon of oxygen's disappearance (which steps into its place, as it were), conversely, oxygen must also be a phenomenon of light's disappearance, or that which steps into the place of light. Oxygen is mutually opposed in *both* spheres of affinity, precisely because it separates both and mediates both. Light must thus disappear where its *antithesis* emerges, and so appears to act as a means to deoxidation (as a combustible body, so to speak). But light, i.e., *that* which *we* call light, does not *deoxidize*, its disappearance only corresponds to the deoxidation.

Light does not *deoxidize*, but the active principle whose phenomenon it is. It is a universal law of this action that it acts on the negative positively, and on the positive negatively (e.g., the oxidized body is negatively electrified, the positive is nonoxidized). So it does not deoxidize, but it makes something *positively charged*. Whether this deoxidation corresponds to a combustion of the light-substrate is another question.—With the assumption of such a [222] deoxidizing cause, much that was previously enigmatic becomes clear; e.g., the quantity of oxygen in the atmosphere always remaining the same *taken as a whole*, which is only explicable by the fact that a *universal*, uniformly acting cause maintains an equilibrium of negative and positive states, and so inhibits the matter from being lost either in one or the other extreme. The universal action acts on the positive in an oxidizing fashion, as on the negative in a deoxidizing fashion, and both effects coexist in Nature just as constantly as positive and negative electricity.

However,

b) concerning the *optical* phenomena that are supposed to prove the materiality of light, we find ourselves the less necessitated to admit this materiality the less those phenomena (e.g., refraction, among others) are of a self-evident nature, and the more certain it is that almost no principle of our optics has an indubitable existence.— —

The same activity that appears on a lower level as *formative drive* appears on a higher as *irritability*, for the fact that both are identical in principle is already certain since the condition of both is heterogeneity; and so now, in order to extend the argument,

2) ELECTRICITY would be that which corresponds to IRRITABILITY in the external world. Of course we will be permitted to supply, in place of all other proofs, the galvanic phenomena.*

a) It is certain that the galvanic phenomena are identical with the electrical in their ultimate principle, although galvanism and electricity are themselves

---

*This has already been demonstrated in the preceding on the basis of the identity of their conditions.

diverse phenomena; for through galvanism electricity is raised, as it were, to a higher function.* Electricity only requires duplicity, and only appears in the contact and separation of heterogeneous bodies. But galvanism requires *triplicity* as its condition, and is active in closed chains and in *rest itself*.† It is the same way with that active principle to the extent that it is cause of irritability, for that action, because its conditions [223] (triplicity) are always present in the organic body,‡ can never *rest*, but its activity is a *uniform* one; it achieves expression by contraction just as it achieves expression through electricity: only by means of a new closure or separation of the *chain*. Thus, the activity in the galvanic *chain* is *not* itself electricity (at least not what one has previously understood by electricity), but is probably conditioned by electricity. It is electricity raised, as it were, into uniform activity, an activity, so to speak, enclosed within a system of bodies, and an effective action only in this circle and affecting nothing outside of it.§ However,

b) it does not follow, therefore, that the agent in the expressions of irritability is *itself* electricity (as little as it follows from the preceding that light is itself agent of the formative drive). Electricity is only that which corresponds to that higher (*organic*) action in universal Nature.‖ Organic action is itself, moreover, a higher power of the galvanic action. Even the contractions of the organ connected in the galvanic chain do not appear to be *immediate* effects of the alteration operative in this chain.—*Electricity* is a wholly *external* appearance in relation to *irritability* (which becomes a seemingly inner activity only under the form of galvanism, because it is effective here only within the chain in which it is enclosed).—Conversely, the cause of the manifestations of irritability is an absolutely inner one, an action absolutely shack-

---

*potency

†So it is not absolute identity, but only identity in ultimate principle. The activity that acts in the galvanic chain is already no longer simple electricity, but electricity raised to a higher power. What is at least certain about the galvanic phenomenon is that what *corresponds* to the phenomena of irritability is *electricity*. This correspondence of organic and universal phenomena of Nature might even come back in the end to the fact that the organic phenomena generally are simply the higher power of the universal phenomena of Nature.

‡one ought to read Fontana's fitting microscopic observations on the structure of the muscle in his *Investigations on the Nature of Animal Bodies*. (Original note.—Trans.)

§Therefore it is conceivable that no electrometer measures it, nor can measure it. (Original note.—Trans.)

‖And that which is operative in the galvanic chain only appears to effect the transition from electrical action to the action of irritability. Even the action which is operative in the galvanic chain is not identical with that which is active in the organ itself when it is contracted. And so galvanism in general seems to be the middle-term which connects the universal phenomena of Nature with the organic, or the bridge over which the universal appearances of Nature cross over into organic, e.g., the action in the galvanic chain is obviously the middle-term between electricity and irritability.

led to the organic.* Electricity is thus only to be seen as a later descendant of that organic force that is only *indirectly* recognizable as cause of formative drive and [224] of irritability in its products, and presents itself directly only where everything organic stops.

Nevertheless, the action that is cause of irritability is joined to the same conditions as electricity, and quite a few unsolved riddles are solved as a result. For the time being it is certain that *oxygen* (as mediator of opposed spheres of affinity) must be the *indirectly determinant* factor in this higher process too (as in the electrical process) and that it cannot *immediately* engage in this process (because otherwise the chemical process would be unavoidable), but only acts upon it through a third body, which is, so to speak, its representative.† The

---

*That electricity itself (whose first conduit would have been the nerves) could not be the cause of irritability is already refuted by Haller's single suggestion that electricity is not a force in and for itself that can be thought to be contained within the nerves (surrounded by conducting substances of all kinds). (Original note to this point, but it continues in the manuscript.—Trans.) That electricity itself is the cause of irritability, whose first conductors would be the nerves (as is usually imagined), is already impossible due to the fact that one cannot conceive how electricity, surrounded by so many conducting substances, could be shackled to the nerves. The cause of irritability is a wholly inner action, chained to the organ, and electricity is that cause intuited only at the lowest layer of appearance—in the first power. Moreover, it is precisely explained here why the process of irritability is so definitely connected to the same conditions to which the chemical and electrical are—without being either one of them: connected to the condition of the chemical process, because it has the electrical in common with it—connected to that of the electrical, because it has the higher power of the electrical in common with it. Virtually all discoveries of *animal chemistry* can be explained by this observation.

†If the process of irritability is only the higher power of the electrical process, then it can be ascertained from this why both processes are joined to the same conditions, as already noted. In order to juxtapose them here still more closely the principle derived from the theory of chemical and electrical processes will be presupposed: oxygen is the one invariable factor of every chemical process—it is indirectly the constant factor of the electrical process. What connection will oxygen have with the process of irritability?

It is well-known that many physicists in recent times have flatly proposed oxygen as principle of stimulability. Quite a few experiments seem to confirm this assumption, many more than are usually supplied, but it is impossible to accept for other reasons that oxygen is actually the direct principle of stimulability. All of those experiments indeed prove that oxygen plays a great role in the phenomena of irritability, but not that it is principle of the manifestations of irritability.—I should, before I investigate the matter more closely, eliminate a few misunderstandings. Very many objections to this, specifically those from Röschlaub, rest on a misunderstanding. He says that oxidized bodies stimulate very little, e.g., plant nutrients—all kinds of vegetables—all vegetable acids—oils, which are employed principally in sthenic diseases with great advantage. Conversely, oxidizable substances stimulate most powerfully, like opium, alcohol, ammonia, and so on.—But these objections rest on the misconception that we suggest that oxygen is the principle of *excitation*. It is asserted, rather, that it is the principle of *excitability*, principle of *stimulability*.

It is false and a complete confusion of concepts when one presents oxygen as a powerful or strong stimulant. It is not, it is precisely the opposite. Oxygen can at most apparently *stimulate*, because it raises the excitability, if it is assumed that the sum total of stimulus is not *decreased* through

third body in the animal life process is the blood, which alone contacts oxygen directly, and steps forward in the life process as its representative.* Because the blood is propelled as a fluid body, and as a substance of variable quality, generally altered (deoxidized) by every contraction, it alone fulfills the condition of the *third factor* in the galvanic life process set out above (p. 119f.); that is, it makes possible a constant becoming and cessation of *triplicity* by its alterability. Without that contact the life process would soon stand still, because its condition (always renewed heterogeneity) is lacking without it. Conversely, while nutrition on the one hand (which occurs in animals[†] by means of combustible materials) and respiration on the other (which [225] changes the blood into an oxidized fluid[‡]) constantly reproduce the condition of all electrical process (namely, an opposite relation of its factors to oxygen), the life process too (as a higher kind of electrical process) must always be rekindled anew.

Just as irritability declines throughout organic nature, and with it the electrical process, so too the conditions of that process will gradually disappear.

---

it—for then the same sum of stimulus will act more strongly on the excitability increased through it, than previously on the lower. All intensity of stimulus is a relative one. Oxygen can thus *seem* to act as stimulant—but only then indirectly. As a rule it acts *weakeningly*; it raises the factor of *aesthenia* or of receptivity, and is thus in the most genuine sense principle of stimulability.

The solution to the contradiction is thus in short this: *Combustible* bodies directly stimulate. Oxygen, on the contrary, must directly suppress the excitation as the opposite factor—and only raise it indirectly, through the raising of stimulability. But if the stimulability is raised beyond a certain boundary, then *aesthenia* follows. Oxygen thus always acts immediately aesthenically, and all of this follows directly from the assertion that it is principle of stimulability. A by far more important question arises: *how* then oxygen does *raise* stimulability, and the answer to this question is one of the most important for all of physiology.

Oxygen cannot engage directly in the life process, just as little as in the electrical, but it plays a role only through a body that is its representative.

*Oxygen thus provides the *negative* term in the electrical processes of life by means of the blood. The oxidized blood does not act insofar as it is oxidized, but insofar as it is *negatively* electrically charged.

†for the most part

‡Incidentally, the blood acts in the animal body as a substance of generally variable quality, since it is again deoxidized through the manifestations of irritability themselves (undoubtedly because *nutrition* is correlated with them). From this perspective, the opposition is particularly remarkable that exists most conspicuously in the contractions of the heart. If the right part of the heart is determined to contraction by the blood returning from the whole body, i.e., for the most part deoxidized blood, then it is inversely the blood coming directly from the lungs, i.e., blood richly endowed with oxygen which stimulates the left part of the heart to contraction, and so the blood (according to its quality, this *lar familiaris* in the galvanism of the life process) seems to have to transform the quality of the remaining factors in the current chain. (Original note.—Trans.)

The plant predominantly has force of reproduction only insofar as the irritability in it is already completely reduced, and since the plant only exists as force of reproduction, its life (and also the degree of irritability that exists with its life, i.e., with this determinate proportion of organic forces) will be promoted through everything that retards irritability. Therefore, the conditions of its life process will already appear as the opposite of those of the animal. The plant will be only *negatively* galvanizable.*

(Galvanism, it is said, does not extend into the plant kingdom. Why not? In plants, it only becomes the negative of *animal* galvanism. It is obvious that stimulability, as far as it is attributed to plants, is promoted by substances which are all *negative* in the electrical conflict, like calciferous metals, water, saltpeter, nitric acid, sulfuric acid, salts of all types, and so on. It becomes clear that here it is not only the oxygen of these substances that is operative, as is usually believed, but also their *negatively electrical* constitution that acts (that *sulfur*, for example, manifests the same effect as the acids).—Now, all of these bodies are *ineffective* in animal galvanism as soon as they cease being liquid (this contributes to the proof that it is not their chemical quality that makes them active).—On the contrary, it is exceedingly clear that just such bodies as are the most effective in animal galvanism (opium, e.g., carbon (according to Ingenhousz), and certainly also metals) depress the stimulability of plants.)

[226] As irritability sinks throughout organic nature, respiration declines along with it (i.e., the influence of oxygen on the organism), and with respiration, circulation. Respiration is the most extensive in the animals where the manifestations of irritability follow upon one another with great rapidity and in close succession (the birds, e.g., in whom the air, through pipelike organs associated with the lungs, penetrates into the hollow and marrowless bones of the bird). And, although more dully and more slowly, it occurs regularly in the same way down to the fishes (now water in the gills may serve them instead of air, according to Vicq' d'Azyr, or they might breathe the air found in the water itself, according to others). But precisely here the whole system of irritability transforms itself at once, one ventricle of the heart disappears and the blood does not return to the heart any longer through a particular channel to the lungs. In the insects the lungs disappear, and in their place air-channels appear. In them, as in the species of worms, the heart is also only a series of nodes which slowly contract one after the other, and that which is called their blood is cold and colorless. Finally, in the polyps there is no more trace of respiration (although it must be presupposed), and every trace of heart or vessels disappears in them

---

*i.e., almost purely negative stimuli must act on it, otherwise it is not a plant

too.—Finally, with the plants (i.e., where irritability sinks the lowest) respiration becomes an *expiration* of pure air, and oxygen, which has the *opposite* function of *nutrition* in the animals, becomes to them nutrient itself (directly or indirectly) as Ingenhousz has shown.

Taking all of this together, it becomes clear how *oxygen* extends its reign throughout the whole of Nature as ground of determination in the dynamical process, and how one can say with Girtanner, in a certain sense, that it is the principle of irritability. It is so in the same way that it is principle of electricity. But the deceptive element in many arguments for and against this opinion is explained.—It can be said in general [227] that the animal, in opposition to the plant, is in a positive state of life (the proof is the constant decomposition of oxygen in the former, and the state of reduction in the latter). Now, since oxidation everywhere induces the *negative* state, since it depresses the phlogistic excitability (increases the heat-capacity) like the electrical, and the negatively electrified is also a negative stimulus for the organism, then it is conceivable how oxygen *increases* the organic receptivity, i.e., the excitability of the animal, and by just this means becomes (indirect) cause of increased activity,* and how, conversely, the substances opposed (positively electrified) to oxygen raise the positive state or depress it indirectly (through exhaustion of excitability). It is conceivable how, conversely, the negative stimulus must act incessantly in the plant (must become habitual), how the plant must be chained to the earth (as combusted substance), how everything deoxidizing (light, combustible substances, and so forth) exhausts its excitability in a moment, and how in contrast negatively electrical bod-

---

*Since life is momentarily extinguished with the lack of respiration, that [negative stimulus, that] influence of the air contrary to life, is actually the force constantly retarding the activity of life, which hinders—through the increase of *excitability* [better: stimulability]—the excitation from reaching its minimum in a single moment (because every stimulus lowers excitability) [since the effect of the stimulus is inexorably propagated, then the stimulability conceived as constantly sinking would sink with accelerated velocity toward the zero-point, unless a never-failing, never-absent stimulus inhibited the consumption of stimulability]. The oxygen, or its representative, the arterial blood, is thus constantly the negative term in the galvanic chain of life (that which, in the chain of mounting stimulability of the individual organ, is the negatively electrical body). (In addition to the bracketed comments above, the original note continues in the manuscript.—Trans.) Pfaff has already proven that when, e.g., zinc, i.e., the positively electrical body, lies constantly in the nerves that are opposite the muscle, the stimulability of the organ is negated more quickly than in the inverse order. Mr. Röschlaub afterward discovered that the organ's stimulability is lost when it is connected-up in a positive chain (I express myself so for the sake of brevity); that conversely the organ, already to a high degree nonstimulable, becomes stimulable again to a high degree when it is brought into the opposite chain. I conclude from this that the negatively electrical body in this chain only augments the stimulability because it acts as a negative stimulus. Now the oxidized blood in the living body constantly has the same function that the negatively electrical body has in such a chain, namely, it functions as the retarding factor of the vital processes, to inhibit the exhaustion of stimulability.

ies alone, while they preserve the plant's weak excitability, indirectly increase their excitability.*

Irritability is only *one* factor of excitability. The external cause of excitability (which we have derived above) indeed produces the *appearances* of excitability (i.e., the manifestations of irritability), but only under the condition of an original duplicity, or, what is the same, the *sensibility* in the organism (see above, p. 112).

[228] So we are driven to a still higher cause in the external world which must be related to electricity in just the same way that sensibility is related to irritability. The *highest* effective cause in Nature that we know *until now*, precisely that universal dynamic action, already presupposes as the condition of its activity a *dynamical juxtaposition*, i.e., an original duplicity. Thus a higher cause *beyond* this cause must be presupposed (as universal dynamic source of activity).

And so†

3) *universal* MAGNETISM will be that which corresponds to SENSIBILITY in the external world, or, the same final cause which in universal Nature is cause of universal magnetism will be cause of sensibility in organic nature.

a) Just as sensibility stands on the boundary of all phenomena in the organic world, so does that which corresponds to sensibility in universal Nature. It must be for universal Nature precisely what sensibility is for organic nature, i.e., *universal, dynamic source of activity*, and just as all organic forces are subordinated to sensibility, all dynamic forces of the universe are subordinated to its correlate.

b) In that which corresponds to sensibility in the whole of nonorganic nature there must truly be only *identity* in *duplicity* and *duplicity* in *identity* (what else is meant by the expression *polarity*?). Precisely this is the decisive feature of all organization. But is not just this identity in duplicity and duplicity in identity the character of the whole universe? If the universe is the absolute totality which comprehends everything within itself, then it is object *for itself*, since it has no object outside of itself, and turns toward itself. The opposites fall in the *interior* of the universe, but all of these opposites are still only various forms into which the one primal opposition, extending itself in infinite branches [229] through the whole of nature, transforms itself—and so the universe is, in its absolute identity, only the product of *one* absolute duplicity.

We have to think the most original state of Nature as a state of universal identity and homogeneity (as a universal sleep of Nature, so to speak).—For the

---

*The thoroughgoing correspondence between the conditions of the process of irritability and the electrical process leaves not a bit of doubt that electricity is the corresponding term to irritability—and this is the proposition that was to be proven.

†Since until now the necessary existence of magnetism in Nature has not been derived like that of light and of electricity, the following for the moment makes claim to a merely hypothetical truth. (Original note.—Trans.)

first and highest causes that we know are active only under the condition of duplicity and already presuppose it. The action of gravity presupposes at least a *mechanical* juxtaposition, the universal dynamical action presupposes a higher *dynamical* juxtaposition. What will be the cause that has been the genuine *source* of their activity, higher than all those subordinate causes?

We can ascertain the following about what kind of cause this is:

—That which is source of all activity is (because activity alone is knowable) itself no longer objectively knowable (as the sensibility in the organism is not). It is something absolutely *nonobjective*. But only that which is itself cause of everything objective, i.e., cause of Nature itself, can be absolutely nonobjective.

What then *is* the organism other than Nature condensed, or the UNIVERSAL *organism in the state of its greatest contraction*? Thus, an *identity of the final cause* must be accepted, by means of which (as through a shared soul of Nature) organic and anorganic, i.e., universal Nature, is ensouled. The same cause which threw the first spark of heterogeneity into Nature has also thrown into it the first germ of life, and that which is the source of activity in Nature overall is also the *source of life* in Nature.

The same cause that inhibits the extremes of Nature from passing into each other and inhibits the universe from disintegrating into one homogeneity, this very cause also inhibits the dissolution of the organism and its transition into a state of identity. Just as *all* activity is conditioned by absolute duplicity, so *organic* activity is conditioned by organic duplicity (a simple modification of the former).

[230] THUS, A COMMON CAUSE OF UNIVERSAL AND OF ORGANIC DUPLICITY IS POSTULATED. The most universal problem, encompassing the whole of Nature, and for this reason the *highest* problem without whose solution nothing is explained by all of the foregoing, is this:

What is the universal source of activity in Nature? What cause brought forth the first dynamic juxtaposition in Nature (of which the mechanical is a mere consequence)? Or what cause first cast the seed of motion into the universal rest of Nature, duplicity into universal identity, the first spark of heterogeneity into the universal homogeneity of Nature?

Appendix to Chapter III.

The problem set out above (p. 53) in which we predicted that all problems of the philosophy of nature may readily be unified is resolved in its complete generality by the preceding section.

Nevertheless, in addition to the perspective that is provided *for the whole of organic nature* by the presentation of that graduated series of organic forces, another is furnished for the organic individual which must be recovered in the form of an appendix. In this respect, all of the individual features of the preceding theory are gathered together in order to indicate, at the same time, the point by means of which another extremely important part of natural history is connected with the universal principles of the philosophy of nature.

Just as a graduated series of functions is established in the whole of organic nature, so too it is established in the individual, and the individual itself is [231] nothing other than *the visible expression of a determinate proportion of organic forces*. Shape, and everything else by means of which the individual is known, is itself only an expression of that higher dynamic proportion; for how the structure conforms itself to that higher proportion, and how an alteration in the latter draws an alteration in the former behind it, has even been shown through a number of examples.

Every organism is *constituted* by this determinate proportion, and exists neither just short of it nor beyond it. A *deviation* from the proportion is at all possible because it is a *determinate* one, and that the whole *existence* of the organism is *limited* by this proportion means that a deviation from it is intolerable for the existence of the whole product—in a word, *both together* make the organism capable of DISEASE.

The concept of disease is a completely relative concept, for *first of all* it has meaning only for the organic product of Nature; that is, in the concept of disease one thinks not only the concept of a deviation from some rule, order or proportion, but also that the deviation does not exist with the existence of the product *as such*; the latter determination really completes the concept of disease.—However, the concept of disease is relative within *this* sphere* as well. With the degree of irritability, e.g., by which the *plant* is sickened, the polyp would perhaps be quite healthy. With the degree of irritability at which you yourself feel ill, an organism lower on the scale would feel quite fit.†—A *determinate* degree of excitability is also part of the constant reproduction of a determinate organism. If the degree of excitability were not relative to every individual, then one could think of it (as intensive magnitude) as decreasing to *infinity*, through infinitely many intermediate degrees approaching to zero. But a determinate *degree* of excitability belongs to every individual in order to

---

*the sphere of the organic

†There is therefore no *absolute* sickness. Every sickness is only a disease in relation to this determinate organism, which this proportion of organic functions cannot tolerate.

preserve this determinate organism against the encroachments of external nature, and to reproduce it in face of its impinging influences.

So much for the *concept* of disease.*—The following principles must be presupposed in the original construction of the concept itself.

1) Disease is produced by the same *causes* through which the phenomenon of life is produced.†

2) Disease must have the same factors as life.‡

The essence of every organism consists in the fact that it is not absolute activity (the likes of which is thought in the concept of vital force, for example), but an *activity mediated* by *receptivity*;§ for the existence of the organism is not a *being*, but a perpetual *being-reproduced*. If the organic activity exhausted itself in its product just as activity is exhausted in its product in the dead object, and unless external, contrary influences inhibited the exhaustion of organic activity in its product and determined the organic to perpetual self-reproduction, organic existence would be a *being*.

The organism as such can exist only under the constant influence of external forces, and the essence of the organic consists in a receptivity by virtue of which activity exists—and in an activity which is conditioned through receptivity—both of which must be summarized in the synthetic concept of *excitability*.‖ This cannot be thought without positing an original duplicity in the organism. That the organism is excitable (or reproduces itself) in conflict with external pressure means that the organism is its own object. It is possible that it never ceases to be its own *object* by constant restoration of the original duplicity in it (by which its sinking back into absolute homogeneity, death, is inhibited). The function of the external causes, i.e., of the stimuli, is the constant restoration that inhibits the organic activity from being lost in its product.

---

*This whole investigation is only presented here as the medium of the principal investigation—in order to prove through the theory of disease that the graduated series, which is found in the whole organic chain, is also expressed in every organic individual.

†It is thus totally nonsensical to call disease an unnatural state, for it is precisely just as natural as life. If disease is an unnatural state, then so is life—and admittedly it is unnatural to the extent that life is really a state extorted from Nature, not favored by Nature, but a state enduring against Nature's will, for it is preserved only by means of struggling against Nature. In this sense one can say that life is a perduring sickness, and death only the recuperation from life.

‡It follows that in the foregoing construction of the phenomena of life, the factors of the construction of the phenomena of disease have also been provided.

§movement

‖The external force would not produce the phenomena of life by itself unless a capacity to be determined to certain functions existed in the organism.

The *factors* (inner conditions) of life are thus contained in the concept of excitability, but its *causes* are found in the uninterrupted influence of *external forces*.

It cannot be thought how the organism is not ruined* by external stimuli (but is instead determined to self-reproduction) unless it occurs through the influence of a higher external cause, a cause which [233] cannot proceed from its immediate external world, but proceeds from a higher dynamical order to which the former is itself submitted. In the construction of the phenomenon of life we distinguish the *first cause of excitability* from the causes of *excitation*. For the latter (Brown's "stimulating potencies") produce the appearance of excitation only under the condition of excitability.†—

Thus, a *cause of excitability, independent of the stimulating potencies, must be accepted* (which is also indirectly cause of excitation‡); *to this extent the original independence of excitability must be presupposed*.

Excitability is recognized only in excitation. It is thus *known* only *insofar as it is determined through the stimulating potencies*, not in its *independence*, for in its self-subsistence or in its independence from the exciting potencies it is dead, without *expression*.—

If excitability is determined only for the APPEARANCE *through the stimulating potencies, then it is* (although originally independent of them) *alterable through nothing other than the stimulating potencies.*§—If it is accepted that it relates inversely to the intensity of the stimulus, then it cannot be increased other than by minimization‖ of stimulus, nor depressed other than by augmentation of the stimulus.

As excitability contains the factors of life, so too of *disease*. The *seat* of disease must be excitability, its *possibility* must be conditioned through the

---

*destroyed

†(According to SW, the last passage in the manuscript reads differently.—Trans.) Now, it cannot be thought how the organism is not destroyed through external stimuli, but is instead determined to self-reproduction, if the final *source* of its activity can be *immediately* or directly affected by external influence. The cause of excitability must therefore be a totally different one from the causes of excitation (Brown's stimulating potencies)—must belong to a higher dynamical order than the former, and generally must be such a cause that can never be directly acted upon, but only indirectly.

‡(According to SW, this parenthesis is struck out in the manuscript, and the sentence continues as follows.—Trans.)—must be accepted. Excitability is originally independent and lies outside of the sphere in which we can directly act. We can thus only act indirectly upon excitability through external influences. But external influences act *only* by stimulating, only by exciting.

§We cannot directly either increase or diminish excitability. If it is diminished or augmented, then this is possible only through the middle-term of excitation.

‖this minimization occurs not directly through *withdrawal* of stimulus (privation), but through negative stimulus (in the genuine sense), through the likes of negative electricity, for example.

alterability of excitability. But excitability is *alterable* only through the stimulating potencies. *The cause of disease cannot lie in excitability insofar as it is self-subsistent,** *but only in its relation to the stimulating potencies.*†

(It also follows directly from this principle that *excitability* cannot be acted upon in any other way than through the *middle-term of excitation*, that the *source* of excitability cannot be *directly* affected, but only indirectly affected through the causes of *excitation*.—The still-dominant theory sees excitability as something *self-subsistent* in theory, but cancels this self-subsistence in practice, for it [234] believes itself able to act directly upon excitability, which is the true meaning of their "soothing," "refreshing," and other specific expedients.‡ This theory views excitability as something still lying within the sphere of our medical means, as something *directly* alterable through the influences of *this* external world of ours. *But the cause of excitability lies outside of the dynamical sphere* within which the means that lie in our power are contained; it must be thought such that it is subordinated to no affinity of the Earth, and can be directly affected through no potency of the Earth. The proof for that proposition can be deduced from the principles of the higher physics.)

It is assumed that no effect from without reaches into the source of excitability itself. The cause of excitability is itself not alterable, but only the

---

*Excitability is a cause always remaining selfsame to the extent that it is self-subsistent, i.e., the cause of excitability itself, which, if it acts nonuniformly, is determinable only through the nonuniformity of the negative conditions.

†The *seat* of disease is excitability, but its *source* the relation to the stimulating potencies. For excitability is only alterable through the relation to the stimulating potencies. Thus disease too can only *arise* through this relation.

‡Example of *opium*—only indirectly soothing (cf. above, p. 18). It is just the same with the refreshing expedients, e.g., *cold*. How is it thinkable that the cold strengthens *in itself*? Is it a particular essence? Is it not mere negation? Thus it can only refresh *through negation*, only indirectly strengthen.—Relative concept of cold—the heat of the freezing-point is still stimulating too (proof: because life is possible in it)—A reviewer of Brownian texts in the *L.Z.* has ascertained quite well where the previous systems are really *sick* when he says that Brown's statement, "all external causes act upon us only as *stimulants*," is slippery, because he has not proven that there are no causes which work *directly* on excitability—able to augment or diminish it directly. Whether Brown snuck in that statement may be left to one side—it must be concluded from such expressions that he at least did not prove it with complete self-evidence. But now the resolution of the major question depends obviously just upon that statement: how can the organism be acted upon at all—whether directly or not—whether to augment or to diminish directly—or whether excitability according to Brown is something *in itself* invariable and is only to be altered indirectly *through* the stimulating potencies. But whether, e.g., medicine could be traced back to its first principles, i.e., could be raised to the level of an actual art, depends again upon this question. It is already won when the disputed issue is correctly construed.

causes of excitation. It is further assumed *that through the mere alteration of these causes* excitability itself will be altered* too.

The proof is the following:

The cause of excitability, whatever it may be, must be thought as a *self-subsistent* cause, as a cause that is independently active anywhere its *conditions* are given† (this has been proven above). There actually exist such self-subsistent causes in Nature that are active from themselves anywhere their conditions are given or are instituted whose DEGREE *of activity* is determined by the degree to which their conditions are given. Such causes are, for example, light, electricity, and so on, whose sources it is indeed not in our power to affect; but it is in our power to institute the conditions for them.‡ Therefore, the cause of excitability must be thought to be like the cause of light, as such a cause that is alterable for us only when its conditions exist. For it is, like the latter, a cause whose principle no longer falls within the dynamical sphere of the Earth, but in a higher sphere, as has been proven above, i.e., it is a self-subsistent cause. The difference between the two causes is [235] only that those universal causes cannot be exhausted, at least in *this* organization of the universe. In contrast, excitability is a determinate cause for every organic individual, and determinate for every moment of its existence. Its source is thus not inexhaustible. *To the extent that the conditions* under which that cause appears as active, i.e., *the stimulating potencies, are* increased, *excitability is necessarily diminished*, and conversely, only *to the extent that* those conditions, i.e., *the stimuli, are diminished* can excitability be *augmented*.

So we have explained how excitability can itself be affected through the middle-term of excitation, without it being necessary to see it as a directly variable magnitude, or to think a hypothetical substrate of excitability that would likely be endowed with chemical affinities which again are not known and upon which chemical means are allowed to act whose mode of action is known once more§ only through randomly performed experiments.‖ If it were even possible, irrespective of the thesis that the cause of excitability would never itself be known, we do know the *conditions* of its appearance, which allow themselves to be investigated on the path of experience and of experimentation and

---

*and in the proportion that the latter are altered

†and which is active *in the degree* to which its conditions are given

‡When I produce light in a dark room through the rubbing of hard substances, I have not produced light that did not exist previously; I have only established the conditions under which it is active of its own accord.

§not according to laws, but

‖i.e., on the grounds of a blind empiricism

lie in our power to alter, and through whose alteration excitability itself is altered, and by whose means the final source of life itself can be affected, not blindly and at random, but through familiar and determinate laws.

Until now we have taken excitability as a *simple* concept. It was accepted that it is alterable through the middle-term of excitation, could be diminished through augmentation of stimuli, and augmented through minimization of stimuli. But it follows from this that excitability always stands in inverse relation to the stimulus; thus the stimulus stands in inverse relation to excitability viewed according to the intensity of its effect.* For it can only *diminish* the excitability through the middle-term of excitation, it† must thus (with identical *absolute* intensity) excite the more, the *higher* excitability stands.‡ Since the same stimulus acts far more strongly [236] on a higher excitability than on a lesser, the *relative* intensity§ of the stimulus increases in direct relation *with* excitability, and conversely, they lose relative intensity in the same proportion that excitability sinks.|| *But excitability is determined through nothing other than the stimulating potencies*; it is only that which the stimuli make out of it. It can therefore only be *raised* in that its stimulus is *withdrawn. In equal proportion as*

---

*i.e., the higher the excitability, the lesser the intensity of the stimulus, and inversely

†the stimulus

‡Let's take the absolute intensity of stimulus = 30° for two individuals, where A has an excitability = 40, B = 50, then the stimulus will act on individual B the more strongly, for its excitability is higher.—It will be completely otherwise with the *relative* intensity of the stimulus.

§The distinction between absolute and relative intensity is very important. We do not know absolute intensity. Were it to be determined, it would be related inversely to excitability. The *relative* intensity, i.e., determined through the degree of excitability, must just for this reason be *inversely* related to the latter. The relative intensity of the stimulus will increase *as* the excitability increases, and decrease as it decreases.

||[Eschenmayer in his "Theses on the Metaphysics of Nature"] has objected to Brown that no excitation that deviates from the mean degree of excitation [thus also no disease] is possible according to his construction of life from stimulus and excitability. This is because one factor cannot rise without the other falling, and conversely; this latter reason, expressed *universally*, is totally false. If one accepts excitability as a variable factor, then the thesis is false, because according to the above the augmentation of the relative intensity of the stimulus runs *parallel* with the augmentation of excitability. (So far original note, but bracketed comments above and what now follows were added.—Trans.) Now the issue is surely that excitability is not the factor *in itself* variable, but since it is only variable through the stimulating potencies the latter must always be assumed as variable factor.—Undoubtedly, something true lies in that objection but the reason is not correctly expressed. Namely, assume that 1) the *relative* intensity of the stimulus rises and sinks in equal proportion *with* excitability; assume that 2) excitability *in itself* is invariable—and is only variable through the stimulating potencies—that these are the single variable factors in the phenomena of life; then we conclude that: excitability is only variable through the stimulus; it can only be raised in that its stimulus is withdrawn.

its stimulus is withdrawn, the RELATIVE *intensity of the remaining* increases,*[†] THE PRODUCT IS THUS THE SAME AND UNCHANGED. Just as little can the excitability be diminished in any other way than by augmentation of the stimulus. *In the same proportion*[‡] *the* RELATIVE *intensity of the remaining decreases,*[§] THE PRODUCT IS NEVERTHELESS UNCHANGED.[||]

It is certain that by taking excitability as a simple concept no variability in the product of excitation can be thought, but such a thing must be thought because excitability is itself only variable through the alteration of this middle-term.

Thus excitability can not be a SIMPLE *factor.*

If it[#] is accepted as *simple*, then there can be disproportion only between excitability and stimulus, but such a thing is impossible, because one cannot

---

*stimulus

[†]namely, because excitability *rises* in the same proportion as its stimulus is withdrawn—and the relative intensity of the stimulus rises in the same way that excitability rises

[‡]as stimulus is added or raised

[§]because in the same proportion as the stimulus is raised, excitability decreases, and the relative intensity of the stimulus falls in equal proportion with excitability

[||]*Example*. Let excitability = 40, the stimulus = 40. Now the question arises, how can excitability be altered, e.g., increased. According to the theory, only by the *minimization* of the stimulus. The question arises whether this is possible. Let the stimulus be lowered from 40 to 20, then the excitability will be increased in the same amount, i.e., it will now be = 60. But a stimulus of 20° acts just as strongly on an excitability of 60° as a stimulus of 40° on a 40°-high excitability. The product, the excitation, is thus the same in both instances. This is because the stimulus recovers in relative intensity what it loses in *absolute* intensity, because the latter rises and falls parallel to excitability.—This is the real vulnerability of the proof, that from *stimulus* and *excitability* alone one cannot construct any variability of excitation, e.g., disease too.—Now conversely, let us posit that excitability is *diminished* to 20°, then this is not possible except by increasing the stimulus by just as much, namely by 20°. Now we have the excitability = 20, the stimulus = 60. But the stimulus = 60 acts on a 20°-high excitability just as strongly as on a 40: 40. The stimuli decrease in relative *intensity* in the same proportion that excitability is diminished, i.e., they lose in relative intensity what they gain in absolute intensity—the excitation, the product, will thus always remain the same, no actual deviation from the mean degree of excitation will be possible—so there cannot be any disease.

It is apparent that we are not able to answer this difficulty from the previously accepted assumptions, and that we are imprisoned with our principles. If the problem is solvable, then a concealed error must lie in the preceding—*or* something has been implicitly accepted. For if the principles are *true*, and they are proven, then nothing false can follow from them; there must have been a mistake made in the conclusion.

The whole argument rests on the thesis: excitability can only be decreased through augmentation of the stimulus, and inversely. How is excitability understood in this thesis? Excitability is understood as a *simple* factor, the *whole* of excitability will be diminished through augmentation of the stimulus. The mistake must therefore lie in this assumption of the simple factor.

[#]excitability

take anything from excitability without giving it to the stimulus, and cannot take from the stimulus without giving it to excitability. Two concealed factors must lie in the concept of excitability itself, and it must be these which make a disproportion in the *excitation* possible. These factors and their relation must be determined.

a) It has been proven through the whole course of our science that in the synthetic concept of excitability the two factors of *sensibility* and of *irritability* are thought together.—It should be noted once again that by *sensibility* we understand nothing [237] other than *organic receptivity, insofar as it is the mediator of organic activity*. Here, as in the whole work, we understand by *irritability* not the mere *capacity* to be stimulated (which is surely the original meaning of the word), but, as a long-established use of language allows, *the organic activity* ITSELF, *insofar as it is mediated by receptivity* (the organic power of reaction).

b) *Both of these factors are themselves opposed to one another.*—It has been proven through a universal induction from the dynamically graduated series of organic nature that *as one of these factors falls, the other rises,* and the converse (III.).

What holds for organic nature overall holds equally for the organic individual (above). *Therefore, such a reciprocal falling and rising of both of these factors can occur in the individual as well.*

c) In the observation of organic nature it is shown that *sensibility* is not permitted to sink infinitely if any degree of irritability is to remain. We see, e.g., in the plant kingdom, where only a weak trace of sensibility exists in only a few individuals, *irritability* also fades at the same time as sensibility.

There is thus a certain boundary within which the law holds that irritability rises as sensibility falls. If this boundary is crossed, if SENSIBILITY SINKS BELOW *a certain point, then the opposing factor does not rise anymore, but it falls simultaneously with it.*

We can explain this law in the following way. All organic activity is mediated by receptivity, according to the first principle of all organic natural history. Receptivity and activity are opposed, one is the negative of the other. The higher the receptivity, the lower the activity, and conversely. Since all organic activity is not *absolute*, but only conditioned through receptivity, *then a* CERTAIN *degree of* RECEPTIVITY *must remain so that a degree of* ACTIVITY *may remain*. To be sure, *within a certain boundary* the rising of activity corresponds to the sinking of receptivity; below this boundary both sink in unison.

[238] (This is the wonderful relation of opposed factors between which organic life balances, as it were, without once being permitted to step out of it, a relation that John Brown first intimated, although never completely developed. It is remarkable to see how the direction peculiar to his whole system of thought is preserved by the observation of this relation in experience. "I saw," he said, "that the increase of strength and of excitation keep in step up to a certain point, but a moment finally comes when the strength and the excitation no

longer keep in step and where the strength passes into indirect weakness."⁹ The discovery of this relation is one of the deepest probings into organic nature. Not only the individual, but also the whole of organic nature oscillates between those bounds.—On the highest level, sensibility has the decisive predominance, but the phenomena of irritability also occur here with greater ease, only with lesser energy than on that level where the preponderance of forces directed toward the outside comes to light with gradually sinking sensibility, in the sthenic natures of the lion, for example, and its co-rulers among the animals. Receptivity becomes narrower and narrower backward in the organic world, and the preponderance of irritability is recognized only in the duration of its appearances. Finally, sensibility disappears completely in appearances, receptivity is near the zero-point, but precisely here those aesthenic natures step forth, the *plants*, where that boundary within which the sinking of receptivity and the rise of activity keep in step is already crossed. The plants are in an indirectly aesthenic state; in an *aesthenic* state because their existence is tolerated only with the lowest degrees of irritability, in an *indirectly* aesthenic state because here their receptivity already stands below the boundary above which its sinking runs parallel with the rise of organic activity.)

---

[239] The conditions of a possible construction of excitability are contained in the three principles just proposed, and through them the construction of excitation as a *variable* MAGNITUDE. If the *whole* of excitability is diminished through increase in stimulus (according to Brown), then the product (the excitation) again loses in excitability what it gains in stimuli; it thus remains the same and unchanged. If *sensibility* (receptivity) *is diminished* by the increase of stimulus, then irritability (or energy) *gains* (at least within the bounds set out above), i.e., the genuine factor of sthenia gains what the opposite factor of aesthenia loses.

Conversely, if by minimization of stimulus the *whole* of excitability is *raised*, then the product again gains in excitability what it loses in stimuli. If, through suppression of the stimuli, *sensibility is raised*, then, according to a universal law of organic nature, irritability will sink in the same proportion, i.e., *aesthenia* will arise.

Universally stated: the law that excitability is inversely related to stimulus does not hold for the whole of excitability, but only for one of its factors, sensibility.

By this separation of excitability variability enters into it, and by means of variability, excitation. *The total product of excitation* (excitation seen as a whole) *is at any rate invariable*, and it even must be so in order that its individual opposite factors can be variable. Suppose that the stimulus rises suddenly from 40° to 60°, then the receptivity (= 40°) must sink by 20°. But receptivity is the converse of organic *energy*, thus the latter will be raised necessarily by just as much *through* the sinking of receptivity by 20° (and so forth, to that determinate boundary for every individual). Now, if one has the receptivity = 20°, the energy or the activity directed outward = 60 (the *whole* of excitability = 80); if one calls the effect on receptivity *sensation* (in the meaning explained above)—the effect on the energy *irritation*, and *both together excitation*, then one has sensation = 20, irritation = 60, the whole of excitation = 80. Thus here excitation as *total* [240] *product* is invariable, and it even must be in order that the individual factors can rise and fall. So a *bipolar positing* of excitation is necessary; the more excitation* directed inwardly, the less excitation directed outwardly, and inversely. Therefore, the *whole* is always equal to itself, but *within* this whole disproportion is possible.

All conditions for the construction *of disease* as a phenomenon of Nature are provided by this construction of excitability and excitation as variable magnitudes. The following are the principles to which the construction may be reduced.

1) In a state without affection from outside (if such a thing can be thought at all), sensibility and irritability would not be at all distinguishable. *In every affection the two are separated.* Now, since *sickness* is provoked (quickly or gradually) only through affection from the outside, like the phenomena of life itself, *these two factors are separated in every disease.*

2) Through every affection from the outside, i.e., through increase of stimulus, sensibility is diminished, so it is necessary that irritability increases in the same proportion (up to a certain boundary), and indeed that *energy* increases.

(We are suggesting that the magnitude of irritability (the power of reaction) has to be gauged not according to the *readiness* of its expressions, but according to their *strength*. The *readiness* of movements stands in direct relation with *sensibility*, as experience also shows by innumerable examples; conversely, the *strength* (at least within the known bounds) always stands in inverse relation to sensibility. On account of the high degree of sensibility, the child, e.g., is very easily determinable, i.e., by means of slight stimulus, but

---

*stimulation

also only to feeble movements. If the organic power of resistance increases, the movements become more forceful, more energetic too—in equal proportion to the sinking sensibility.—Or, one might observe the difference of the sexes, or the climatic [241] differences of peoples, or finally the increase of the forces directed outwardly in Nature, which also happen in a certain (inverse) relation to sensibility.)

3) This thesis must be proposed as a principle of the construction of all disease: *the two factors of excitability are opposed to one another, such that within a certain limit* (which is a determinate one for every organic individual, and which one must investigate through experience) *the irritability or the energy rises as the sensibility or the receptivity falls, and conversely, and all disease is conditioned by this reciprocal sinking and falling of the two factors of excitability.*

According to Brown, disease is conditioned by the disproportion between stimulus and excitability (but it has been shown that such a thing is unthinkable)—according to us, it is conditioned by the disproportion *between the factors of excitability itself* (produced, of course, by means of the unrelenting or sudden effect of stimuli). According to Brown the stimulus is itself a factor of disease, according to us merely a *cause*.

4) The possibility of a disproportion is only presented in the organism by the fact that the two factors of excitability are posited as mobile and in an inverse relation—this possibility is exhibited as the *energy* (or the factor of *sthenia*) is raised, while the *receptivity* (or the factor of aesthenia) is diminished, and inversely. It is not yet explained thereby, however, *how the rising of the one factor and the sinking of the other produces* DISEASE.—Also supposing that Brown had actually constructed sthenia and aesthenia—then are sthenia and aesthenia *disease*? The question still remains, how do the two of them become *disease*?

Disease is only present where the organism as *object* is altered. As long as the organism as object does not appear to be an *other* it is not ILL. This is the question: How does a disproportion in the factors of excitability produce alterations in the organism as *object*?—

The organism as object only falls within a determinate proportion of the factors of excitability, for the whole circle of the organism is sealed by receptivity and activity. Since the whole [242] manifoldness of organic nature is itself conditioned with respect to its structure by the sinking and rising of those higher factors of life, it is conceivable how, according to the same mechanism, the whole organization can be altered—and how even the structure of the individual can be altered. *Every individual requires for its existence* (which is nothing other than a perpetual reproduction) *a certain degree of receptivity, and a certain degree of energy standing in inverse relation to it.* It is evident that a certain *range* must be admitted here, within which that reciprocity of the two factors does not produce any alteration in the organism as object. A degree of one or the other surpassing the limit is intolerable for the existence of the whole *product*, and it is *this intolerability for the existence of the whole product* that is felt as *sickness*.

5) *The diseases must be divided into diseases of raised sensibility* (receptivity) *and diminished irritability* (power of action) on the one hand, *and into diseases of diminished sensibility and raised irritability* on the other hand. A third class includes those where the increase of irritability no longer runs parallel with the sinking of sensibility, the diseases of the *indirect* weakness of the power of reaction. Since all organic functions are subordinated to sensibility and disease is possible only through (indirect) affection of the final source of life itself, to that extent *sensibility is the seat of all diseases* (in the proper sense of the word, since it signifies nothing other than the mediator of all organic activity).

Since sensibility is not at all directly recognizable, but only indirectly in its object (the expressions of irritability), and a diminishing of the former is recognizable only in a rising of the latter, and conversely, *then the diseases are all, on the first level of their appearance, diseases of irritability*.

All *phenomena of reproduction* are also determined by the higher factors of life and of disease. *An alteration in their relation must* [243] *be propagated all the way to the force of reproduction*. Only after the disease has propagated itself from its original seat of sensibility through irritability to force of reproduction does it take on a *visibly specific* character, and thus the whole manifoldness of disease forms *emerges* from two original basic diseases. Irritability is not *the same* throughout all systems of the organism (according to degree); its identity only means that it cannot be raised or diminished otherwise than *uniformly*. Irritability passes into force of reproduction (above p. 149) in the *same* proportion as the degree to which it is diminished (e.g., in force of secretion), thus it can produce a merely *gradual* alteration of the same altered phenomena of reproduction, or altered phenomena of secretion, e.g., without any *specific* affections of irritability (of which the *nerve pathologist* dreams).

Practical physicians, in their usual stupidity,[10] see disease only on this lowest level of its appearance; for example, in the spoiling of the humors (*humoral pathology*), but this already presupposes disease.*

6) The proposition must be established *as the principle of all medicine* that the force of reproduction can only be acted upon by means of the higher factors to which it is subordinated, but sensibility (the ultimate source of life) can be acted upon only through the middle-term of irritability. *Irritability is the single middle-term by means of which the organism can be acted upon at all*, thus all external forces must be directed upon it. But *how* the ultimate source of all move-

---

*Even in every infection (a concept which makes sense only for the organic natural product) something higher occurs than the average humoral pathologist imagines. The product is a uniform one, the affection of the formative drive is the same as in higher operations. (Original note.—Trans.)

ments could be acted upon through irritability is only conceivable by virtue of the inverse relation in which it stands with sensibility.

The conditions of the process of irritability are familiar and can be investigated experimentally (its conditions are identical with those of the chemical process, as well as with those of the electrical process, although it is itself not chemical). Thus it is to be expected, assuming the principle that the source of life can only be acted upon [244] through the middle-term of excitation, that when the theory of excitation is first reduced to the fundamental principles of physics, medicine too will be brought back to more secure principles, and its exercise to infallible rules.

General Remark.

The concept of disease, like that of life, drives us necessarily to the assumption of a physical cause which, outside of the organism, contains the ground of its excitability, and indirectly through it, of all the alterations taking place in it. For how could we believe that the organism itself has the sufficient reason for its life and its duration in itself, since, with respect to all alterations (in particular those of disease), we observe it to be dependent upon a uniformly acting external force only variable through its conditions, which must act uninterruptedly upon the first source of life of the organized body,* and which seems to sustain the life of universal Nature just as much as it sustains the individual life of every organic being (as the life of Nature is exhibited in universal alterations).

Now, when we look back on the preceding to see which forces have been proposed as corresponding to the organic in universal Nature, we find just those which, by universal agreement, must be seen as the causes of those natural alterations, and whose connection with the phenomena of life the natural historians of all times have in part intimated, and in part actually asserted.

All of these assertions concerning the physical causes of life and the theories built upon them (whose founders have seen farther, at bottom, than those who posit life in excitability, and who find it impossible or superfluous to explain it further) express a fundamental lack (aside from the fact that none actually constructs life from them), namely, that the basic character of all [245] theories, their inner necessity, escapes them all. This lack cannot be remedied otherwise than by the demonstration, from the possibility of Nature in general, of the necessary existence of those causes in Nature, and from the possibility of an organism generally, as well as from the necessary existence of conditions under which alone they are effective in the organism, all of which we believe

---

*Schäffer on sensibility as life-principle in organic nature. (Note in original—Trans.)

has been achieved in the preceding. We have not only proven that the conditions under which those causes are active are *necessary* in the organism by virtue of its essence and its nature, by which it *is* an organism at all, but we have also presented the existence of those causes themselves and their uninterrupted effectiveness in universal Nature (as conditioned by the existence of a universe generally), and we have thus joined the organism and life, even the most innocuous plant, to the eternal order of Nature by means of their final causes.

IV.

The highest function of the organism (sensibility) drives us to the question of the first origin of *universal heterogeneity* (see above, p. 158). And in the organic world *formative drive* is what *chemical* PROCESS is in the anorganic world.

The condition of the *chemical* process is also a *universal heretogeneity*, and to that extent it has the same conditions as the force of reproduction. The solution to that problem of heterogeneity is to be seen as a theory of the chemical process, and conversely, the theory of the chemical process is simultaneously to be viewed as the solution to that problem.

---

[246] General Theory of the Chemical Process.

A.
Concept of the Chemical Process.

§.1.
The cause that we recognized above as operative in irritability and force of reproduction was characterized as one that appears under the condition of duplicity. But a cause whose activity is conditioned by duplicity can only be such that it tends toward intussusception, because the latter is not thinkable without two individual bodies that pass into one indentical subject. The tendency of that cause must thus be intussusception, and if intussusception exists only in the chemical process, it *must be the cause of every chemical process.*

§.2.
Between organic and anorganic nature a like gradation exists as exists between higher and lower forces in organic nature itself. That which is irritability

in organic nature has already faded into electricity in anorganic nature, and what is force of reproduction in the organic has already dissipated in the chemical process.

§.3.

The cause of the chemical process tends toward the cancellation of all duality. Thus, there must be absolute intussusception in the chemical process, i.e., transition of two heterogeneous bodies into one identical occupation of space.* An identical occupation of space does not arise where [247] one body is only augmented by the other, for such an expansion would still leave us with two bodies; it must arise only where the individuality of each individual is absolutely canceled and a new body is formed as common product.

§.4.

Intussusception is simply *mechanically* impossible, as in the way the atomists represent dissolution, and according to whose concepts it is always only partial, i.e., it extends only down to the smallest parts of solid bodies which are spread out infinitely near to one another in the solute. Aside from the fact that this theory rests on the concept of matter as a mere aggregate of parts whose bond is unbreakable by physical force (for why otherwise should the force of the solute have its limits?)—further, aside from the fact of the unnatural representations to which the concept of a mechanical dissolution leads, such a thing absolutely should not be called intussusception since it deals only with *surfaces*, and if it does go further it is no longer thinkable mechanically.

§.5.

Rather, since the impenetrability of matter can only be thought as the standstill of expansion and contraction, penetrability can not be thought otherwise than as the restoration of that oscillation (i.e., through the disturbed equilibrium of expansive and compressive forces of matter). Since two materials cannot interpenetrate without becoming *one* material, each individual must cease, as it were, to be *matter*, i.e., a homogeneous occupation of space; matter must be restored to the state of primal BECOMING.

[248] §.6.

Assuming that the chemical process exists only where heterogeneous bodies absolutely pass into one another, the question arises, how can such an absolute transition of one into another be mathematically constructed?—However, this question belongs in the *formal* part of natural philosophy or universal mechanics, where it will also be answered.

§.7.

It evidently follows from this concept of the chemical process that its cause is not a cause again subordinated to the chemical process (at least to the

---

*Chemical division is always the correlate of chemical combination.

chemical process of the Earth); since a complete penetration occurs in it, and since individual bodies of the same sphere necessarily form a mechanical juxtaposition, it must be a cause from a higher sphere for which substances of the lower sphere are penetrable, and not impenetrable, as they are for one another.*

B.
Material conditions of the Chemical Process.

§.8.

The first consequence of the principles deduced is that the chemical process is possible only between *heterogeneous* bodies (for only where there is heterogeneity is there duplicity), and that if there is an intussusception between *homogeneous* bodies it cannot be of a chemical kind. The first problem of a theory of the chemical process is thus to deduce heterogeneity in Nature, which is its condition.

[249]
§.9.

For the time being it is understood that since every heterogeneity is necessarily a determinate one, and this concept is merely a relational concept, there must be certain fixed points of reference of all quality; therefore, the chemical process is a necessarily limited one, i.e., it has an extreme point beyond which it cannot go. If the chemical process had no such extreme by means of which it is limited, then neither would it have a point from which it could begin. Only the fact that the chemical process begins somewhere makes *determinate* chemical products possible. If it began nowhere and ceased nowhere, then a general intermixture of all qualities in one another would take place, i.e., no determinate quality would come to light in Nature.

§.10.

We have explained for the moment how the chemical process in the universe generally becomes limited through our theory of world-formation; that is, according to our theory the organization of the cosmos into systems of gravity is at once a *dynamical* (thus also chemical) organization of the universe, and a certain limit to the univeral evolution is determined through universal gravity.

§.11.

How the chemical process of the individual heavenly body (e.g., the Earth) is limited must be capable of demonstration through the indication of the points of inhibition at which all evolution of the Earth stands still.

---

*Baader on the pythagorean square, or the four world-regions in nature, 1798 [*sic*].—A text that will be discussed later on. (Original note.—Trans.)

§.12.

Since being inhibited is something solely negative, a solely negative presentation of those points of inhibition must also be possible. They will be indicated by that whose composition can be overcome by no chemical potency of this determinate sphere, i.e., by the indecomposable. The indecomposable is at all possible in Nature, [250] following the above (p. 31), only because it is the most composable, for otherwise matter would be lost in pure extremes. The most composable is recognized just in that it is not presentable in isolation—but only in combination with others. The genuine concept for that negative factor is thus the concept of *impresentability*, and no more nor less is to be thought under the concept of *matter*.

§.13.

The concept of a *simple* thing cannot be thought in the concept of matter. What matter is for chemistry is a material like every other, but it is composable to a higher degree, and just for that reason it is not matter presentable in isolation.—Therefore, it is evident at the same time that the concept of the "simple" in Nature is a false concept. Since a mechnically simple thing is unthinkable (like the atom of the mechanistic physicist), then only a dynamically simple factor can be thought, something that is no longer product but is solely *productive*. Such a thing has been designated above (pp. 20–22) by the concept of the "simple actant," and since an infinite product is evolving in Nature, one can only think an infinite multiplicity of simple actants as elements of Nature if the evolution is thought to be actually completed. Nevertheless, an absolute evolution in Nature can never come to pass, and the assumption of the simple is thus *false*, as well as its conclusion; no simple thing exists in Nature, and, because everything is *product* to infinity, neither does the dynamically simple, or the purely *productive*. The most original points of inhibition in Nature are not exhibited by simple actants, but by real products that are not able to evolve further (at least at this stage of Nature), and the simple actants are only the ideal factors of matter. [251]

§.14.

The *most composable* is thought in the concept of matter (§.12). But every composition requires *two* factors. There has to exist a *composable factor of the opposite kind* in Nature. The question arises how this is possible.

§.15.

The absolutely composable (which, just for that reason, is the *indecomposable*) must limit the process of the Earth. It has to be limited in opposite directions.

§.16.

For the time being, we cannot think any limit other than that of the evolution of the Earth overall. There is an absolutely inhibiting factor in the determination of qualities of the Earth. This *inhibiting* factor generally is the single truly insoluble, chemically inconquerable factor. Thus, what makes individual substances that stand at this limit insoluble is not their *simplicity*, but that *negative principle* of the Earth which is communicated to all of them in common and which one can call *phlogiston* (according to the original concept of its inventor), "earth-principle"; i.e., since earth is only the sensuous image of the insoluble, it may preferably be called the "insoluble."

§.17.

Since this negative factor must be seen as *determining* quality, and also as cause of the *chemical intertia* of bodies, then the purely indecomposable will come forth where this negative principle attains preponderance—(e.g., in the metals).

Since the indecomposable can only exist as composable in Nature, a positive principle must act on Nature, acting against that principle which, because the *shape* is fixed by that negative principle, will present itself as a principle inimical to all shape (because [252] it is in conflict with the negative forces of the Earth, those favorable to *formation*), i.e., heat. This principle will be a principle awakened only by an alien (positive) influence (light).

Remark. It is self-evident from the foregoing why this positive principle acts most strongly on those parts of Nature where the negative principle of the Earth attains greatest preponderance. This is why the capacity for heat, e.g., of the metals, is the least; why they increase in equal proportion with *oxidation*; and finally why the force of cohesion of such bodies is destroyed by every chemical process, and, while their absolute weight increases, their specific weight is reduced; why, conversely, the chemical function of a body is also altered by increased cohesion (why, e.g., ice becomes *positively* charged, and breaks the light *more weakly* than water, and so on).

§.18.

Another type of indecomposable (or composable) factor must be opposed to this indecomposable for the reasons set out above (§.14). It will become clear from the following observations what type of factor this is.

§.19.

If every chemical product is an *association* of heterogeneous parts, then the factors of the product must be *opposed* in relation to the product. Now, all material of the Earth is really only *one* factor of a single higher product, which follows necessarily from the theory of universal world-formation set out above. If the universe has been formed by an infinite differentiation of one primal product into an increasing number of novel factors, then every *individual* factor can only = one of these factors, and what belongs to it must be *homogeneous* within itself (all material of the Earth, for example). But the condition of the chemical process is *heterogeneity*.—If all materials of the Earth = one matter (their diversity merely a diversity of variety), then there is no real opposition [253] between them, and thus no chemical process is possible.

§.20.

If a chemical process is to be possible, then one factor must be a material which is opposed *to all* materials of the Earth, and in relation to which all materials of the Earth count as only *one* factor. If it is this one factor opposed *to all* matter of the Earth that makes all chemical processes possible, then no chemical process must be possible between materials of the Earth, other than one mediated by that one constant factor, i.e., only insofar as any body from the sphere of affinity of the Earth is a *representative* of that principle.

§.21.

That principle must be the middle-term of all chemical affinity and of all chemical processes, and also, just for that reason, the ideal point of reference for all determination of quality.

§.22.

Since that one factor of all matter on the Earth is collectively opposed, it will intervene directly or indirectly into every dynamical process of the Earth, but it cannot again be a product from the sphere of affinity of the Earth; it must be a product of the higher sphere of affinity, i.e., a product of the *Sun*, and to that extent the *Sun* (or rather the relation of the Earth to it) is the final cause of all chemical processes of the Earth.

§.23.

It obviously follows that this principle must not be *reducible* by any substance of the Earth, since it is not a product of the Earth; thus, it must be an absolutely *insoluble* factor, but for that reason an absolutely *composable* one.

[254] §.24.

We know from the theses on the chemical influence of the Sun (set out in the preceding) *which* material of the Earth that product is. The necessary existence of such a principle is deduced here *a priori* as a condition of all chemical processes, and in experience it presents itself as that which our chemistry calls *oxygen*. It will become clear from the following what *function* this principle has in the dynamical process.

§.25.

Oxygen can only be the indirectly or the directly determining factor in the dynamical process of the Earth. In the first case, a body must come forth as its *representative*, by virtue of its relation to oxygen; it has been deduced above that this happens in the *electrical* process. In the other case, oxygen would *engage* in the process itself, either indirectly through a body with which it identifies, or directly. In the latter case, since it is only the middle-term that separates the two opposed spheres of affinity of the Earth and the Sun, as soon as it disappears the higher sphere of affinity has to step forth in its phenomenon, light (as Sun), which oxygen represents in opposition to the Earth, i.e., a *process of combustion* must take place.—It cannot be conceived how an earthly body can become its own source of light (like the Sun) except by virtue of the breaking up or opening which separates both spheres of affinity effected by the middle-term.

§.26.

Thus, oxygen is the condition of the electrical process because electricity is possible only under the condition of the separation of opposite spheres of affinity, and *oxygen* is just the separating element. It is the condition of the process of combustion because the latter presupposes a transition of both into one another. But there is no transition without separation. Both processes therefore rest on the same opposition, but this opposition, which in the former is a mediated process, becomes an unmediated one in the latter.

[255]   §.27.

Since oxygen only represents the higher sphere of affinity to the Earth, it has at bottom the same function in the process of combustion as the *positive* body has in the electrical process. Just as the latter is only a representative of oxygen, the former is only representative of a higher affinity (of the Earth to the Sun). Just as oxygen is the determining factor in the electrical process, in the process of combustion it is the higher affinity of the Sun.

§.28.

Since this higher affinity in the process of combustion manifests itself as light, just as it must come forth as oxygen in the electrical process (before it can pass over into the process of combustion), one might say that oxygen is only itself again the representative of a higher principle, that is, of *light*.

§.29.

It is only possible that oxygen is opposed to all substances of the Earth collectively (i.e., that all of them burn with it while it does not burn with any other), because in the circle of affinity of the Earth it has no higher factor with which it burns. It is necessary that the *absolutely* incombustible substance in relation to a higher system is either a combusted one, or the substance in the subordinate system flammable to the highest degree. Thus all substances of the Earth, while they combine themselves with oxygen, combust indirectly through it with a higher principle.

§.30.

The process of combustion drives us to a heterogeneity receding to infinity. What will finally be the absolutely incombustible factor in the universe [256] with which everything ultimately burns and itself burns with nothing else?—It is readily seen that this chain recedes to infinity through constant mediation, and since all chemical processes are reducible to the process of combustion, every chemical process is conditioned by the final factors of the universe whose transition into one another would promote absolute homogeneity.

The chemical phenomena, like the organic, drive us to the question of the ultimate origin of all duplicity.* One factor of the chemical process always falls outside of the individual product (e.g., the Earth), it lies in a higher product; but for the chemical process of this higher sphere, its one, invariable factor again lies in a higher order, and so on to infinity.[†]

There is thus ONE universal dualism which runs throughout the whole of Nature, and the individual antitheses that we see in the universe are only shoots of that one primal opposition, between which the universe itself exists.[‡]

What has that primal opposition itself called forth, beckoned from the universal identity of Nature? If Nature is to be thought as absolute totality, then nothing can be opposed to it, for everything falls within its sphere and nothing outside of it. It is impossible that this unlimited (from the outside) change itself into a finite being for intuition except insofar as it becomes *object to itself*, i.e., in its infinitude.[§]

That antithesis must be assumed to have sprung from a universal identity. By this means we[||] find ourselves driven to a cause which no longer *presupposes* heterogeneity,[#] but itself *produces* it.

To produce heterogeneity means to create duplicity in identity. But duplicity is only cognizable in identity. Identity must again proceed from duplicity itself.[**]

[257] Unity in diremption only exists where the heterogeneous attracts, and diremption in unity only where the homogeneous repels. Both necessarily coexist; the homogeneous flees itself only insofar as the heterogeneous seeks

---

*The three stages that can be distinguished in the construction of the organic product must also be distinguishable in universal nature, just as in anorganic nature—there will be a *universal* dynamically graduated sequence of stages. This graduated series should now be established.

We begin with that which corresponds to sensibility in universal and in anorganic nature.

Sensibility is for us nothing other than *organic duplicity* and the first condition of the construction of a product in general. Now, just as sensibility is source of all organic activity, so duplicity *in general* is source of all activity in nature.—The chemical phenomena, e.g., rest on an opposition which recedes into infinity, as was proven in the theory of the chemical process.

[†]The electrical phenomena are also conditioned by the same opposition by which the chemical phenomena of Nature are conditioned. Further, the phenomena of gravity presuppose at least a mechanical juxtaposition, and this a higher juxtaposition.

[‡]If the most extreme ends of this opposition could pass into one another, then all dynamical phenomena would disappear and Nature would sink back into universal inactivity.

[§]turned toward itself—is divided.

[||]thus

[#]duplicity

[**]If there were not once more identity in the opposition, again reciprocal relation, then it could not at all endure *as* opposition. Thus, there is no duplicity where there is no identity.

itself, and the heterogeneous seeks itself only insofar as the homogeneous flees itself. We observe this production of the heterogeneous from the homogeneous and the homogeneous from the heterogeneous most originally in the phenomena of MAGNETISM. Thus, the *cause* of UNIVERSAL MAGNETISM would also be the cause of *universal heterogeneity in homogeneity and of homogeneity in heterogeneity*.*

Since heterogeneity is source of activity and of movement, the cause of universal magnetism would be the final cause of all activity in Nature; *original* magnetism being for universal Nature what sensibility is for organic nature—*dynamical* source of activity: for in the domain of mechanism one sees movement spring from movement. What then is the first source of *all* movement? It cannot *again* be movement. It must be the opposite of movement. Movement must well forth from rest. It is the same as in the chemical process, where it is not the moved body that moves the resting or moved, but the resting body moves the resting body. It happens likewise in the organism where no movement directly produces movement, but where every movement is mediated by rest (by sensibility).†

If one compares the features of that which ought to correspond to sensibility in universal Nature proposed above (p. 156f.) the following agreement is found.

---

*If we know that the cause of *magnetism* is cause of universal duplicity in identity, then we surely do not come to know the cause itself *more closely* (which is also impossible since it is the condition of everything objective and thus recedes into the most intimate reaches of Nature—is absolutely nonobjective)—but we can yet demonstrate its *effectiveness* in Nature, we can establish the stage of Nature at which it is *itself* still distinguishable.

I want to repeat the proof for this thesis once again. The thesis is "that we distinguish in the phenomenon of magnetism *alone* the universal duplicity in its *first origin*." The proof can be reduced to the following major premisses.

1) Nature is absolute identity within itself—absolutely equal to itself—and yet in this identity it is opposed to itself once more, object to itself.—The general expression of Nature is therefore "identity in duplicity and duplicity in identity."

2) All opposition in Nature reduces itself to *one* original opposition. If there were not again *unity* in this opposition, then Nature would not be a whole subsisting in itself. Conversely, if there were not again duplicity in this *unity*, then Nature would be absolute rest—absolute inactivity.—Therefore, neither unity without diremption nor diremption without unity can be thought in Nature. One must constantly proceed from the other.

3) How can it be thought that unity proceeds from diremption and diremption from unity unless the heterogeneous seeks itself and the homogeneous flees itself? Thus, this is the law reigning throughout Nature, in this *inner contradiction* lies the ground of all of its activity.

4) But this inner contradiction cannot be known originally, it is known only in the *phenomenon of magnetism*; in the latter alone do we distinguish universal duplicity in its *first* origin.

†Just as sensibility is the organic source of activity, so magnetism is *universal* source of activity. So that which corresponds to sensibility in universal and in anorganic nature is magnetism.—Now, one can reach the same result in another way.

a) Everyone will admit that magnetism stands at the boundary of the universal phenomena of Nature, just as sensibility stands at the bounds of the organic; i.e., that no phenomenon of Nature exists from which it could be deduced. The only phenomena from which anyone could attempt to deduce it, the electrical phenomena, have nothing in common with the magnetic phenomena aside from action through division, and this is precisely the higher factor in the electrical process—incidentally, *one can indeed match up every magnetic* [258] *phenomenon with an electrical one, but not a magnetic phenomenon with every electrical one.* That every magnetic body is electrical, but not every electrical body magnetic, proves that magnetism is a much more limited force as regards its *breadth*, and that just for this reason magnetism is also not subordinate to electricity, but electricity is subordinate to magnetism.

b) It is too clear to be proven at great length that in magnetism, in universal as well as in specific individual substances (which seem to step forth from the universal), the most original identity in duplicity, and the converse, exists (which is the character of the whole of Nature).*

Granted this identity of *sensibility* and of *magnetism* with respect to their cause, *magnetism must be the determining factor of all dynamical forces, just as sensibility is the determining factor of all organic forces.*

In order to bring this thesis to full self-evidence it is required of the proof not only that the same gradation of forces in universal Nature exist as in organic nature (for this is already certain), but also that this gradation follows the same *proportion* and the same *laws* in universal and in organic nature.

For the gradation of forces in organic nature (above, Division III.) the following proportion was found.

That which Nature has distributed most widely in the organic world is force of reproduction. Certainly more sparingly, but still very richly, it has handed out irritability, but most sparsely the highest force, sensibility.

What could be distributed more meagerly in the inorganic world than *magnetic force*, which we only perceive in a few substances?† The number of bodies *electrical* to some degree already increases extraordinarily, and there is no body that is absolutely nonelectrical, just as no organism is absolutely nonirritable. Conversely, the chemical property belongs to all bodies (perhaps in a certain inverse proportion to its electrical property, not yet discovered).

---

*In magnetism, we see in the whole of *nonorganic* nature that which is really the character of Nature as a whole—namely, identity in duplicity and duplicity in identity (which, said otherwise, is the expression of polarity). It should be said that every magnet is a symbol of the whole of Nature.

†Indeed, in many more than has been believed until now. Quite a few crystals, e.g., of the iron-rich and magnet-rich island Elba, also show phenomena of polarity.

[259] Further, every magnetic body is also electrical and chemical, just as no organism that has any part of sensibility lacks irritability or force of reproduction. But not every electrical body is magnetic, just as not every organism that shows traces of irritability also has sensibility.

But reproductive force is also irritability, irritability is also sensibility.* For example, that which is still irritability in the animal has already faded into reproductive force in the plant *for appearance*, and that which in the higher animal is distinguished as sensibility is distinguished in the lower animal only in irritability *for appearance*. Likewise, that which in the electrical body is still electricity has already faded in the chemical body into the chemical process *for appearance*, and in the electrical body, that which in the magnetic body is still magnetism is diffused in electricity. Magnetism is as universal in universal Nature as sensibility is in organic nature, which also belongs to plants. It is only *canceled* in individual substances *in appearance*; in the nonmagnetic substances that which in the magnetic substances is still distinctly magnetism dissipates (by contact) directly into electricity, just as in plants that which is still distinctly sensation in the animal dissipates directly into contractions.†

Only the *means* are lacking in order to recognize the magnetism of the so-called nonmagnetic substances, and to prevent that which appears on a higher level as magnetism from being lost in electricity or the chemical process.‡

If one looks further into the *mechanism* of that graduated series as it was determined for organic nature, the following results:

[260] There is one cause that fades gradually from one function to the other. Sensibility passes into irritability; this is not possible unless both have at least *one* factor in common. But do they not?—In appearance this is the system of the nerves, the organs are both sensible and irritable at once. Where the higher factor of sensibility (the brain) gradually disappears and the lower grad-

---

*as has been expressly ascertained

†What is remarkable is that that which is favorable to the chemical process or electricity, e.g., *heat*, weakens the magnetic force. Indeed, it is not true that oxidized iron ceases to be drawn by the magnet. [Yet the passive attraction in iron decreases in the same proportion as the iron is oxidized; Gehler, p. 94, and here: "quite complete iron-oxide is no longer drawn."] But only the superficially oxidized (rusted) magnet loses force.—Electrical sparks can rob it of this force (for whether they reverse the magnetic poles is uncertain). (Original note, excepting the bracketed sentence.—Trans.)

‡Reminders are not often needed, but in this case it is necessary to say that the discussion was not about this special magnetism (recognizable in the individual), but of original magnetism, with which the former is connected only by infinitely many intermediate steps. (Original note.—Trans.)

ually attains preponderance, sensibility also begins to fade into irritability (therefore, Sömmering's law that sees sensibility in the inverse relation of nerves to the brain).

In the same way, irritability and reproductive force must have at least one factor in common, otherwise, how could the former pass into the latter? And so they do. One factor of irritability, that oscillation of expansion and contraction, is also the condition of reproductive force, and just where irritability passes into reproductive force, one factor of irritability—the higher—is seen to disappear as well.—It is a universal law that the reproductive force of, e.g., individual parts, is viewed in inverse relation to its dependence upon the *nerves*. If irritability is to become reproductive force, then its higher factor must disappear; and conversely, where only the lower factor of irritability remains (e.g., contractility in cell membranes), it will become force of reproduction.

It can be established as a universal law for this graduated sequence that *the higher function is lost in the lower because its higher factor disappears, and the lower function becomes the higher factor of the subordinate force.*

Transferring this law to the dynamically graduated series in universal Nature, magnetism is the *producer* of heterogeneity acting by means of division (as perhaps the brain does). And that which is the oscillation of contraction and expansion in the phenomena of irritability is the oscillation of attraction and repulsion in the phenomena of electricity. Attraction occurs by virtue of the higher factor of electricity (action through dissociation), repulsion works by virtue of the lower factor, namely the communication of homogeneous electricity. (And who knows whether or not a similar sequence of dissociation occurs by means of the brain, and if the communication of homogeneous electricity by the nerves produces the appearances [261] of contraction and expansion of the organ?) *Precisely this oscillation of expansion and contraction is also the condition of possibility of all chemical processes.* Only by means of an oscillation of expansive and compressive forces can two different bodies pass into one identical occupation of space. If one supposes that that higher factor disappears (the *oscillation* of expansion and contraction), then the movement will either stand still in contraction (with the formation of *solid* bodies, crystallization, and so on), or in expansion (with the formation of *fluid* bodies)—and the *caput mortuum* is a homogeneous filling of space = *dead* matter.

We thus observe the final exertions of organic force in the chemical movements of bodies, and there is one force that binds together the most aggregated animal body just like the chemical body.

If general analogies have any demonstrative force there is no doubt that the same function must be ascribed to magnetism for universal Nature that we ascribe to the unknown cause of sensibility for organic nature. First of all, all duality comes into Nature by the power of magnetism. Since universal duality withdraws into the organism as into its narrowest sphere (the reason for its powerful and concentrated effects), the final cause of all duality is the same for the organism as for universal Nature.

Since the universal organism appears only in the state of its greatest expansion in the world system, magnetism will be that which inhabits the universe,* which makes it such that† every effect on the part propagates itself to the whole, just as in the individual organism. The impacts that the universe constantly sustains in this universal reciprocal action fade into movements that are recognizable only in reactive substances—(although the uninterrupted falling of the planets toward certain central points may be a movement mediated by universal sensibility).—But why is the magnetic needle sensitive to every considerable alteration in Nature, sensitive to the electrical light that illumines at the opposite pole, or sensitive to a vulcanic eruption [262] in the other hemisphere?—When one member in the great dynamical organization is disturbed, the whole *reacts*; Lichtenberg says that a solar storm that erupts in the Sun can descend upon us within eight minutes; but what is the so-called igniting of a fire other than such a descent of the Sun's solar storm?—

It probably cannot be doubted as a result of our discussion *that* magnetism has the same function for universal Nature that sensibility has for organic nature. It is proven THAT it is cause of universal heterogeneity, and as such is the determining factor in all activity conditioned by heterogeneity, but not shown *how* it is such. This must be shown.

It is conceivable how an original opposition is brought into Nature through magnetism. But the question is: *How have all individual antitheses in Nature developed from this one original opposition?*

It really is our position that all antitheses have evolved from one original.—That which has been proven by induction elsewhere,‡ *that one and the same universal dualism diffuses itself from magnetic polarity on through the electrical phenomena, finally even into chemical heterogeneities, and ultimately crops up again in organic nature,* is here to be deduced *a priori*.—The question is, *how* has that one opposition been broadened into such manifold antitheses?

If magnetism brought the first antithesis into Nature, then at the same time the seed of an infinite evolution was sown by this means, the seed of that

---

*as it were

†e.g.,

‡*On the World-Soul* (Original note. See AA I,6 179.—Trans.)

infinite dissociation into ever new products in the universe. Assuming the evolution that has been postulated above to be completed—or also as progressively occurring—then that original opposition is also posited as enduring (by the postulate of this evolution); the factors that are separated in it are posited as separate to infinity and always [263] separating anew. Where should the progressive action of that cause, not presupposing *heterogeneity* but *producing* it, be recognized? We know of no *production* of heterogeneity other than through that which is called *division*. If the universe is evolved, then that cause of heterogeneity will preserve the self-propagating *division* from product to product through the universal heterogeneity. The division which is reciprocally exercised will not only be the condition of gravitation in every system, but also the universally determining factor of the dynamical process.

Opposed forces are awakened by every action through division. But these, since they maintain equilibrium, produce a state of indifference, and actually all materials of the Earth find themselves in this state of indifference before they are subjected to the effect of (special) magnetism, or are brought into electrical or chemical conflict. The state of indifference will appear as a state of homogeneity. With respect to its qualities, such a homogeneous state also exists in every dynamical sphere (for, like the material of the Earth, the material of every other sphere must be posited as homogeneous in itself). This homogeneous state is not, however, a state of *absolute* homogeneity, it is only a state of *indifference*. The constant action of division from without, while it sustains this homogeneous state of quality, makes possible a removal of the condition of indifference, i.e., the dynamical process, and particularly the chemical process. Every body that is subjected to the chemical process must be *divided* IN ITSELF; without this diremption in the homogeneous itself no solution can be thought at all—that alternating play of expansion and contraction without which no chemical process is possible could not be thought. In order to be able to construct the chemical process the homogeneity of quality presupposed above must be resolved into duplicity. This homogeneity is only magnetic *indifference*. *Therefore*, magnetism must be posited as abolished *universally* and only *in appearance*. If that action from without could cease then the substances of the Earth would become completely inactive in the dynamical process, just as iron is (magnetically) inactive before the magnet has acted upon it—thus no difference of quality would be *recognizable* [264] either.—(The universal action of division can only be analogically compared with that which we see the magnet exercise. The latter always awakens the same poles—to infinity; for it, and every substance upon which it acts, is embraced in the universal sphere of earthly magnetism. The magnet can communicate no polarity, nor can the substance receive any, that would not be homogeneous with the universal polarity of the Earth. Conversely, the Earth is, e.g., *outside* of the Sun, therefore the magnetism of the Sun must awaken outside of itself a polarity distinct from its magnetism.)

The action of the Sun indeed produces polarity* through division in the dynamical sphere of the Earth, but the product of this polarity† is a universal state of indifference (the universal *point* of indifference is presented as center of *gravity*). There is indeed a universal heterogeneity in the *universe*, but every *individual* product is homogeneous within itself. If there is to be a dynamical process (whose condition is difference), then matter must be posited outside of the state of indifference. The question arises, posited by what means?—Will the higher product act on the subordinate one only through *division?*—Another manner of acting is still possible—through *communication*. If a communication really takes place between Sun and Earth (of which Light is at least the *phenomenon*), then the Sun will communicate something *homogeneous* to the Earth, as an electrically charged body communicates electricity to the nonelectrically charged.—*Heterogeneity* comes into the subordinate product by this communication, and with it the condition of the electrical and chemical processes.

Every dynamical process begins with a conflict of the originally heterogeneous. *Where the homogeneous contacts its heterogeneous factor it is displaced from the point of indifference* (the dynamical intertia in it is disturbed). *Throughout the whole of Nature homogeneity is only the expression of a* STATE OF INDIFFERENCE, *because homogeneity can only proceed from heterogeneity.* By this means, the dynamical process is grounded, and it cannot rest before the *absolute intussusception* of the heterogeneous, i.e., with the absolute abolition of its condition.‡

[265] Thus, there is ONE *cause that brought the original antithesis into Nature and we can designate this cause the* (unknown) *cause of original magnetism.*

An effect stretching itself throughout the universe to infinity by means of DIVISION *is conditioned by this cause, through the latter a state of indifference for every individual product is conditioned, and through this state of indifference the possibility of a difference in the homogeneous; through this the cause conditions the* POSSIBILITY *of a dynamical process* (to which the life process also belongs), *and in particular the chemical process, as a dissolution of the heterogeneous into the heterogeneous.*

The ACTUALITY *of the dynamical process for every individual product is conditioned by* COMMUNICATION, *which takes place in the universe to infinity, and whose universal medium is* LIGHT, *for the part of the universe known to us.*

---

*i.e., duplicity in identity

†universal duplicity

‡or with the restoration of indifference.

Not only are the conditions of the construction of every dynamical process contained in the theses proposed up to this point, but we have also deduced how all other oppositions, even those presented in the chemical heterogeneities, are determined by *one* original antithesis.

V.
The Theater of the Dynamic Organization of the Universe

The dynamical organization of the universe has been deduced, but not its theater. This organization presupposes an evolution of the universe from one original product, a dissociation of this product into ever new products. The ground of this infinite dispersion in Nature must have been laid by one original duality, and this diremption must be seen as having emerged in an originally identical being; but this is not thinkable unless that identical being is posited as an *absolute involution*, as a dynamical infinity,* for then an infinite tendency to development was cast into the product with [266] one duplicity.†—For intuition, this infinite tendency will be a tendency toward an evolution with infinite *velocity*. Thus nothing would be distinguishable in this evolution, i.e., no moment of *time* would be filled in a determinate way unless there were a *retarding* element in this series which kept the balance with that tendency. Therefore, the evolution of Nature‡,§ with finite velocity presupposes, as ultimate factors, an accelerating and a retarding force which are both infinite in themselves, and which are only mutually limited by one another. By virtue of the reciprocal restriction of these forces no absolute evolution will occur in any given *moment* (of time).

If an absolute evolution did occur, then Nature would offer itself as nothing other than an *absolute juxtaposition*. Since that *absolute* juxtaposition is only

---

*The dynamically infinite is here opposed to the mechanically infinite, i.e., the infinite *juxtaposition*.—In another sense, dynamical infinity is predicated of the organic product, and probably also of the art product, to the extent that if such a thing should emerge through *aggregation* (mechanically) no *inception* of aggregation could be found, because every individual presupposes an infinitely other, and everything other presupposes that individual. (Original note.—Trans.)

†The question still to be answered is this: what will the original opposition in the product be found to be from the standpoint of analysis? We now place ourselves wholly on the empirical standpoint, where Nature is merely *product* to us, in order to see what is to be found in it through *analysis*. Nature as mere product will appear as development from *one* original synthesis. But the *universal opposition* will appear as *condition of evolution*. That is, if Nature is an absolute synthesis, then the tendency to an infinite development was sown in it with *one* duality.

‡insofar as it occurs

§happens

absolute space, the *accelerating* force thought in its unrestrictedness leads to the idea of infinite space.*

Conversely, if the *retarding* force was unrestricted, then there would only be an absolute *interpenetration* for intuition, i.e., the *point* would arise, which as mere *limit* of space is symbol of *time* in its independence from space.

Nature cannot be either of these alone; it is a juxtaposition in interpenetration, and an interpenetration in juxtaposition—for the moment, a being only *conceived* in evolution—a being oscillating between evolution and involution.

Since the tendency to evolution[†] is an originally infinite one *ex hypothesi*, it must be thought as a force that would fill an infinitely large space in an infinitely small time. If one allows space to increase to infinity, or time to decrease to infinity, then one has $\frac{\infty}{1}$ in the one case, in the other $\frac{1}{0}$, i.e.,[‡] the *infinitely large*.[§]

The retarding force, as the antithesis, must be thought as the one that preserves the expansive force through a *finite* time in a finite space.

Neither of the two for itself would generate a real occupation of space. If the force of expansion could permeate an infinitely large space in an infinitely short time, then it would not linger even for a moment [267] in any part of space, thus it would nowhere fill space. The more the counterweight of the retarding force increases, the more the expansive tendency will linger in each point of space for a longer period of time, thus filling space to a higher degree.—In this way *diverse degrees of density* are possible.

Matter is not so much an occupation of space as a *filling up of space*, and indeed a filling of space with *determinate velocity*. Since the measure of one of those forces is space filled, the other time filled, then their proportion is $= \frac{S}{T} = C$, and the various degrees of density are only various velocities of the filling of space.

The absolutely elastic is what fills space with infinite velocity; the absolutely dense is what fills space with infinite slowness; neither of the two exists in Nature.

The finite velocity of evolution overall is also deduced along with the two deduced forces, i.e., it is explained how Nature is a determinate product for every single moment of time, but not how it is so for every moment of *space*. However, evolution should not occur only with a finite velocity, it must be absolutely inhibited—i.e., be inhibited at determinate *points*, for otherwise evolution (at finite velocity) would only be completed in an infinite time; evolution

---

*of infinite expansion; it will have to appear as *expansive* force

[†]the force of expansion

[‡]in both

[§]The force of expansion tends toward the infinitely large according to its nature, and therefore everywhere represents the positive factor.

would be progressive indeed, and Nature an infinitely polymorphic being for every *moment* of time, but not a fixed and determinate product for *all time*.

Thus the force by which an absolute *boundary* of evolution enters, a determination of the product in Nature for every moment of *space*, must be a force different from and independent of the force which only determines the *velocity* of evolution and the determination of the product for every moment of *time*.

There is no force through which an *original* limit in space is posited other than universal *gravity*. The latter must thus be conjoined to those two forces as the third force through which Nature first becomes a permanent product and fixed for all time.

---

[268] Nature can be seen as a *product* only from this standpoint, the standpoint that Kant adopted in his *Metaphysical Foundations of Natural Science*. What we have called "accelerating" force corresponds to Kant's *repulsive* or *expansive force*; what we call *retarding* force corresponds to his *attractive force*, with the difference that Kant conceives *gravity* under the latter too, and so believes the construction of matter to be complete with two forces.*—He thinks it is completed

1) insofar as he seems to see all *difference of quality* as reducible to a variable proportion of those forces,† which indeed *mechanism* recognizes, and which knows matter at all only as occupation of space, but not the higher dynamics—(according to the former, every material must be changeable into every other at least through dynamical, e.g., chemical, alteration of the original proportions of those forces. However, all quality is determined through something far higher than the mere degree of density. See above.);

---

*Expansive and retarding force show themselves here as necessary factors of every occupation of space to a determinate degree.—Since matter, from the merely empirical standpoint, is nothing other than occupation of space, the opposition from the standpoint of analysis too can appear only as an opposition between repulsive and attractive force. This is the point from which Kant begins the dynamical philosophy—the same point at which our theory stops.

If Kant's expansive and attractive forces (he names "attractive" what we have called "retarding" up to this point) represent nothing other than the original opposition, then he cannot complete the construction of matter from *two forces alone*. He still requires the third force which fixes the opposition, and which, according to us, is to be sought in the universal striving toward indifference, or in *gravity*.

†For what constitutes the quality of a body is not the *proportion* of the two factors, but the relative preponderance of the one over the other. The body cannot act in the dynamical process with a force immanent and directed to its construction, but only with one reaching out beyond the product.

2) insofar as Kant assumes that that which he calls force of attraction is identical with gravity, and which is the determinant of density in his construction of matter; but the identity of these is clearly impossible because the force of attraction for every body is already expended in its mere construction (see above p. 77f.).*

—(This reason also holds against the construction of chemical effects out of those two forces. In the chemical conflict too the substance can only act with a force directed toward the outside, but those two forces are only immanent forces contributing to the construction of their product.)—

[269] The transcendental proof of those two forces, gravity and retarding force, *as forces independent of one another* is, in brief, the following:

There must be for every finite being a limited intuition of the world; this original restrictedness is, for the intellectual world, just that which *gravity* is for the physical world, that which binds the individual to a determinate system of things and assigns to it its position in the universe. The intuition of the world is determined, with respect to every individual object, within a determinate system. By this means limitation comes *into* limitation. But the individual object, since its position in the universe is already determined for it by gravity, can only be further determined with respect to the *degree* to which it occupies space. The *degree* of its occupation of space is only determinable through the form of time, by the inverse ratio of time in which the space is filled to the occupied space. The existence of the object for time is thus limited through a force that is as little identical with gravity as time is identical with space. Conversely too, through this force (the retarding force) the velocity of the filling of space is lessened, but the evolution itself is not inhibited; the latter must occur through a force different from it.

It is to be expected in advance that since both forces are of a negative nature, i.e., are limiting forces, both stand in some relation to one another, one will determine the other. The following is self-evident:

The greater the preponderance of retarding force, the slower the evolution. The further evolution progresses, the more must the retarding force gradually decline. Every product of Nature must be inhibited at a determinate point of evolution in order to be a determinate product. If one supposes that the product will be inhibited at a point where the retarding force still has a great preponderance, then the expansive tendency must act more strongly on this point (because it is inversely related to the space in which it [270] expands). In

---

*Other, quite profound reasons against the identity of the two forces can be found in Mr. Baader's text mentioned above, exceedingly important for the whole dynamical philosophy, which came too late into my hands in order to make use of it earlier. (Original note.—Trans.)

order to keep it in equilibrium gravity must act on those parts of Nature the most strongly where the retarding force still has the greatest preponderance.

The body of greatest mass lies closer to the dynamic center *in itself* than the body of lesser mass. The mass is thus determined by gravity, but not, as one generally says, as if the weight were proportional to the mass.—Is mass then a magnitude known *in itself*? Is it recognized, perhaps, through the multitude of its parts? But this multitude is infinite. So no determination of mass through the multitude of its parts is possible; there is no ground of determination of mass aside from the effect of gravity. The product is determinate for every moment of time, but it does not act *outside* itself, it *fills* only *its* sphere; gravity first gives to it the tendency proportional to the degree of its occupation of space, which by this means first becomes a fixed and as such cognizable degree.

Matter only manifests itself through gravity; there may be an imponderable matter, but it does not manifest itself. Therefore, the *unity* of a material is known only through the unity of its gravity, a multitude of material is organized into a unity because it gives itself a common center of gravity.—Kant understands the essence of solidity as the parts not being able to be pushed into one another without immediately being separated, which means, in other words, that the part has no movement independent of the whole. In the fluid the part is shed from the whole *by its sheer weight*; the reason for this difference lies in the fact that the fluid body has no common center of gravity and every particle forms its preferred center of gravity. (Therefore, the preferential adoption of the spherical form in the formation of droplets.)—The *unity* of the center of gravity is that which organizes matter into one, it is the forming, binding, determining element of all formation.*

---

[271] The two forces, the expansive and the retarding, are the forces of evolution itself; gravity already presupposes evolution, thus gravity can have *conditions*. It can only be found, for example, at a certain degree of universal evolution; if it is conditioned, then it will be conditioned by the most original reciprocal relation in the universe, i.e., that universal, mutually exercised effect through (magnetic) division. Although it is originally *one*, it will, in the proportion that the universe evolves itself, split itself into multiple gravities as individual rays. So this force constitutes, as it were, the binding mediator of forces which subtend Nature as a *theater*, and those which sustain it as a *dynamical organization*.

---

*Baader in the aforementioned text. (Original note.—Trans.)

Only after the stage is set, as it were, by the higher dynamical forces, can the merely mechanical forces take possession; the consideration of these forces and their laws no longer falls within the limits of the philosophy of nature, which is nothing other than a higher dynamics, and whose spirit is expressed in the principle that views the dynamic as the single positive and original aspect of Nature, the mechanical only as the negative and derived aspect of the dynamical.

It was assumed that Nature is a development from one original involution. This involution cannot be anything real, however, according to the above: thus it can only be thought as *act*, as ABSOLUTE SYNTHESIS, which is only ideal, and signifies the turning point, as it were, of transcendental philosophy and philosophy of nature.

Introduction to the Outline of a System of the Philosophy of Nature, or, On the Concept of Speculative Physics and the Internal Organization of a System of this Science (1799)

[271] §.1.
What we call Philosophy of Nature is a Necessary Science in the System of Knowledge.

The intelligence is productive in two modes: either blindly and unconsciously, or freely and consciously; it is unconsciously productive in external intuition, consciously in the creation of an ideal world.

Philosophy removes this distinction by assuming the unconscious activity to be originally identical with, and, as it were, sprung from the same root as the conscious. This identity is *directly* proved in the case of an activity at once clearly conscious and unconscious, which manifests itself in the productions of *genius*; *indirectly*, *outside* of consciousness, in the products of *Nature*, so far as in all of them the most complete fusion of the ideal and the real is perceived.

Since philosophy assumes the unconscious, or as it may likewise be termed, the real activity to be identical with the conscious or ideal, its tendency will be to bring back everywhere the real to the ideal—a process which gives rise to what is called transcendental philosophy. The regularity displayed in all the movements of Nature—for example, the sublime geometry which is exercised in the motions of the heavenly bodies—is not explained by saying that Nature is the most perfect geometry. Rather conversely, [272] it is explained by saying that the most perfect geometry is the productive power in Nature; a mode of explanation whereby the real itself is transported into the ideal world, and those motions are changed into intuitions which take place only in ourselves, and to which nothing outside of us corresponds. Again, the fact that Nature, wherever it is left to itself, in every transition from a fluid to a solid state, produces of its own accord, as it were, regular forms (a regularity which, in the higher species of crystallization, namely the organic, seems even to become purposefulness); or the fact that in the animal kingdom (that product of the blind forces of Nature) we see actions arise which are equal in regularity to those that take place with consciousness, and even external works of art, perfect in their kind—all of this is explained in our view by saying that it is an unconscious productivity in its origin akin to the conscious, whose mere reflection we see in Nature, and which from the standpoint of the natural view must appear as one and the same blind drive that exerts its influence from crystallization upward to the highest point of organic formation (in which, on one side, through the technical drive, it returns again to mere crystallization) only acting on different planes.

According to this view, since Nature is only the visible organism of our understanding, Nature *can* produce nothing but what shows regularity and purpose, and Nature is *compelled* to produce it. But if Nature *can* produce only the regular, and produces it from necessity, it follows that the origin of such regular and purposeful products must again be capable of being proved to be necessary in the relation of its forces, in Nature thought as independent and real—it follows *that therefore, conversely, the ideal must arise out of the real and admit of explanation from it.*

Now if it is the task of transcendental philosophy to subordinate the real to the ideal, it is, on the other hand, the task of the philosophy of nature to explain the ideal by the real. The two sciences are therefore but one science, differentiated only in the opposite orientation of their tasks. Moreover, as the [273] two directions are not only equally possible, but equally necessary, the same necessity attaches to both in the system of knowledge.

§.2.
Scientific Character of the Philosophy of Nature.

Philosophy of nature, as the opposite of transcendental philosophy, is distinguished from the latter chiefly by the fact that it posits Nature (not, indeed, insofar as it is a product, but insofar as it is at once productive and product) as the self-existent; therefore it can most concisely be designated the *Spinozism of physics*. It follows naturally from this that there is no place in this science for idealistic methods of explanation, such as transcendental philosophy is fitted to supply, since for it Nature is nothing more than the organ of self-consciousness, and everything in Nature is necessary merely because it is only through the medium of such a Nature that self-consciousness can take place. This mode of explanation, however, is as meaningless for physics (and for our science which occupies the same standpoint) as were the old teleological modes of explanation, and the introduction of a universal reference to final causes into the science of nature, which was adulterated as a result. For every idealistic mode of explanation, dragged out of its own proper sphere and applied to the explanation of Nature, degenerates into the most adventurous nonsense, examples of which are well-known. The first maxim of all true natural science, to explain everything by the forces of Nature, is therefore accepted in its widest extent in our science, and even extended to that region at the limit of which all interpretation of Nature has until now been accustomed to stop short: for example, to those organic phenomena which seem to presuppose an analogy with reason. For, granted that there really is something which presupposes such analogy in the actions of animals, nothing further would follow on the principle of realism than that what we call "reason" is a mere [274] play of higher and necessarily unknown natural forces. For, inasmuch as all thinking is at last reducible to a producing and reproducing, there is nothing impossible in the thought that the same activity by which Nature reproduces itself anew in each successive phase, is reproductive in thought through the medium of the organism (very much in the same manner in which, through the action and play of light, Nature, which exists independently of it, is really created *immaterially*, and as it were for a second time), in which case it is natural that what forms the limit of our intuitive faculty no longer falls within the sphere of our intuition itself.

§.3.
Philosophy of Nature is Speculative Physics.

Our science, as far as we have gone, is thoroughly and completely realistic; it is therefore nothing other than physics, it is only *speculative* physics. In its tendency it is exactly what the systems of the ancient physicists were, and what, in more recent times, the system of the restorer of Epicurean philosophy is, i.e., *Lesage's* mechanical physics, by which the speculative spirit in physics, after a long scientific sleep, has again for the first time been awakened. It cannot be shown in detail here (for the proof itself falls within the sphere of our science) that on the mechanical or atomistic basis that has been adopted by Lesage and his most successful predecessors, the idea of a speculative physics cannot be realized. For, inasmuch as the first problem of this science, that of inquiring into the *absolute* cause of motion (without which Nature is not in itself a finished whole), is absolutely incapable of a mechanical solution. Because mechanically motion results only from motion to infinity, there remains for the real construction of speculative physics only one way open, the dynamic, with the presupposition that motion arises not only from motion, but even from rest; we suppose, therefore, that there is motion in the rest of Nature, [275] and that all mechanical motion is the merely secondary and derivative motion of that which is solely primitive and original, and which wells forth from the very first factors in the construction of a Nature overall (the fundamental forces).

In making clear the points of difference between our undertaking and all those of a similar nature that have hitherto been attempted, we have at the same time shown the difference between speculative physics and so-called empirical physics; a difference which may principally be reduced to the fact that the former occupies itself solely and entirely with the original causes of motion in Nature, that is, solely with the dynamical phenomena; the latter on the contrary, inasmuch as it never reaches a final source of motion in Nature, deals only with the secondary motions, and even with the original ones only as mechanical (and therefore likewise capable of mathematical construction). The former, in fact, aims generally at the inner clockwork and what is *nonobjective* in Nature; the latter, on the contrary, only at the *surface* of Nature, and what is *objective* and, so to speak, *outside* in it.

§.4.
On the Possibility of Speculative Physics.

Insofar as our inquiry is directed not so much upon the phenomena of Nature as upon their final grounds, and our business is not so much to deduce the latter from the former as the former from the latter, our task is simply this: to erect a science of Nature in the strictest sense of the term; and in order to find out whether speculative physics is possible, we must know what belongs to the possibility of a doctrine of Nature viewed as science.

(a) The idea of knowledge is here taken in its strictest sense, and so it is easy to see that, in this use of the term, we can be said *to know* objects only when they are such that we see the principles of their possibility, for without this insight my whole knowledge of an object, e.g., of a machine [276] with whose construction I am unacquainted, is a mere seeing, that is, a mere conviction of its existence, whereas the inventor of the machine has the most perfect knowledge of it, because he is, as it were, the soul of the work, and because it preexisted in his head before he exhibited it as a reality.

Now, it would certainly be impossible to get a glimpse of the internal construction of Nature if an invasion of Nature were not possible through freedom. It is true that Nature acts openly and freely; its acts however are never isolated, but performed under the concurrence of a host of causes which must first be excluded if we are to obtain a pure result. Nature must therefore be compelled to act under certain definite conditions, which either do not exist in it at all, or else exist only as modified by others.—Such an invasion of Nature we call an experiment. Every experiment is a question put to Nature, to which it is compelled to give a reply. But every question contains an implicit *a priori* judgment; every experiment that is an experiment, is a prophecy; experimenting itself is a production of phenomena. The first step toward science, therefore, at least in the domain of physics, is taken when we ourselves begin to produce the objects of that science.

(b) We *know* only the self-produced; knowing, therefore, in the *strictest* sense of the term, is a *pure* knowing *a priori*. Construction by means of experiment is, after all, not an absolute self-production of the phenomena. There is no question that much in the science of Nature may be known comparatively *a priori*; as, for example, in the theory of the phenomena of electricity, magnetism, and even light. There is such a simple law recurring in every phenomenon that the results of every experiment may be told beforehand; here my knowing follows immediately from a known law without the intervention of any particular experience. But whence then does the law itself come to me? We suggest that all phenomena are correlated in one absolute and *necessary* law, from which they can all be deduced; in short, that [277] in natural science all that we know, we know absolutely *a priori*. Now, that experimentation never leads to such a knowing is plainly manifest from the fact that it can never get beyond the forces of Nature, of which it makes use as means.

Since the final *causes* of natural phenomena are themselves not phenomenal, we must either give up all attempt ever to arrive at a knowledge of them, or else we must altogether put them into Nature, endow Nature with them. However, that which we put into Nature has no other value than that of a presupposition (hypothesis), and the science founded upon it must be equally as hypothetical as the principle itself. It would be possible to avoid this only in one case, i.e., if that presupposition itself were involuntary, and as necessary as Nature itself. Assuming, for example, what must be assumed, that the sum of phenomena is not a mere world, but of necessity a Nature (that is, that this whole is not merely a product, but at the same time productive), it follows that in this whole we can never arrive at absolute identity, because this would bring about an absolute transition of Nature as productive into Nature as product, that is, it would produce absolute rest. Such a wavering of Nature, therefore, between productivity and product, will necessarily appear as a universal duplicity of principles, whereby Nature is maintained in continual activity, and prevented from exhausting itself in its product; and universal duality as the principle of explanation of Nature will be as necessary as the idea of Nature itself.

This absolute hypothesis must bear its necessity within itself, but it must, besides this, be brought to an empirical test; *for, inasmuch as all the phenomena of Nature cannot be deduced from this hypothesis as long as there is in the whole system of Nature a single phenomenon which is not necessary according to that principle, or which contradicts it, the hypothesis is thereby at once shown to be false*, and from that moment ceases to have validity as a hypothesis.

[278] By this deduction of all natural phenomena from an absolute hypothesis, our knowing is changed into a construction of Nature itself, that is, into a science of Nature *a priori*. If, therefore, such deduction itself is possible, a thing which can be proved only by the deed, then too a doctrine of Nature is possible as a science of Nature; a system of purely speculative physics is possible, which was the point to be proved.

Note. There would be no necessity for this remark if the confusion that still prevails in regard to ideas perspicuous enough in themselves did not render some explanation with regard to them requisite.

The assertion that natural science must be able to deduce all its principles *a priori* is in a sense understood to mean that natural science must dispense with all experience, and, without any intervention of experience, be able to spin all its principles out of itself; an affirmation so absurd that the very objections to it deserve pity.—*Not only do we know this or that through experience, but we originally know nothing at all except through experience, and by means of experience*, and in this sense the whole of our knowledge consists of the judgments of experience. These judgments become *a priori* principles when we become conscious of them as necessary, and thus every judgment, whatever its content may be, may be raised to that dignity, insofar as the distinction between *a priori* and *a posteriori* judgments is not at all, as many people may have imagined, one originally cleaving to the judgments themselves, but is a distinction made solely *with respect to our knowing*, and the *kind* of our knowledge of these judgments, so that every judgment which is merely historical for me—i.e., a judgment of experience—becomes, notwithstanding, an *a priori* principle as soon as I arrive, whether directly or indirectly, at insight into its internal necessity. Now, however, it must in all cases be possible to recognize every natural phenomenon as absolutely necessary; for, if there is no chance in Nature at all, then likewise no original phenomenon of Nature can be fortuitous; on the contrary, for the very reason that Nature is a system, there must be [279] a necessary connection, in some principle embracing the whole of Nature, for everything that happens or comes to pass in it.—Insight into this internal necessity of all natural phenomena becomes, of course, still more complete, as soon as we reflect that there is no real system which is not, at the same time, an organic whole. For if, in an organic whole, all things mutually bear and support each other, then this organization must have existed as a whole previous to its parts; the whole could not have arisen from the parts, but the parts must have arisen out of the whole. It is *not, therefore*, that WE KNOW Nature as *a priori*, but Nature IS *a priori*; that is, everything individual in it is predetermined by the whole or by the idea of a

Nature generally. But if Nature *is a priori*, then it must be possible to *recognize* it *as* something that is *a priori*, and this is really the meaning of our affirmation.

Such a science, like every other, does not deal with the hypothetical or the merely probable, but depends upon the evident and the certain. Now, we may indeed be quite certain that every natural phenomenon, through whatever number of intermediate links, stands in connection with the last conditions of Nature; the intermediate links themselves, however, may be unknown to us, and still lying hidden in the depths of Nature. To find out these links is the work of experimental research. Speculative physics has nothing to do but to show the need of these intermediate links;* but since every new discovery throws us back upon a new ignorance, and while one knot is being loosed a new one is being tied, it is conceivable that the complete discovery of all the intermediate links in the chain of Nature, and therefore also our science itself, is an infinite task.—Nothing, however, has more impeded the infinite progress of this science than the arbitrariness of the fictions by which [280] the lack of profound insight was so long doomed to be concealed. The fragmentary nature of our knowledge becomes apparent only when we separate what is merely hypothetical from the pure outcome of science, and then set out to collect the fragments of the great whole of Nature again into a system. It is, therefore, conceivable that *speculative* physics (the soul of true experimentation) has, throughout all time, been the mother of all great discoveries in Nature.

§.5.
On a System of Speculative Physics in General.

Up to this point the idea of speculative physics has been deduced and developed; it is another business to show how this idea must be realized and actually carried out. The author, for this purpose, would at once refer to his *Outline of a System of the Philosophy of Nature*, if he had no reason to suspect that many even of those who might consider that *Outline* worthy of their attention, would come to it with certain preconceived ideas, which he has not presupposed, and which he does not desire to have presupposed by them. The causes which may render an insight into the tendency of that *Outline* difficult, are (exclusive of defects in presentation) mainly the following:

1) That many persons, perhaps misled by the phrase "philosophy of nature," expect to find transcendental deductions from natural phenomena, of the

---

*Thus, for example, it becomes very clear through the whole course of our inquiry, that, in order to render the dynamic organization of the Universe evident in all its parts, we still lack that *central phenomenon* of which Bacon already speaks, which certainly lies in Nature but has not yet been extracted from it by experiment. (Original note.—Trans.)

sort that exist elsewhere in various fragments, and will regard natural philosophy generally as a part of transcendental philosophy; whereas it forms a science altogether peculiar, altogether different from, and independent of, every other.

2) That the notions of dynamical physics popularized until now are very different from, and partially at variance with, those which the author lays down. I do not speak of the modes of representation which several persons, whose business is really merely experiment, have made up in this connection; for example, where they suppose it to be a dynamical explanation [281] when they reject a galvanic fluid, and accept instead certain vibrations in the metals; for these persons, as soon as they observe that they have understood nothing of the matter, will revert of their own accord to their previous representations, which were made for them. I speak of the modes of representation which have been put into philosophic heads by Kant, and which may be mainly reduced to this: that we see in matter nothing but the occupation of space in definite degrees, and in all variety of matter, therefore, only mere difference of occupation of space (i.e., density), in all dynamic (qualitative) changes only mere changes in the relation of the repulsive and attractive forces. Now, according to this mode of representation, all the phenomena of Nature are seen only on their lowest level, and the dynamical physics of these philosophers begin precisely at the point where they ought properly to leave off. It is indeed certain that the last result of every dynamical process is a changed degree of occupation of space, that is, a changed density; now, since the dynamical process of Nature is one, and the individual dynamical processes are only fragments of the one fundamental process, even magnetic and electrical phenomena, viewed from this standpoint, will not be actions of particular materials, but changes in *the constitution of matter itself*; and as this depends upon the mutual action of the fundamental forces, finally, will be changes in the relation of the fundamental forces themselves. We do not indeed deny that these phenomena at the extreme limit of their manifestation are changes in the relation of the principles themselves; we only deny that these changes are *nothing more*. On the contrary, we are convinced that this so-called dynamical principle is too superficial and defective a basis of explanation for all of Nature's phenomena in order to reach the real depth and manifoldness of natural phenomena, since by means of it, in fact, no qualitative change of matter *as* such is constructible (for change of density is only the external phenomenon of a higher change). To adduce proof of this assertion is not incumbent upon us, until, from [282] the opposite side, that principle of explanation is shown by actual fact to exhaust Nature, and the great chasm is filled up between that kind of dynamical philosophy and the empirical attainments of physics (for example, in regard to the very different kinds of effects exhibited by simple substances, a thing which, let us say at once, we consider to be impossible).

We may therefore be permitted, in place of the dynamic mode of representation prevailing until now, to put our own without further ado, a gesture which will no doubt clearly show how the latter differs from the former, and by which of the two the doctrine of Nature may most certainly be raised to a science of Nature.

§.6.
Internal Organization of the System of Speculative Physics.

I.

An inquiry into the *principle* of speculative physics must be preceded by inquiries into the distinction between the speculative and the empirical generally. This distinction depends mainly upon our conviction that between empiricism and theory there is such a complete opposition that there can be no third thing in which the two may be united; that, therefore, the idea of "experimental science" is a mongrel idea that implies no consistent thought, or rather, is an idea which cannot be thought at all. What is pure empiricism is not science, and conversely, what is science is not empiricism. This is not said for the purpose of at all deprecating empiricism, but is meant to exhibit it in its true and proper light. Pure empiricism, be its object what it may, is history (the absolute opposite of theory), and conversely, history alone is empiricism.*

[283] Physics, as empiricism, is nothing but a collection of facts, of accounts of what has been observed, what has happened under natural or artificial circumstances. In what we at present call physics, empiricism and science run riot together, and for that very reason they are neither one thing nor another. Our aim, in view of this object, is to separate science and empiricism as soul and body, and by admitting nothing into science which is not susceptible of an *a priori* construction, to strip empiricism of all theory, and restore it to its original nakedness.

The opposition between empiricism and science rests therefore upon this: that the former regards its object in *being*, as something already prepared and accomplished; science, on the other hand, views its object in *becoming*, and as something that has yet to be accomplished. As science cannot set out from anything that is a product, that is, a thing, it must set out from the unconditioned;

---

*If only those warm panegyrists of empiricism, who exalt it at the expense of science, did not, true to the idea of empiricism, try to palm off upon us their own judgments as empiricism, and what they have put into Nature and imposed upon objects. Though many people think they can talk about it, there is a great deal more belonging to empiricism than many imagine—to eliminate purely the product from Nature, and to render it with the same fidelity with which it has been eliminated. (Original note.—Trans.)

the first inquiry of speculative physics is that which relates to the unconditioned in natural science.

II.

As this inquiry is, in the *Outline*, deduced from the highest principles, the following may be regarded as merely an illustration of those inquiries. Inasmuch as everything of which we can say that it *is*, is of a conditioned nature, it is only *being itself* that can be the unconditioned. But seeing the individual being, as a conditioned thing, can only be thought as a particular limitation of the productive activity (the sole and ultimate substrate of all reality), *being itself* is *thought* as the same productive activity *in its unlimitedness*. For the science of nature, therefore, Nature is originally only productivity, and from this as its principle science must set out.

[284] As long as we only know the totality of objects as the sum total of all being, this totality is a mere *world*, that is, a mere product for us. It would certainly be impossible in the science of nature to rise to a higher idea than that of being if all permanence (which is thought in the idea of being) were not deceptive, and really a continuous and uniform reproduction.

Insofar as we regard the totality of objects not merely as a product, but at the same time necessarily as productive, it becomes *Nature* for us, and this *identity of the product and the productivity*, and this alone, is implied by the idea of Nature, even in the ordinary use of language. *Nature* as a mere *product* (*natura naturata*) we call Nature as *object* (with this alone all empiricism deals). *Nature as productivity* (*natura naturans*) we call *Nature as subject* (with this alone all theory deals).

As the object is never unconditioned, something absolutely nonobjective must be put into Nature; this absolutely nonobjective factor is nothing else but the original productivity of Nature. In the conventional view productivity vanishes in the product; conversely, in the philosophic view the product vanishes into the productivity.

Such an identity of the product and the productivity in the *original* conception of Nature is expressed by the ordinary view of Nature as a whole, which is at once the cause and the effect of itself, and is in its duplicity (which runs through all phenomena) again identical. Furthermore, with this idea the identity of the real and the ideal agrees, an identity which is thought in the idea of every product of Nature, and with respect to which only the nature of art can be placed in contrast. For whereas in art the idea precedes the act or the execution, in Nature idea and act are rather contemporary and one; the idea passes immediately over into the product, and cannot be separated from it.

[285] This identity is canceled by the empirical perspective, which sees in Nature only the *effect* (although on account of the continual wandering of

empiricism into the field of science, we have, even in purely empirical physics, maxims which presuppose an idea of Nature as subject; such as, for example, "Nature chooses the shortest way"; "Nature is sparing in causes and lavish in effects"); the identity is also canceled by speculation, which looks only at *cause* in Nature.

III.

We can say of Nature as object that it *is*, not of Nature as subject; for this is being or productivity itself. This absolute productivity must pass over into an empirical nature. In the idea of absolute productivity is the thought of an *ideal* infinity. The ideal infinity must become an empirical one. But empirical infinity is an infinite becoming.—Every infinite series is but the exhibition of an intellectual or ideal infinity. The original infinite series (the ideal of all infinite series) is that wherein our intellectual infinity evolves itself, i.e., time. The activity which sustains this series is the same as that which sustains our consciousness; consciousness, however, is continuous. Time, therefore, as the evolution of that activity, cannot be produced by composition. Now, as all other infinite series are only imitations of the originally infinite series, time, no infinite series can be otherwise than continuous. In the original evolution the inhibiting agent (without which the evolution would take place with infinite rapidity) is nothing but *original reflection*; the necessity of reflection upon our acting in every organic moment (continued duplicity in identity) is the secret stroke of art whereby our being receives *permanence*.—Absolute continuity, therefore, exists only for intuition, but not for [286] *reflection*. Intuition and reflection are opposed to each other. The infinite series is continuous for productive *intuition*, interrupted and composite for reflection. It is upon this contradiction between intuition and reflection that those sophisms are based, in which the possibility of all motion is contested, and which are solved at every successive step by the productive activity. For intuition, for example, the action of gravity takes place with perfect continuity; for reflection, by fits and starts. Hence all the laws of mechanics, whereby that which is properly only the object of the productive intuition becomes an object of reflection, are really only laws for reflection.—Hence those fictitious notions of mechanics, the atoms of time in which gravitation acts, the law that the moment of solicitation is infinitely small because otherwise an infinite rapidity would be produced in finite time, etc., etc. Hence, finally, the assertion that in mathematics no infinite series can really be represented as continuous, but only as advancing by fits and starts.

The whole of this inquiry into the opposition between reflection and the productivity of intuition serves only to enable us to deduce the general statement that in all productivity, and in productivity alone, is there absolute continuity; a statement of importance in the consideration of the whole of Nature.

For example, when the law that in Nature there is no leap, that there is a continuity of forms in it, etc., is confined to the original productivity of Nature, in which certainly there must be continuity, and where from the standpoint of reflection all things must appear *disconnected* and *without* continuity, placed beside each other, as it were, we must therefore admit that both parties are right. Those who assert continuity in Nature (for example, in organic Nature) are correct, no less than those who deny it, when we take into consideration the difference of their respective standpoints; and we thus at the same time come upon the distinction between dynamical and atomistic physics; for, as will soon become apparent, the two are distinguished only by the fact that the former occupies the standpoint of *intuition*, the latter that of *reflection*.

[287] IV.

Assuming these general principles, we shall be able, with more certainty, to reach our aim and provide an exposition of the internal organization of our system.

(a) In the idea of becoming, we think the idea of gradualness. But an absolute productivity will exhibit itself empirically as a becoming with infinite rapidity, whereby nothing real results for the intuition. (Since Nature must in reality be thought as engaged in infinite evolution, the permanence, the resting of the products of Nature (the organic ones, for instance), is not to be viewed as an absolute resting, but only as an evolution proceeding with infinitely small rapidity or with infinite tardiness. However, at this point evolution, with even finite rapidity, not to speak of infinitely small rapidity, has not been constructed.)

(b) It is not thinkable that the evolution of Nature should take place with finite rapidity, and thus become an object of intuition, without an original limitation (a being limited) of the productivity.

(c) But if Nature is absolute productivity, then the ground of this limitation cannot lie *outside* of it. Nature is originally *only* productivity; there can, therefore, be nothing determined in this productivity (all determination is negation) and so products can never be reached by it.—If products are to be reached, the productivity must pass from being undetermined to being determined, that is, it must, *as pure* productivity, be canceled. If the ground of determination of productivity lay outside of Nature, Nature would not originally be absolute productivity. Determination, that is, negation, must certainly come into Nature; but this negation viewed from a higher standpoint must again be positivity.

(d) But if the ground of this limitation lies *within Nature itself*, then Nature ceases to be *pure identity*. (Nature, in so far as it is *only* productivity, is pure identity, and there [288] is absolutely nothing in it capable of being distinguished. In order for anything to be distinguished in it, its identity must be canceled; Nature must not be identity, but duplicity.)

Nature must originally be an object to itself; this change of the *pure subject* into an *object to itself* is unthinkable without an original diremption in Nature itself.

This duplicity cannot therefore be further deduced physically; for, as the condition of all Nature generally, it is the principle of all physical explanation, and all physical explanation can only have for its aim the reduction of all the antitheses which appear in Nature to that original antithesis in the heart of Nature, *which does not, however, itself appear.*—Why is there no original phenomenon of Nature without this duplicity, if in Nature all things are not mutually subject and object to each other to infinity, and Nature even, in its origin, is not at once product and productive?—

(e) If Nature is originally duplicity, there must even be opposite tendencies in the original productivity of Nature. (The positive tendency must be opposed by another, which is, as it were, antiproductive, retarding production; not as the contradictory, but as the negative, the real opposite of the former.) It is only then that, in spite of its being limited, there is no passivity in Nature, even when that which limits it is again positive, and its original duplicity is a contest of real antithetical tendencies.

(f) In order to arrive at a product, these opposite tendencies must encounter one another. But since they are supposed *equal* (for there is no ground for supposing them unequal), wherever they meet they will annihilate each other; the product is therefore = to 0, and once more no product is reached.

This inevitable, though hitherto not very closely remarked contradiction (namely, that a product can arise only through the concurrence of opposite tendencies, while at the same time these opposite tendencies mutually annihilate each other) can be solved only in the following manner.

Absolutely no *subsistence* of a product is thinkable *without* [289] *a continual process of being reproduced*. The product must be thought *as annihilated at every step*, and *at every step reproduced anew*. We do not really see the subsisting of a product, but only the continual process of being reproduced. (It is of course quite conceivable how the series $1-1+1-1\ldots$ on to infinity is thought as equal neither to 1 nor to 0. The reason why this series is thought as $= \frac{1}{2}$ lies deeper. There is one absolute magnitude ($= 1$) which, though continually annihilated in this series, continually recurs, and by this recurrence produces, not itself, but the mean between itself and nothing.—Nature, as object, is that which comes to pass in such an infinite series, and is = a fraction of the original unit, to which the never canceled duplicity supplies the numerator.)

(g) If the subsistence of the product is a continual process of being reproduced, then all *persistence* also only exists in Nature as *object*; in Nature as *subject* there is only infinite *activity*. The product is originally nothing but a mere point, a mere limit, and it is only through Nature's battling against this point that it is, so to speak, raised to a full sphere, to a product. (Suppose, for illustration, a stream; it is *pure identity*; where it meets resistance, a whirlpool is formed; this whirlpool is not an abiding thing, but something that vanishes at every moment, and every moment springs up anew.—Originally, in Nature there is nothing distinguishable; all products are, so to speak, still in solution, and invisible in the universal productivity. It is only when retarding points are given that they are thrown off and advance out of the universal identity.—At every such point the stream breaks (the productivity is annihilated), but at every step there comes a new wave which fills up the sphere).

The philosophy of nature does not have to explain the productive power of Nature; for if it does not posit this as originally in Nature it will never bring it into Nature. It has to explain the permanent. But the fact *that* anything should become permanent in Nature, can itself [290] only be explained by that contest of Nature *against all permanence*. The products would appear as mere points if Nature did not give them extension and depth by its own pressure, and the products themselves would last only an instant if Nature did not at every moment shove into them.

(h) This seeming product, which is reproduced at every step, cannot be a really infinite product; for otherwise productivity would actually be exhausted in it. In like manner it cannot be a finite product; for the force of the whole of Nature itself surges into it. It must therefore be at once infinite and finite; it must be only seemingly finite, but in infinite *development*.

The point at which this product originally enters is the universal point of inhibition in Nature, the point from which all evolution in Nature begins. But in Nature, as it is evolved, this point lies not here or there, but everywhere where there is a product.

This product is a finite one, but as the infinite productivity of Nature concentrates itself in it, it must have a drive toward infinite development.— And thus gradually, and through all the foregoing intermediate links, we have arrived at the construction of that infinite becoming, the empirical exhibition of an ideal infinity.

We behold in what is called Nature (i.e., in this collection of individual objects), not the primal product itself, but its evolution (hence the point of inhibition cannot remain *one*).—It has not yet been explained by what means *this* evolution is again absolutely inhibited (which must happen if we are to arrive at a fixed product).—

Through this product an original infinity evolves itself; this infinity can never decrease. The magnitude that evolves itself in an infinite series is still [291] infinite at every point of the line, and thus Nature will be still infinite at every point of the evolution.

There is only one original point of inhibition to productivity; but any number of points of inhibition to evolution may be thought. Every such point is marked for us by a product. Nature is still infinite at every point of the evolution, however; therefore Nature is still infinite in every product, and the germ of a universe lies in every one.*

(The question, by what means the infinite striving is retarded in the product, is still unanswered. The original inhibition in the *productivity* of Nature explains only why the evolution takes place with finite rapidity, but not why it takes place with infinitely small rapidity.)

(i) The product evolves itself to infinity. In this evolution, therefore, nothing can happen which is not already a product (synthesis) and which might not divide up into new factors, each of these again having its factors. Thus even by an analysis pursued to infinity, we could never arrive at anything in Nature which would be absolutely simple.

(k) If, however, we *suppose* the evolution to be completed (although it *never* can be completed), the evolution could not stop at anything which was a product, but only at the purely *productive*.

The question arises whether a final term—one that is no longer a substrate, but the cause of all substrate, no longer a product, but absolutely productive—we will not say "occurs," for that is unthinkable, but can at least be proved in experience.

(l) Since it bears the character of the unconditioned, it would have to exhibit itself as something which, although itself not in space, is still the principle of all occupation of space. (See the *Outline*, p. 19.)

[292] What *occupies* space is not matter, for matter is the occupied space itself. That, therefore, which occupies space cannot be matter. Only that which is, is in space, but *being itself* is not.

It is self-evident that no positive external intuition of that which is not in space is possible. It would therefore have to be capable of being exhibited at least negatively. This happens in the following manner. That which is in space, is, as such, mechanically and chemically destructible. That which is not

---

*A traveler in Italy makes the remark that the whole history of the world may be demonstrated on the great obelisk at Rome—so, likewise, in every product of Nature. Every mineral body is a fragment of the annals of the Earth. But what is the Earth?—Its history is interwoven with the history of the whole of Nature, and so passes from the fossil through the whole of inorganic and organic Nature, until it culminates in the history of the universe—one chain. (Original note—Trans.)

destructible either mechanically or chemically must therefore lie beyond space. It is only the final ground of all *quality* that has anything of this nature; for although one quality may be extinguished by another, this can nevertheless only happen in a third product, C, for the formation and maintenance of which A and B (the opposite factors of C), must continue to act.

But this indestructible factor, which is thinkable only as *pure intensity*, is, as the cause of all substrate, at the same time the principle of divisibility to infinity. (A body divided to infinity still occupies space to the same degree as its smallest part.)

That, therefore, which is purely productive without being a *product* is but the ultimate ground of *quality*. But every quality is a determinate one, whereas productivity is originally indeterminate. In the qualities, therefore, productivity appears as already inhibited, and since it appears most originally in them overall, it appears in them *most originally inhibited*.

This is the point at which our mode of conception diverges from that of conventional so-called dynamical physics. Our assertion, briefly stated, is this:

If the infinite evolution of Nature were *completed* (which is impossible) it would separate out into original and simple *actants*, or, if we may so express ourselves, into simple productivities. Our assertion therefore is not that *there are* such simple *actants* in Nature, [293] but only that they are the *ideal* grounds of the explanation of quality. These *entelechies* cannot actually be shown, they do not *exist*. We therefore do not have to explain anything more than is asserted here, namely, that such original productivities must be *thought* as the grounds of the explanation of all quality. This proof is as follows:

The affirmation that nothing which *is* in space is mechanically simple, that is, that nothing at all is mechanically simple, requires no demonstration. That, therefore, which is in reality simple, cannot be thought as in space, but must be thought as outside of space. But beyond space only *pure intensity* is thought. This idea of pure intensity is expressed by the idea of the actant. It is not the product of this action that is simple, but the *actant itself* abstracted from the product, and it must be simple in order that the product may be infinitely divisible. For although the parts are near vanishing, the intensity must still remain. And this pure intensity is what, even in infinite divisibility, sustains the substrate.

If, therefore, the assertion that affirms something simple as the basis of the explanation of quality is atomistic, then our philosophy is atomistic. But, inasmuch as it places the simple in something that is only productive without being a product, it is *dynamical atomism*.

It is clear that if we admit an absolute division of Nature into its factors, the last factor that remains over must be something that absolutely defies all division, that is, the simple. But the simple can only be thought as dynamical, and as such it is *not in space at all* (it designates only what is thought as altogether beyond occupation of space); therefore, no intuition of it is possible, except through its *product*. In like manner, no measure for it is given other than its product. To pure thought it is the mere *inception* of the product (as the point is only the origin of the line), in a word, pure *entelechy*. But that which is known, not in itself, but only in its product, is known altogether *empirically*. If, therefore, every original quality, *as* quality [294] (not as substrate, in which quality merely inheres), must be thought as pure intensity, pure *action*, then qualities generally are just the absolutely empirical factors in our knowledge of Nature, of which no construction is possible, and in respect to which there remains nothing for the philosophy of nature except the proof that they are the absolute limit of its construction.

The question in reference to the ground of quality posits the evolution of Nature as completed, that is, it posits something merely thought, and therefore can be answered only by an ideal ground of explanation. This question adopts the standpoint of reflection (on the product), whereas genuine dynamics always remains on the standpoint of *intuition*.—

(However, it must be at once remarked here that if the ground of the explanation of quality is conceived as an *ideal* one, the question only regards the explanations of quality, insofar as it is thought as *absolute*. There is no question of quality, for instance, insofar as it shows itself in the dynamical process. There is certainly a* ground of explanation and determination for quality, so far as it is relative; quality in that case is determined by its opposite, with which it is placed in conflict, and this antithesis is itself again determined by a higher antithesis, and so on back into infinity; so that, if this universal organization could dissolve itself, all matter would likewise sink back into dynamical inactivity, that is, into absolute absence of quality. Quality is a higher power of matter, to which the latter elevates itself by reciprocity.) It is demonstrated below that the dynamical process is a limited one for each individual sphere, because it is only through limitation that definite points of relation for the determination of quality arise. This limitation of the dynamical process, that is, the proper *determination* of quality, takes place by means of no other force than that by which the evolution is universally and absolutely limited, and this negative element in things is the only one that is indivisible, and mastered by nothing. —The [295] absolute relativity of all quality may be shown from the electric relation of bodies, inasmuch

---

*not merely ideal, but actually real

as the same body that is positive with one is negative with another, and conversely. But we might from now on abide by the statement (which is also laid down in the *Outline*) that *all quality is electricity*, and conversely, *the electricity of a body is also its quality* (for all difference of quality is equal to difference of electricity, and all* quality is reducible to electricity).—Everything that is sensible for us (sensible in the narrower sense of the term, as colors, taste, etc.), is doubtless sensible to us only *through* electricity, and the only *immediately* sensible factor would then be electricity,† a conclusion to which the universal duality of every sense leads us independently, since in Nature there is properly only one duality. In galvanism, sensibility, as a reagent, reduces all quality of bodies for which it is a reagent to an original difference. All bodies which, in a chain, at all affect the sense of taste or that of sight, be their differences ever so great, are either alkaline or acid, excite a negative or positive shock, and here they always appear as active in a higher than the *merely* chemical potency.

Quality considered as *absolute* is inconstructible, because quality generally is not anything absolute, and there is no other quality at all except that which bodies show mutually in relation to each other, and all quality is something by virtue of which the body is, so to speak, *raised above itself*.

All previous attempts at the construction of quality are reducible to two: to express qualities by *figures*, and so to assume a particular figure in Nature for each original quality; or else, [296] to express quality by *analytical formulae* (in which the forces of attraction and repulsion supply the negative and positive magnitudes). To convince oneself of the futility of the latter attempt, the shortest method is to appeal to the emptiness of the explanations to which it gives rise. Hence we limit ourselves here to the single remark that through the construction of all matter out of the two fundamental forces, different degrees of density may indeed be constructed, but certainly never different qualities *as* qualities; for although all dynamical (qualitative) changes appear, at their lowest stage, as changes of the fundamental forces, yet we see at that stage only the product of the process, not the *process itself*, and those changes are *what require explanation*, and the ground of explanation must therefore certainly be sought in something higher.—

The only possible ground of explanation for quality is an ideal one; because this ground itself presupposes something purely ideal. Whoever inquires into the final ground of quality is transported back to the starting point of Nature. But where is this starting point? And does not all quality consist in this,

---

*chemical

†Volta already asks, with reference to the affection of the senses by galvanism: "Might not the electric fluid be the immediate cause of all flavors? Might it not be the cause of sensation in all the other senses?" (Original note.—Trans.)

that matter is prevented by the general concatenation from reverting into its originality?

From the point at which reflection and intuition separate (a separation which is possible only on the hypothesis of the completed evolution), physics divides into two opposite directions, into which the two systems, the atomistic and the dynamical, have been divided.

The *dynamical* system *denies* the absolute evolution of Nature, and passes from Nature as synthesis (= Nature as subject) to Nature as evolution (= Nature as object); the *atomistic* system passes from the evolution, as the original, to Nature as synthesis; dynamics passes from the standpoint of intuition to that of reflection; atomistics from the standpoint of reflection to that of intuition.

Both directions are equally possible. If only the analysis is correct, then the synthesis must be capable of being found again through analysis, just as [297] the analysis in its turn can be found through the synthesis. But whether the analysis is correct can be tested only by the fact that we can pass from it again to the synthesis. The synthesis therefore is, and continues to be, the absolutely presupposed.

The problems of the one system turn exactly around into those of the other; that which, in atomical physics, is the cause of the *composition* of Nature is, in dynamical physics, *that which inhibits evolution*. The former explains the composition of Nature by the force of cohesion, by means of which, however, no continuity is ever introduced into it; the latter, on the contrary, explains cohesion by the continuity of evolution. (All cohesion is originally only in the productivity.)

Both systems set out from something purely ideal. Absolute synthesis is as much purely *ideal* as absolute analysis. The real occurs only in Nature as *product*; but Nature is not product, neither when thought as absolute involution or as absolute evolution; *product* is what is contained between the two extremes.

The first problem for both systems is to construct the product, i.e., that in which the opposites become real. Both reckon with purely *ideal* magnitudes so long as the product is not constructed; it is only in the *directions* in which they accomplish this that they are opposed. Both systems, as far as they have to deal with merely ideal factors, have the same value, and the one forms the test of the other.—That which is concealed in the depths of productive Nature must be reflected as product in Nature as Nature, and thus the atomistic system must be the continual reflection of the dynamical. In the *Outline*, of the two directions, that of atomistic physics has been chosen intentionally. It will contribute not a little to the understanding of our science if we here demonstrate in the *productivity* what was there shown in the *product*.

(m) *In the pure productivity of Nature absolutely nothing is distinguishable without diremption;* [298] *it is only productivity dualized in itself that gives the product.*

Since the absolute productivity arrives only at producing per se, not at the producing of a determinate something, the tendency of Nature, by virtue of which a product is arrived at, must be the *negative* of productivity.

In Nature, insofar as it is real, there can no more be productivity without a product than a product without productivity. Nature can only approximate to the two extremes, and it must be demonstrated *that* it approximates to both.

(α) *Pure productivity originally passes into formlessness.*

Wherever Nature loses itself in formlessness, productivity is exhausted in it. (This is what we express when we talk of a "becoming latent.")—Conversely, wherever the form predominates, i.e., wherever the productivity is *limited*, the productivity manifests itself; it appears, not as a (representable) product, but *as* productivity, although passing over into one product, as in the phenomena of heat. (The idea of imponderables is only a *symbolic* one.)

(β) *If productivity passes into formlessness, then, objectively considered, it is the absolutely formless.*

(The boldness of the atomical system has been very imperfectly comprehended.—The idea which prevails in it, that of an absolutely formless element everywhere incapable of manifestation as determinate matter, is nothing other than the symbol of Nature approximating to productivity.—The nearer to productivity the nearer to formlessness.)

(γ) *Productivity appears as productivity only when limits are set to it.*

That which is everywhere and in everything, is, for that very reason, nowhere.—Productivity is fixed only by limitation.—*Electricity exists* only at that point at which limits are given, and it is only a poverty of conception that would look for anything else in its phenomena [299] beyond the phenomena of (limited) productivity.—The condition of *light* is an antithesis in the electric and galvanic processes, as well as in the chemical process, and even light which comes to us without our cooperation (the phenomenon of productivity exerted all around by the Sun) presupposes that antithesis.*

---

*According to recent *experiments*, it is at least not impossible to regard the phenomena of light and those of electricity as one, since in the prismatic spectrum the colors may at least be considered to be opposites, and the white light, which regularly falls in the middle, *can* be regarded as the point of indifference; and for reasons of *analogy* one is tempted to consider *this* construction of the phenomena of light as the real one. (Original note.—Trans.)

(δ) *It is only limited productivity that gives rise to the product.*

(The explanation of the product must begin at the origination of the fixed point at which the start is made.)—*The condition of all formation is duality.* (This is the more profound signification that lies in Kant's construction of matter from opposite forces.) Electrical phenomena are the general scheme for the construction of matter universally.

(ε) *In Nature, neither pure productivity nor pure product can ever exist.*

The former is the negation of all product, the latter the negation of all productivity. (Approximation to the former is the absolutely decomposable, to the latter the absolutely indecomposable substance of the atomists. The former cannot be thought without, at the same time, being the absolutely incomposable, the latter without, at the same time, being the absolutely composable.)

Nature will therefore originally be the mean factor arising out of the two, and thus we arrive at the idea *of a productivity engaged in a transition into product, or of a product that is productive* to infinity.—We hold to the latter definition. The idea of the product (the fixed) and that of the productive (the free) are mutually opposed.—Since what we have postulated is already [300] product, it can be productive, if it is productive at all, only in a *determinate way*. But determined productivity is (active) *formation*. That third factor must therefore be *in the state of formation*.

But the product is supposed to be productive to infinity (that transition is never to take place absolutely); it will therefore be productive at every stage in a determinate way; the productivity will remain, but not the product.

(The question might arise how a transition from form to form is possible at all here, when *no* form is fixed. Still, that *momentary* forms should be reached has already been rendered possible by the fact that the evolution cannot take place with infinite rapidity, in which case, therefore, for every moment at least, the form is certainly a determinate one.) The product will appear to be gripped in *infinite metamorphosis*.

(From the standpoint of reflection, it will appear to be continually in the leap from fluid to solid, without ever reaching, however, the form sought.— Organisms that do not live in the cruder element at least live on the deep ground of the aerial sea—many pass over, by metamorphoses, from one element into another; and what does the animal, whose vital functions almost all consist in contractions, appear to be, other than such a leap?)

The metamorphosis will not possibly take place *without rule*. For it must remain within the original antithesis, and is thereby confined within limits.*

---

*Hence wherever the antithesis is canceled or deranged, the metamorphosis becomes irregular.— For what is even disease but metamorphosis? (Original note.—Trans.)

(This accordance with rule will express itself solely by an internal relationship of forms, a relationship which again is not thinkable without an archetype which lies at the basis of all—and which, with however manifold divergences, they nevertheless all express.)

But even with such a product we do not have that which we [301] were searching for, a product which, while productive to infinity, remains *the same*. That this product should remain the same seems unthinkable, because it is not thinkable without an absolute inhibition or cancellation of the productivity.—The product would have to be inhibited as the productivity was inhibited, for it is still productive, inhibited by diremption and the limitation resulting from it. But it must at the same time be explained how the productive product can be inhibited at each individual stage of its formation, without its ceasing to be productive, *or how, by diremption itself, the permanence of the productivity is secured*.

In this way we have brought the reader as far as the problem of the fourth section of the *Outline*, and we leave him to find in it for himself the solution, along with the corollaries which it brings up.—Meanwhile, we shall endeavor to indicate how the deduced product would necessarily appear from the standpoint of *reflection*.

The product is the synthesis wherein the opposite extremes meet, which on the one side are designated by the absolutely decomposable, on the other the indecomposable.—How continuity comes into the absolute discontinuity with which the atomic philosopher sets out, he endeavors to explain by means of cohesive, plastic power, etc. But he does so in vain, for *continuity* is only productivity itself.

The manifoldness of the forms which such a product assumes in its metamorphosis was explained by the difference in the stages of development, such that parallel with every step of development goes a particular form.—The atomic philosopher posits in Nature certain fundamental forms, and since in it everything strives toward form, and everything which does form itself also has its *particular* form, so the fundamental forms must be conceded, but certainly only as *indicated* in Nature, not as *actū* existent.

From the standpoint of reflection, the becoming of this product must appear as a continual striving of the original actants toward the production of a determinate form, and a continual annihilation of those forms again.
[302] Thus, the product would not be the product of a simple tendency—it would only be the visible expression of an internal proportion, of an internal equipoise of the original actants, which neither reduce themselves mutually to absolute formlessness, nor yet do they allow the production of a determinate and fixed form, on account of the universal conflict.

Until now (so long as we have had to deal merely with ideal factors), opposite directions of investigation have been possible; from this point, since we have to pursue a real product in its developments, there is only one direction.

(m) By the unavoidable separation of productivity into opposite directions at every single stage of development the product itself is separated into *individual products* by which, however, for that very reason, only different stages of development are marked.

That this is so may be shown *either* in the products themselves, as is done when we compare them with each other with regard to their form and seek a continuity of formation. This is an idea which, from the fact that continuity is never in the *products* (for the reflection), but always only in the *productivity*, can never be perfectly realized.

In order to find continuity in productivity, the successive steps of the *transition of productivity into product* must be more clearly exhibited than they have been until now.—By the fact that the productivity gets *limited* (see above), we have in the first instance only the inception of a product, only the fixed point for the productivity overall.—It must be shown how the productivity gradually materializes itself and changes itself into products ever more and more fixed, so as to produce a *dynamically graduated scale in Nature*, and this is the real subject of the fundamental problem of the whole system.

(In advance, the following may serve to throw light on the subject.—In the first place, a diremption of the productivity is demanded; the cause through which this diremption is effected remains in the first instance altogether outside of the investigation. [303]—An alternation of contraction and expansion is perhaps conditioned by diremption. This alternation is not something in matter, but is *matter itself*, and the first stage of productivity passing over into product.—*Product* cannot be reached except through a stoppage of this change, that is, through a *third* factor which *fixes* that change itself, and thus matter in its lowest stage—in the *first* power—would be an object of intuition; that change would be seen in rest, or in equilibrium, just as, conversely again, by the cancellation of the third factor, matter might be raised to a higher power.—Now it might be possible that those products just deduced stood upon *quite different levels* of materiality, or *of that transition*, or that those different levels were more or less *distinguishable* in the one than in the other—that is, a *dynamically graduated scale of those products* would thereby have to be demonstrated.)

(n) In the *solution* of the problem itself we shall continue in the direction hitherto taken, for the time being, without knowing where it may lead us.

Individual products are brought into Nature; but in these products productivity, *as* productivity, is still held to be always distinguishable. Productivity has not yet absolutely passed over into product. The subsistence of the product is supposed to be a continual self-reproduction. The problem arises, by what is this absolute transition—exhaustion of the productivity in the product—prevented? Or by what means does its subsistence become a continual self-reproduction?

It is absolutely unthinkable how the activity that everywhere tends toward a product is prevented from going over into it *entirely*, unless that transition is prevented *by external influences*, and the product, if it is to subsist, is compelled at every moment to reproduce itself *anew*.

Up to this point, however, no trace has been discovered of a cause opposed to the product (to organic nature).—Such a cause can, therefore, at present, only be postulated. (We thought [304] we saw the whole of Nature exhaust itself in that product, and it is only here that we note that in order to comprehend such a product *something else* must be presupposed, and a new antithesis must come into Nature. Nature has been for us absolute *identity* in duplicity—here we come upon an antithesis that must again take place *within* the other.—This antithesis must be capable of being shown in the deduced product itself, if it is capable of being deduced at all.)

The deduced product is an activity *directed outward*—this cannot be distinguished *as* such without an activity *directed inward from without* (i.e., directed upon itself), and this activity, on the other hand, cannot be thought unless it is *counteracted* (reflected) from without.

In the opposite directions, which arise through this antithesis, lies the principle for the construction of all the phenomena of life—upon the cancellation of those opposite directions, life remains over either as *absolute activity* or *absolute receptivity*, since it is only possible as the perfect *reciprocal determination* of receptivity and activity.

We therefore refer the reader to the *Outline* itself, and merely call his attention to the higher stage of construction which we have here reached.

We have above (g) explained the origin of a *product generally* by a struggle of Nature against the original point of inhibition, through which this point is raised to a full sphere, and thus receives permanence.—Here, since we are deducing a struggle of *external* Nature, not against a *mere* point, but against a *product*, the first construction rises for us to a *second* power, as it were; we have a doubled product (and thus it might well be shown subsequently that organic nature generally is only the higher power of the inorganic, and that it rises above the latter for the very reason that in it precisely that which was already product *again* becomes product).

[305] Since the product, which we have deduced as the most primary, drives us to a side of Nature that is opposed to it, it is clear that our construction of the origin of a product generally is *incomplete*, and that we have not yet, by far, satisfied our problem—(the problem of all science is to construct the origin of a fixed product). A productive product, as such, can subsist only under the influence of external forces, because it is only thereby that productivity is interrupted, prevented from being extinguished in the product.—There must now be again a particular sphere for these external forces; these forces must lie

in a world which is *not productive*. But that world, for this very reason, would be a world fixed and undetermined in every respect. The problem of how a product comes to exist in Nature has therefore received a onesided solution through all that has preceded. "The product is inhibited by diremption of the productivity at every single stage of development." But this is true only for the *productive* product, whereas we are here dealing with a *nonproductive* product.

The contradiction that we encounter here can be resolved only by finding a *general* expression for the construction of a *product generally* (regardless of whether it is productive or has ceased to be so).

Since the existence of a world that is *not productive* (inorganic) is for the time being merely postulated in order to explain the productive one, its conditions can be laid down only hypothetically; and as we do not in the first instance know it at all except through its opposition to the productive, those conditions likewise must be deduced only from this opposition.—(From this it is of course clear (also referred to in the *Outline*) that this second section, as well as the first, contains throughout merely hypothetical truth, since neither organic nor inorganic nature is explained without our having reduced the construction of the two to a common expression, which, however, is possible only [306] through the synthetic part.—This must lead to the highest and most universal principles for the construction of a *Nature* generally; hence we must refer the reader who is concerned about a knowledge of our system altogether to that part.)—The hypothetical deduction of an inorganic world and its conditions we may pass over here all the more readily, since they are sufficiently detailed in the *Outline*, and hasten to the most universal and highest problem of our science.

The most universal problem of speculative physics may now be expressed thus: *to reduce the construction of organic and inorganic products to a common expression*. We can only provide the main principles of such a solution, and of these, for the most part, only such as have not been completely educed in the *Outline* itself (third principal division).

A.

Here at the very beginning we lay down the principle that *since the organic product is the product in the second power, the* ORGANIC *construction of the product must be, at least, the sensuous image of the* ORIGINAL *construction of* EVERY *product*.

(a) In order that the productivity may be at all fixed at a point, *limits must be given*. Since *limits* are the condition *of the first phenomenon*, the *cause* through which limits are produced *cannot be a phenomenon*, it withdraws into the interior of Nature, or the interior of each respective product. In organic nature, this limitation of productivity is shown by what we call *sensibility*, which must be thought as the first condition of the construction of the organic product.

(b) The immediate effect of confined productivity is an *alternation of contraction and expansion* in the matter already given, and as we now know, constructed, as it were, for the second time.

[307] (c) Where this alternation ceases, productivity passes over into product, and where it is again restored, product passes over into productivity.— For since the product must remain productive to infinity, *those three stages of productivity must be capable of being* DISTINGUISHED in the product; the absolute transition of the latter into product is the canceling of product itself.

(d) Just as these three stages are distinguishable in the *individual*, they must be distinguishable *in organic nature as a whole*, and the graduated series of organizations is nothing more than a graduated scale of *productivity itself*. (Productivity exhausts itself to degree *c* in the product A, and can begin with the product B only at the point where it left off with A, that is, with degree *d*, and so on downward to the *vanishing* of all productivity. If we knew the absolute *degree* of productivity of the *Earth* for example (which is determined by the Earth's relation to the Sun) the limit of organization upon it might be more accurately determined by this means than by incomplete experience—which must be incomplete for this reason, if for no other, that the catastrophes of Nature have, beyond doubt, swallowed the last links of the chain.—A true system of natural history, which has for its object not the *products* of Nature but *Nature itself*, follows the *one* productivity that battles, so to speak, against freedom, through all its windings and sinuosities, to the point at which it is at last compelled to perish in the product.)

It is upon this dynamical graduated scale in the individual, as well as in the whole of organic nature, that the construction of all organic phenomena rests.

B.*

These principles, stated universally, lead to the following fundamental principles of a universal theory of Nature.

[308] (a) Productivity must be *primarily* limited. Since *outside* of limited productivity there is[†] *pure identity*, the limitation cannot be established

---

*(From this point onward, there are, as in the *Outline*, additions in notes, included in the SW text.—Trans.)

[†]only

by a difference already existing, and therefore must be furnished by *an opposition arising in productivity itself*, to the existence of which we here revert as a first postulate.*

(b) This difference thought *purely* is the first condition of all[†] activity, the productivity is attracted and repelled[‡] between opposites (the primary limits); in this alternation of expansion and contraction there necessarily arises a common element, but one which exists only *in change*.—If it is to exist *outside* of change, then the *change itself* must become fixed.—The active factor in change is the productivity sundered within itself.

(c) It is asked:

(α) By what means such alternation can be fixed at all.—It cannot be fixed by anything that is contained as a *member* in the alternation itself, and must therefore be fixed by a *tertium quid*.

(β) But this *tertium quid* must be able to *prehend* that original antithesis; however, *outside* of that antithesis nothing *exists*[§]—it[‖] must therefore be originally contained in it, as something that is mediated by the antithesis, and by which in turn the antithesis is mediated; for otherwise there is no reason why it should be originally contained in that antithesis.

[309] The antithesis is dissolution of identity. But Nature is *primarily* identity.—In that antithesis, therefore, there must again be a striving toward identity. This striving is immediately conditioned *through* the antithesis; for if there was no antithesis, there would be identity, absolute rest, and therefore no striving toward identity.[#]—If, on the other hand, there were not identity in the antithesis, the antithesis itself could not endure.

Identity produced from difference is indifference; that *tertium quid* is therefore a *striving toward indifference*, a striving which is conditioned by the difference itself, and by which it, on the other hand, is conditioned.—(The

---

*The first postulate of natural science is an antithesis in the pure identity of Nature. This antithesis must be thought quite purely, and not with any other substrate besides that of activity; for it is the condition of all substrate. The person who cannot think activity or opposition without a substrate cannot philosophize at all. For all philosophizing only concerns the deduction of a substrate.

[†]natural

[‡]The phenomena of electricity show the scheme of nature oscillating between productivity and product. This condition of oscillation or change, attractive and repulsive force, is the real condition of formation.

[§]For it is the only thing that is given us to derive all other things from.

[‖]that *tertium quid*

[#]That *tertium quid* 1) must be directly determined through the antithesis; 2) the antithesis must likewise be conditioned through that third factor. Now by what is the antithesis conditioned? It is antithesis only by virtue of that *striving* toward identity. For where there is no striving toward unity, there is no antithesis.

difference must not be looked upon as a difference at all, and is nothing for intuition, except through a third that sustains it—to which change itself adheres.)

This *tertium quid*, therefore, is the exclusive substrate in that primal alternation.—But substrate posits change as much as change posits substrate—and there is here no first and no second, but difference and striving toward indifference, are, as far time is concerned, one and contemporary.

Axiom. No identity in Nature is absolute, but all is only indifference.*

Since that *tertium quid* itself *presupposes* the primary antithesis, the antithesis itself cannot be *absolutely* canceled by it; *the condition of the continuance of that tertium quid*† *is the perpetual continuance of the antithesis*, just as, conversely, *the continuance of the antithesis is conditioned by the continuance of the tertium quid.*

But how, then, shall the antithesis be thought as enduring?

We have one primary antithesis, between the limits of which all Nature must lie; if we assume that the factors of this antithesis [310] can really pass over into each other, or come together absolutely in some *tertium quid* (some individual product), then the antithesis is removed, and along with it the *striving*, and so all the activity of Nature.—But that the antithesis should endure is thinkable only by its being *infinite*—by the extreme limits being held asunder to infinity, *so that always only the mediating links of the synthesis, never the last and absolute synthesis itself, can be produced*, in which case it is only *relative points of indifference* that are always attained, never absolute ones, and every successively originated difference leaves behind a new and still unremoved antithesis, and this again passes into indifference, which, in its turn, *partially* removes the primary antithesis. By virtue of the original antithesis and the striving toward indifference there arises a product, but the product partially does away with the antithesis; *through* the doing away with that part, that is, through the origination of the product itself, there arises a new antithesis, different from the one that has been done away with, and through it, a product different from the first; but even this leaves the *absolute* antithesis in place, therefore duality, and through it a product will arise anew, and so on to infinity.

Let us say, for example, that by the product A, the antitheses *c* and *d* are united; the antitheses *b* and *e* still lie outside of that union. This latter antithesis is done away with in B, but this product also leaves the antithesis *a* and *f* unremoved—if we say that *a* and *f* mark the extreme limits, then the union of these will be that product which can never be reached.

---

*Nature is an activity that constantly *strives* toward identity, an activity, therefore, which in order to endure *as* such, constantly presupposes the antithesis.

†of that third activity, or of Nature

Between the extremes a and f lie the antitheses c and d, b and e; but the series of these intermediate antitheses is infinite; all these intermediate antitheses are included in the one absolute antithesis.—In the product A, of a only e, and of f only d is canceled; let what remains of a be called b, and of f, e; these will indeed, by virtue of the absolute striving toward indifference, become again united, but they leave a new antithesis uncanceled—and so there remains between a and f an infinite series of intermediate antitheses, and the product in which those absolutely cancel themselves never *is*, but only *becomes*.

This infinitely progressive formation must be thus represented.—The original antithesis would necessarily be canceled in the primal product A. The product would necessarily fall at the point of indifference of a [311] and f, but inasmuch as the antithesis is an absolute one, which can be canceled only in an infinitely continued, never actual, synthesis, A must be thought as the center of an infinite periphery (whose diameter is the infinite line af). Since in the product of a and f, only e and d are united, there arises in it the new division b and e, and the product will therefore divide up into opposite directions; at the point where the striving toward indifference attains preponderance, b and e will combine and form a new product different from the first, but between a and f there still lie an infinite number of antitheses; B, the point of indifference, is therefore the center of a periphery which is comprehended in the first, but is itself again infinite, and so on.

The antithesis of b and e in B IS MAINTAINED through A, because it[*] leaves the antithesis *disunified*; so[†] the antithesis in C *is maintained* through B, because B, in its turn, cancels *only a part* of a and f. But the antithesis in C is maintained through B, only insofar as A maintains the antithesis in B.[‡] What therefore in C and B results *from* this antithesis[§] is *occasioned* by the common influence of A, so that B and C, and the infinite number of other products that come as intermediate links between a and f—are, in relation to A, only *one* product.—The *difference*, which remains over in A after the union of c and d, is only *one*, into which then B, C, etc., again divide.

[312] But the endurance of the antithesis is, in the case of every product, the condition of the striving toward indifference, and thus a striving toward indifference is maintained through A in B, and through B in C.—But

---

[*] A

[†] in like manner

[‡] The whole of the uncanceled antithesis of A is carried over to B. But again, it cannot entirely cancel itself in B, and is therefore carried over to C. The antithesis in C is therefore maintained by B, but only insofar as A maintains the antithesis which is the condition of B.

[§] suppose, for example, the result of it were universal gravitation

the antithesis which A leaves uncanceled, is only one antithesis, and therefore also this tendency in B, in C, and so on to infinity, is only conditioned and maintained through A.

The organization thus determined is none other than the organization of the universe in the system of gravitation.—*Gravity is simple*, but its *condition* is duplicity.—Indifference arises only out of difference.—The canceled duality is matter, inasmuch as it is only *mass*.

The *absolute* point of indifference exists nowhere, but is, as it were, divided among several *single* points.—The universe which forms itself from the center toward the periphery *seeks* the point at which even the extreme antitheses of Nature cancel themselves; the impossibility of this canceling guarantees the infinity of the universe.

From every product A, the uncanceled antithesis is carried over to a new one, B, the former by this means becoming the cause of duality and gravitation for B.—(This *carrying over* is what is called "action by distribution," the theory of which becomes clear only at this point.)*—Thus, for example, the Sun, being only *relative* indifference, maintains, as far as its sphere of action reaches, the antithesis that is the condition of weight upon the subordinate planetary bodies.†

[313] The indifference is canceled at every step, and at every step it is restored. Hence, weight acts upon a body at rest as well as upon one in motion.—The universal restoration of duality, and its recanceling at every step, can‡ appear only as a *nisus* toward a third factor. This third factor§ abstracted from tendency is nothing,‖ therefore purely *ideal* (marking only direction)—a point.#

---

*That is, distribution exists only when the antithesis in a product is not absolutely but only *relatively* canceled.

†The striving toward indifference attains the preponderance over the antithesis, at a greater or lesser distance from the body which exercises the distribution (as, e.g., at a certain distance the action by distribution which an electric or magnetic body exercises upon another body, appears as canceled). The difference in this distance is the ground of the difference of planetary bodies in one and the same system, inasmuch, namely, as one part of the matter is subjected to indifference more than the rest. Since, therefore, the condition of all product is difference, difference must again arise at every moment as the source of all existence, but must also be thought as again canceled. By this continual reproduction and resuscitation creation takes place anew at every step.

‡that is

§is therefore the pure zero

‖$= 0$

#It is precisely zero to which Nature continually strives to revert, and to which it would revert if the antithesis were ever canceled. Let us suppose the original condition of Nature $= 0$ (lack of reality). Now zero can certainly be thought as dividing itself into $1 - 1$ (for this $= 0$); but if we posit that this division is not infinite (as it is in the infinite series $1 - 1 + 1 - 1...$), then Nature will, as it were, oscillate continually between zero and unity—and this is precisely its condition.

Gravity* is in the case of every total product only *one*,† and so also the relative point of indifference is only *one*. The point of indifference of the *individual* body marks only the line of direction of its tendency toward the universal point of indifference; hence this point may be regarded as the only one at which gravity acts; just as that by whose means alone bodies attain consistency for us is simply this tendency outward.‡

Vertical falling toward this point is not a simple, but a compound motion, and it is to be wondered at that this has not been perceived before.§

Gravity is not proportional to mass (for what is this mass but an abstraction of the specific gravity which you have hypostatized?); but, conversely, the mass of a body is only the expression of the momentum with which the antithesis in it cancels itself.

[314] (d) With the preceding, the construction of matter in general is completed, but not the construction of specific difference in matter.

That which all the matter of B, C, etc., in relation to A has in *common*, is the difference which is not canceled by A, and which again cancels itself *in part* in B and C—hence, therefore, the gravity mediated by that difference.

What *distinguishes* B and C from A, therefore, is the difference which is not canceled by A, and which becomes the condition of gravity in the case of B and C.—Similarly, what distinguishes C from B (if C is a product subordinate to B), is the difference which is not canceled by B, and which is again carried over to C. Gravity, therefore, is not the same thing for the higher and for the subaltern planetary bodies, and there is as much variety in the central forces of attraction as in their conditions.

The means by virtue of which another difference of individual products is possible (in the products A, B, and C, which, insofar as they are opposed to *each other*, represent products absolutely *homogeneous*‖), is the possibility of a difference of relation between the factors in the *canceling*, so that, for example, in X, the positive factor, and in Y, the negative factor has preponderance (thus rendering the one body positively, and the other negatively electrified).—All difference is difference of electricity.#

---

*the center of gravity

†for the antithesis is one

‡*Baader* on the *Pythagorean Square*, 1798. (Original note.—Trans.)

§Except by the thoughtful author of a review of my work on the *World-Soul* in the Würzburg *Gelehrte Anzeiger*, the only review of that work that has since come to my attention. (Original note.—Trans.)

‖because the antithesis is the same for the *whole product*,

#It is here taken for granted that what we call the quality of bodies, and what we are wont to regard as something homogeneous and the ground of all homogeneity, is really only an expression for a canceled difference.

(e) That the identity of matter is not *absolute* identity, but only *indifference*, can be proved from the possibility of again canceling the identity, and from the accompanying phenomena.*—We may be allowed, for brevity's sake, to include this recanceling and its resultant phenomena under the expression [315] *dynamical process*, without, of course, affirming decisively whether anything of the sort is everywhere actual.

Now there will be exactly as many stages in the dynamical process as there are stages of transition from difference to indifference.

(α) The first stage will be marked by objects *in which the reproduction and recanceling of the antithesis at every moment is still itself an object of perception.*

The whole product is reproduced anew at every moment.† That is, the antithesis which is canceled in it springs up afresh every moment; but this reproduction of difference loses itself immediately in *universal* gravity.‡ This reproduction, therefore, can be perceived only in *individual* objects, which seem to gravitate *toward each other*; since, if to the one factor of an antithesis its opposite is offered (in another) *both factors* become *heavy with reference to each other*, in which case, therefore, the universal gravity is not canceled, but a special one occurs *within* the universal.—An instance of such a mutual relation between two products is that of the Earth and the magnetic needle, in which the continual recanceling of indifference in gravitation toward the poles is ascertained.§ It is the continual sinking back into identity‖ in gravitation toward the universal point of indifference.—Here, therefore, it is not the *object*, but the *reproduction of the object itself* that becomes object.#

[316] (β) At the first stage, the duplicity of the product again appears *in the identity*; at the second, the antithesis will divide up and distribute itself among different objects (A and B). From the fact that the one factor of the antithesis attained a *relative* preponderance in A, the other in B, there will arise, according to the same law as in (α), a gravitation of the factors *toward each other*, and so a new difference, which, when the relative equilibrium is restored

---

*(According to SW, the last part of this sentence reads as follows.—Trans.) The construction of quality ought necessarily to be capable of experimental proof, by recanceling of the identity, and of the phenomena which accompany it.

†Every body must be thought as reproduced at every step, and therefore also every total product.

‡The *universal*, however, is never perceived, for the simple reason that it is universal.

§By which what was said above is confirmed, that falling toward the center is a compound motion.

‖The reciprocal canceling of opposite motions.

#Or the object is seen in the first stage of becoming, or of transition from difference to indifference. The phenomena of magnetism even serve, so to speak, as an incentive to transport us to the standpoint beyond the product, which is necessary in order for the construction of the product.

in each, results in *repulsion*.*—(Alternation of attraction and repulsion, *second stage* in which matter is seen)—Electricity.

(γ) At the second stage the one factor of the product had only a *relative* preponderance.† At the third it will attain an *absolute* one—in the two bodies A and B, the original antithesis is again completely represented—matter will revert to the *first stage* of becoming.

At the *first stage* there is still PURE *difference*, without substrate;‡ at the second stage it is the *simple* factors of two PRODUCTS that are opposed to each other; at the third it is the PRODUCTS THEMSELVES that are opposed; here is difference in the *third* power.

If two *products* are absolutely opposed to each other,§ then in each of them singly indifference *of gravity* (by which alone each *is*) must be *canceled*, and they must gravitate *toward each other*.‖ (In the second stage there was only a mutual [317] gravitating of the *factors* to each other—here there is a gravitating of the products.)#—This process, therefore, first assails the *indifferent element of the* PRODUCT, that is, the products themselves dissolve.

Where there is equal difference there is equal indifference; difference of *products*, therefore, can end only with *indifference* of *products*.—(All indifference deduced until now has been only indifference of substrateless, or at least simple, factors.—Now we come to speak of an indifference of products.) This striving will not cease until a joint product exists. The product, in forming itself, passes, from both sides, through all the intermediate links that lie between the two products,** until it finds the point at which it succumbs to indifference, and the product is fixed.

---

*There will result the opposite effect—a negative attraction, that is, repulsion.—Repulsion and attraction stand to each other as positive and negative magnitudes. Repulsion is only negative attraction, attraction only negative repulsion; as soon, therefore, as the maximum of attraction is reached, it passes over into its opposite, into repulsion.

†If we designate the factors as + and − electricity, then, in the second stage, + electricity had a relative preponderance over − electricity.

‡for it was only out of it that a substrate arose

§If the individual factors of the two products are no longer opposed, but the whole products themselves are absolutely opposed to each other

‖For a product is something in which antithesis cancels itself, but it cancels itself only through indifference of gravity. When, therefore, two products are opposed to each other, the indifference in each *individually* must be absolutely canceled, and the whole products must gravitate toward each other.

#In the electric process, the *whole product is not* active, but only the one factor of the product, which has relative preponderance over the other. In the chemical process in which the *whole product* is active, it follows that the indifference of the whole product must be canceled.

**for example, through all the intermediate stages of specific gravity

General Remark.

By virtue of the first construction, the product is posited as identity; this identity, it is true, again resolves itself into an antithesis, which is no longer an antithesis cleaving to *products* however, but an antithesis in the *productivity* itself.—The product, therefore, *as* product, is* identity.—But even in the sphere of products, there again arises a duplicity in the second stage, and it is only in the third that even the duplicity of the *products* again becomes *identity* of the products.†—There is therefore here too a progress from thesis to antithesis, and thence to synthesis.—The final synthesis of [318] matter concludes in the chemical process; if composition is to proceed yet further in it, then this circle must open again.

We must leave it to our readers themselves to make out the conclusions to which the principles here stated lead, and to consider the universal interdependence which is introduced by them into the phenomena of Nature.—Nevertheless, to give one instance: when in the chemical process the bond of gravity is loosed, the phenomenon of *light* which accompanies the chemical process in its greatest perfection (in the process of combustion) is a remarkable phenomenon which, when followed out further, confirms what is stated in the *Outline*, page 100: "The action of light must stand in secret interdependence with the action of gravity which the central bodies exercise."—For, is not the indifference dissolved at every step, since gravity, as ever active, presupposes a continual canceling of indifference?—It is thus, therefore, that the Sun, by the distribution exercised on the Earth, causes a universal separation of matter into the primary antithesis (and hence gravity). This universal canceling of indifference is what appears to us (who are endowed with life) as *light*; wherever, therefore, that indifference is dissolved (in the chemical process), there light *must* appear to us.—According to the foregoing, there is *one* antithesis which, beginning at magnetism and proceeding through electricity, finally dissipates in the chemical phenomena.‡ In the chemi-

---

*was

†We have therefore the following scheme of the dynamical process. First stage: Unity of the product—magnetism. Second stage: Duplicity of the products—electricity. Third stage: Unity of the products—chemical process.

‡The conclusions which may be deduced from this construction of dynamical phenomena are partly anticipated in the preceding. The following may serve for further explanation. The chemical process, for example, in its highest perfection, is a process of combustion. Now I have already shown on another occasion that the condition of light in the body undergoing combustion is nothing else but the maximum of its positive electrical condition. For it is always the positively electrical condition that is also the combustible. Might not, then, this coexistence of the phenomenon of light with the chemical process in its highest perfection give us information about the ground of *every* phenomenon of light in Nature? What happens, then, in the chemical process? Two whole products gravitate toward each other. The *indifference* of the *individual* is therefore *absolutely* canceled. This absolute canceling of indifference puts the whole body into the condition of light, just as the partial cancel-

cal process, that is, [319] *the whole product* becomes +E or −E (the *positively* electric body, in the case of absolutely *uncombusted* bodies, is always the *more combustible*.* Whereas the *absolutely incombustible* is the cause of every *negatively* electric condition). And if we may be allowed to invert the case, what else are bodies themselves but condensed (confined) electricity?—In the chemical process the whole body dissolves into +E or −E. Light is everywhere the appearing of the *positive* factor in the primary antithesis; hence, wherever the antithesis is restored, *light* is there for us, because generally only the positive factor is beheld, and the negative one is only felt.—Is the connection of the diurnal and annual deviations of the magnetic needle with light now conceivable—and, if in every chemical process the antithesis is dissolved—is it conceivable that light is the cause and beginning of all chemical processes?†

---

ing in the electric process puts it into a partial condition of light. Therefore, the light too that seems to stream to us from the sun is nothing else but the phenomenon of indifference canceled at every step. For as gravity never ceases to act, its condition—the antithesis—must be regarded as springing up again at every step. We should thus have in light a continual, visible phenomenon of gravitation, and it would be explained why, in the planetary system, it is exactly those bodies which are the principal seat of gravity that are also the principal source of light. We should then, also, have an explanation of the connection *in which* the action of light stands to that of gravitation.

The manifold effects of light on the deviations of the magnetic needle, on atmospheric electricity, and on organic nature, would be explained by the very fact that light is the phenomenon of indifference continually canceled—therefore, the phenomenon of the dynamical process continually rekindled. There is, therefore, one antithesis that prevails in all dynamical phenomena—in those of magnetism, electricity and light; for example, the antithesis that is the condition of the electrical phenomena must already enter into the first construction of matter; for all bodies are certainly electrical.

*Or rather, conversely, the more combustible is always also the positively electric; whence it is manifest that the body which burns has merely reached the maximum of + electricity.

†And indeed it is so. What then is the absolutely incombustible? Doubtless, simply that by means of which everything else burns—oxygen. But it is precisely this absolutely incombustible oxygen that is the principle of negative electricity, and thus we have a confirmation of what I have already stated in the *Ideas for a Philosophy of Nature*, i.e., that oxygen is a principle of a negative kind, and therefore the representative, as it were, of the power of attraction; whereas phlogiston, or, what is the same thing, positive electricity, is the representative of the positive, or of the force of repulsion. There has long been a theory that the magnetic, electric, chemical, and, finally, even the organic phenomena, are interwoven into one great interdependent whole. This must be established.—It is certain that the connection of electricity with the process of combustion may be shown by numerous experiments. One of the most recent of these that has come to my knowledge I will cite. It occurs in Scherer's *Journal of Chemistry*. If a Leyden jar is filled with iron filings, and repeatedly charged and discharged, and if after the lapse of some time this iron is taken out and placed upon an isolator-paper, for example, it begins to get hot, becomes incandescent, and changes into an oxide of iron.—This experiment deserves to be frequently repeated and more closely examined—it might readily lead to something new.

This great interdependence, which a scientific system of physics must establish, extends over the whole of Nature. It must, therefore, once established, shed a new light on the *history* of the

[320]   (f) *The dynamical process is nothing but the second construction of matter, and as many stages as there are in the dynamical process, there are the same number in the original construction of matter.*

[321] This axiom is the converse of axiom (e).* That which, in the dynamical process, is perceived in the product, takes place *beyond* the product with the simple factors of all duality. The first inception of original production is the limitation of productivity through the primitive antithesis, which, *as* antithesis (and as the condition of all construction), is distinguished only in *magnetism*; the second stage of production is the *alternation* of contraction and expansion, and *as* such becomes visible only in *electricity*; finally, the third stage is the transition of this change into indifference, a change which is recognized as such only in *chemical* phenomena.

MAGNETISM, ELECTRICTY, AND CHEMICAL PROCESS are the *categories* of the original construction of Nature†—the latter escapes us and lies outside of intuition, the former are what of it remains behind, what stands firm, what is fixed—the general schemata for the construction of matter.

And (in order to close the circle at the point where it began), just as in organic nature, where in the graduated series of sensibility, irritability, and formative drive the secret of the production of the *whole of organic nature* lies in each individual, so in the graduated series of magnetism, electricity, and chemical process, so far as the series of powers can be distinguished in the individual body, is to be found the secret of the production *of Nature from itself*‡,§

---

whole of Nature. Thus, for example, it is certain that all geology must start from terrestrial magnetism. But terrestrial electricity must again be determined by magnetism. The connection of North and South with magnetism is shown even by the irregular movements of the magnetic needle.—But again, with universal electricity, which, no less than gravity and magnetism, has its indifference point—the universal process of combustion and all volcanic phenomena stand in connection.

Therefore, it is certain that there is one chain going from universal magnetism down to the volcanic phenomena. Still these are all only scattered experiments. In order to make this interdependence *fully* evident, we need the central phenomenon, or central experiment, of which Bacon speaks oracularly—I mean the experiment wherein all those functions of matter, magnetism, electricity, etc., so run together in one phenomenon that the *individual* function is distinguishable—proving that the one does not lose itself immediately in the other, but that each can be exhibited separately, an experiment which, when it is discovered, will stand in the same relation to the *whole* of Nature, as galvanism does to organic nature.

Proof: All dynamical phenomena are phenomena of transition from difference to indifference, but it is in this very transition that matter is primarily constructed.

†of matter

‡of the whole of Nature

§Every individual is an expression of the whole of Nature. As the existence of the *single* organic individual rests on that graduated series, so does the whole of Nature. Organic nature maintains

[322]   C.

We have now approached nearer the solution of our problem, which was to reduce the construction of organic and inorganic nature to a common expression.

Inorganic nature is the product of the *first* power, organic nature of the *second** (this was demonstrated above; it will soon appear that the latter is the product of a still higher power).—Hence the latter, in view of the former, appears contingent; the former, in view of the latter, necessary. Inorganic nature can take its origin from *simple* factors, organic nature only from *products*, which again become factors. Hence an inorganic nature generally will appear as having been from all eternity, organic nature as having *originated*.

In organic nature, indifference can never come to be in the same way in which it comes to exist in inorganic nature, because life consists in nothing more than a continual *prevention of the attainment of indifference*,† through which there manifestly comes about a condition which is only, so to speak, extorted from Nature.

By organization, matter, which has already been composed for the second time by the chemical process, is once more thrown back to the initial point of formation (the circle above described is again opened); it is no wonder that matter always thrown back again into formation at last returns as a perfect product.

[323] The same stages through which the production of Nature originally passes, are also passed through by the production of the organic product; only that the latter, even *in the first stage*, at least begins with products of the *simple* power.—Organic production also begins with limitation, not of the *primary* productivity, but of the *productivity of a product*; organic formation also takes place through the alternation of expansion and contraction, just as primary formation does; but in this case it is a change taking place, not in the simple productivity, but in the compound.

---

the whole wealth and variety of its products only by continually changing the relation of those three functions.—In like manner inorganic Nature brings forth the whole wealth of its products only by changing the relation of those three functions of matter to infinity; for magnetism, electricity, and chemical process are the functions of matter generally, and on that ground alone are they categories for the construction of all matter. The fact that those three factors are not phenomena of special kinds of matter, but *functions of all matter universally*, gives its real, and its innermost sense to dynamical physics, which, by this circumstance alone, rises far above all other kinds of physics.

*That is, the organic product can be thought only as subsisting under the hostile onslaught of an external nature.

†in prevention of the absolute transition of productivity into product

But there is all of this, too, in the chemical process,* and yet in the chemical process indifference is attained. The vital process, therefore, must again be a higher power of the chemical; and if the schema that lies at the base of the latter is duplicity, the schema of the former will of necessity be *triplicity*.† But the schema of triplicity is‡ that§ of the galvanic process (Ritter's demonstration, etc., p. 172); therefore, the galvanic process (or the process of excitation) stands a power higher than the chemical, and the third element, which the latter lacks and the former has, prevents indifference from being arrived at in the organic product.||

As excitation does not allow indifference to be arrived at in the individual product, and since the antithesis is still there (for the primary antithesis still pursues us),# there remains for Nature no alternative [324] but separation of the factors into *different* products.**—The formation of the individual product, for that very reason, cannot be a completed formation, and the product can never cease to be productive.††—The contradiction in Nature is that the product must be *productive*,‡‡ and that, notwithstanding, the product, *as* a product of the third power, must pass over into indifference.§§ This contradiction Nature tries to re-

---

*The chemical process, too, does not have substrateless or simple factors; it has *products* for factors.

†the former will be a process of the third power

‡in reality

§the fundamental schema

||The same deduction is already furnished in the *Outline*, p. 118.—What the dynamical action is, which according to the *Outline* is also the cause of excitability, is now surely clear enough. It is the *universal action* which is everywhere conditioned by the cancellation of indifference, and which at last tends toward intussusception (indifference of products) when it is not continually prevented, as it is in the process of excitation. (Original note.—Trans.)

#The abyss of forces down into which we gaze here opens up with the single question: in the *first* construction of our Earth, what can have been the ground of the fact that no genesis of new individuals is possible upon it, otherwise than under the condition of opposite powers? Compare an utterance of Kant on this subject, in his *Anthropology*. (Original note.—Trans.)

**The two factors can never be *one*, but must be separated into different *products*—in order that thus the difference may be permanent.

††In the product, indifference of the first and second powers is arrived at (for example, by excitation itself comes an origin of *mass* [i.e., indifference of the first order], and even chemical *products* [i.e., indifference of the second order] are reached), but indifference of the third power can never be reached, because it is a contradictory idea. (Original note, excluding bracketed additions.—Trans.)

‡‡i.e., a product of the third power

§§The product is productive only from the fact of its being a product of the third power. But the idea of a productive product is itself a contradiction. What is productivity is not product, and what is product is not productivity. Therefore a product of the third power is itself a contradictory idea. From this it is even manifest what an extremely artificial condition life is—wrenched, as it were, from Nature—subsisting against the will of Nature.

solve by mediating *indifference* itself through *productivity*, but even this does not succeed, for the act of productivity is only the kindling spark of a new process of excitation; the product of productivity is a *new productivity*. The productivity of the *individual* now indeed passes over into this as its product; the individual, therefore, ceases to be productive more rapidly or slowly, and Nature reaches the point of indifference with it only after the latter has descended to a product of the second power.*

[325] And now the result of all this?—The condition of the inorganic (as well as of the organic) product, is duality. In any case, however, the organic *productive product is so* only from the fact *that the difference* NEVER *becomes indifference*.

It is† therefore *impossible* to reduce the construction of organic and of inorganic product to a *common* expression, and the problem is incorrect, and therefore the solution impossible. The problem presupposes that organic product and inorganic product are mutually *opposed*, whereas the latter is only the *higher power* of the former, and is produced only by the higher power of the forces through which the latter also is produced.—Sensibility is only the higher power of magnetism; irritability only the higher power of electricity; formative drive only the higher power of the chemical process.—But sensibility, irritability, and formative drive are all only included in that *one* process of excitation. (Galvanism affects them all).‡ But if they are only the higher

---

*Nothing shows more clearly the contradictions out of which life arises, and the fact that it is altogether only a heightened condition of *ordinary* natural forces, than the contradiction of Nature in what it tries, but tries in vain, to reach through the *sexes*.—Nature *hates* sex, and where it does arise, it arises against the will of Nature. The separation into sexes is an inevitable fate, with which, after Nature is once organic, it must put up, and which it can never overcome.—By this very hatred of separation it finds itself involved in a contradiction, inasmuch as what is odious to Nature it is compelled to develop in the most careful manner, and to lead to the summit of existence, as if it did so on purpose; whereas it is always striving only for a return into the identity of the genus, which, however, is enchained to the (never to be canceled) duplicity of the sexes, as to an inevitable condition.—That Nature develops the individual only from compulsion, and for the sake of the genus, is manifest from this, that wherever in a genus it *seems* desirous of maintaining the individual longer (though this is never really the case), it finds the genus becoming more uncertain, because it must hold the sexes farther asunder and, as it were, make them flee from each other. In this region of Nature, the decay of the individual is not so visibly rapid as it is where the sexes are nearer to each other, as in the case of the rapidly withering flower, in which, from its very birth, they are enclosed in a calix as in a bridal bed, but in which for that very reason the *genus* is *better secured*.

Nature is the *laziest of animals* and curses separation because it imposes upon it the necessity of activity; Nature is active only in order to rid itself of this compulsion. The opposites must forever shun, in order forever to seek each other; and forever seek, in order never to find each other; it is only in *this* contradiction that the ground of all the activity of Nature lies. (Original note.—Trans.)

†insofar

‡Its effect upon the power of reproduction (as well as the reaction of particular conditions of the latter power upon galvanic phenomena) is less studied still than might be needful and useful. See the *Outline*, p. 120. (Original note.—Trans.)

functions of magnetism, electricity, etc., there must again [326] be a higher synthesis for these in Nature.* And this, however, it is certain, can be sought for only in Nature, insofar as, viewed as a whole, it is *absolutely* organic.

And this, moreover, is also the result to which the genuine science of nature must lead, i.e., that the difference between organic and inorganic nature is only in Nature as object, and that Nature as originally productive soars above both.†

There remains only one remark which we may make, not so much on account of its intrinsic interest, as in order to justify what we said above in regard to the relation of our system to the current so-called dynamical system.—If it were asked, for instance, in what form our original antithesis, canceled, or rather fixed, in the product, would appear from the standpoint of reflection, we cannot better designate what is found in the product by analysis than as *expansive* and *attractive* (retarding) *force*, to which then, however, *gravitation* must always be added as the *tertium quid*, by virtue of which those opposites become what they are.

Nevertheless, the designation is valid only for the standpoint of reflection or of *analysis*, and cannot be applied for *synthesis* at all; and thus our system leaves off exactly at the point where the dynamical physics of *Kant* and his successors begins, namely, at the antithesis as it presents itself in the *product*.

And with this the author delivers over these *Elements of a System of Speculative Physics* to the thinking heads of the age, begging them to make common cause with him in this science—which opens up views of no mean order—and to make up by their own powers, knowledge and external relations, for what, in these respects, he lacks.

---

*Compare above note, p. 199. (Original note.—Trans.)

†That it is therefore the same Nature, which, by the same forces, produces organic phenomena, and the universal phenomena of Nature, and that these forces are in a heightened condition in organic nature.

Appendix: Scientific Authors

Franz Xaver von Baader (1765–1841). Philosopher, theologian, and mystic, Baader first studied medicine and sciences in Ingolstadt and Vienna (1781–85). In 1788 he joined the mining college in Freiburg and became a mining official. While traveling and studying mining in England he discovered the work of Jacob Böhme. In 1808 he became a member of the Bavarian Academy of Sciences, and from 1826 until his death was professor of philosophy and theology in Munich. His aim was the unification of catholic theology with philosophy.

Johann Friedrich Blumenbach (1752–1840). He received his MD in 1775 at the University of Göttingen. In Göttingen he was one of the first scientists to view human beings as an object of natural history. His dissertation on the topic, *De generis humani varietate nativa liber*, became world-famous and was translated into several different languages. He shared the belief with other early scientists such as George-Louis Leclerc Comte de Buffon that an organism's morphology was capable of being modified by the environment and that the resultant changes were inherited. In another work, *Handbuch der Naturgeschichte* (1779), he presented a compelling argument that zoological classifications could and should be based on structures associated with an animal's specific functions.

Joachim Brandis (1762–1846). Physician and pharmacist, health official at Hildesheim and spa physician at Driburg, he later became professor of medicine at Kiel (1803) and royal physician in Copenhagen (1810). His conception of vital force was popularized in his *Versuch über die Lebenskraft* (1795).

John Brown (1735–88). Brown received his MD from St. Andrews (1779) and developed the theory that all living tissues are "excitable" and postulated that the state of life is dependent on certain internal and external "exciting powers," or stimuli, that operate on it. The normal excitement produced by all the agents which affect the body constitutes the healthy condition, while all diseases arise either from deficiency or from excess of excitement, and must be treated with stimulants or sedatives. In 1780 he published the reknowned exposition of his doctrine, *Elementa Medicinae*.

Erasmus Darwin (1731–1802). Charles Darwin's grandfather was one of the leading intellectuals of eighteenth-century England, a respected physician, a well-known poet, philosopher, botanist, and naturalist. As a naturalist, he formulated one of the first formal theories on evolution in *Zoonomia, or, The Laws of Organic Life* (1794–96). He also presented his evolutionary ideas in verse, in particular in the posthumously published poem *The Temple of Nature*. He discussed ideas that his grandson elaborated on sixty years later, such as how life evolved from a single common ancestor.

Karl August Eschenmayer (1768–1852). Trained as a doctor, he became professor of philosophy and medicine at Tübingen (1811). An important commentary and response to Schelling's philosophy of nature entitled "Spontaneität = Weltseele" (Spontaneity = World-Soul) appeared in the *Journal for Speculative Physics*, edited by Schelling.

Leonhard Euler (1707–83). A Swiss professor of mathematics and physics in St. Petersburg, his interests and contributions in mathematics and the sciences ranged from number theory and calculus to hydrodynamics, optics, and astronomy. He is considered by many to be one of the most important mathematicians of all time.

Felice Fontana (1730–1805). An Italian naturalist, physiologist, and court physician, he was a follower of Albrecht von Haller and wrote a series of letters in confirmation of the latter's views on irritability. He made a special study of the eye and in 1765 carried on a series of experiments on the contractile power of the iris. He investigated the physiological action of poisons, particularly of serpents and of the laurel berry. He also devoted some attention to the study of the physical and chemical properties of gases.

Benjamin Franklin (1706–90). The first internationally known American scientist, printer, publisher, and diplomat, he conducted various electrical experiments that led him to the law of charge conservation and what we now view to be a basically accurate theory of electricity. The invention of the lightning rod resulted from the famous kite experiment that established the identity of atmospheric and laboratory electricity.

Franz Joseph Gall (1759–1828). Beginning in 1796 he lectured on phrenology and practiced medicine in Vienna from 1785 to 1807 and in Paris from 1807 to 1828. Gall believed that the mind could be divided into separate faculties that were discretely localized in the brain, and that the exercise of or innate prominence of a faculty would enlarge the appropriate brain area that, in turn, would show up as a cranial prominence.

Johann Gehler (1751–95). German mathematician, physicist, translator, and editor of a dictionary of natural science, "Physikalisches Wörterbuch oder Versuch einer Erklärung der vornehmsten Begriffe und Kunstwörter der Naturlehre" (1787–96).

Christof Girtanner (1760–1800). A Swiss-born, Brunonian physician, he wrote on John Brown and Erasmus Darwin, was an adherent of Antoine Lavoisier's chemistry, and published *Anfangsgründe der antiphlogistische Chemie* in 1792.

Albrecht von Haller (1708–77). Physician, poet, and natural scientist, he was trained and graduated in medicine at Tübingen (1723) and worked as professor of anatomy, botany, and surgery at the University of Göttingen (1736–53).

William Harvey (1578–1657). Schooled in medicine and anatomy and appointed personal physician to James I and subsequently to King Charles I, he was the first to present a reasonably accurate theory of the circulation of the blood and the operation of the heart ("On the Motion of the Heart and Blood in Animals" [1628]), as well as a theory of reproduction via egg and sperm.

Frederick William Herschel (1738–1822). A German-born musician, he emigrated to England in 1757 and took up astronomy and telescope making. He discovered Uranus (from which his fame derives), two moons of Saturn, and infrared radiation. Schelling refers to his discussions of star clusters or nebulae where Herschel suggests the approximate shape of the Milky Way.

John Hunter (1728–93). A Scottish physician, he is considered one of the greatest anatomists of all time and the founder of experimental pathology in England. Hunter put the practice of surgery on a scientific foundation and laid the framework for twentieth-century developments. "Hunter's Lightning" is the term for the light effects that result when one presses on the closed eye.

Jan Ingenhousz (1730–99). A Dutch physician, he emigrated to England where he met Joseph Priestley and Benjamin Franklin. His *Experiments upon Vegetables* (1779) developed a theory of the chemical nature of photosynthesis, and his interest in electricity led him to an explanation of Alessandro Volta's electrophore.

Karl Friedrich Kielmeyer (1765–1844). A professor of chemistry, pharmacy, and medicine in Tübingen, he directed the construction of the old botanic garden there in 1804. Schelling likely knew his "Über die Verhältnisse der organischen Kräfte unter einander" (1793).

Johann Heinrich Lambert (1728–77). He published on logic, mathematics, physical measurement, philosophical method, and cosmology. Remembered as an important correspondent of Immanuel Kant, he was also a pioneer in non-Euclidean geometry.

George-Louis Lesage (1724–1803). Lesage studied medicine and physics and developed the mechanical theory of gravitation that Schelling both admires and attacks here. Schelling often refers to his atomistic *Lucrèce Newtonien* (1784) and *Attempt at a Mechanical Chemistry* (1758).

Georg Christoph Lichtenberg (1742–99). A German physicist, satirical writer, and philosophical aphorist, he became professor at Göttingen in 1769. He wrote on various topics including vulcanology, electricity, and the shape of the Earth, and was one of the first to propose a particle-and-wave theory of light.

Peter Simon Pallas (1741–1811). A German naturalist and physician, aside from traveling extensively in Russia and recording his observations, he did experiments on planaria, hydra, and other flatworms. Species of hydra or polyp now bear his surname, for example, *Hydra oligactis Pallas*.

Christoph Heinrich Pfaff (1773–1852). With interests in chemistry, medicine and pharmacy, he worked with Alessandro Volta on electricity in animals, and published *Ueber thierische Electricität und Reizbarkeit* (1795).

Johann Christian Reil (1759–1813). After medical studies at Göttingen and Halle, where he later became professor and physician, in 1795 he founded the first journal dealing with physiology in Germany, *Archiv für die Physiologie*, which presented works in physics, chemistry, histology, biology, and comparative anatomy. In it he published his essay "Von der Lebenskraft" (1795).

Hermann Samuel Reimarus (1694–1768). German philosopher and Enlightenment deist, he was appointed professor of Hebrew and Oriental languages at the Hamburg Gymnasium in 1727 and made his house a cultural center and meeting place for learned and artistic societies. His first important philosophical work was "Treatise on the Principal Truths of Natural Religion" (1754), a deistic discussion of cosmological, biological, psychological, and theological problems. In "Doctrine of Reason" (1756) he combated traditional Christian belief in revelation.

Henry Ridley (1653–1708). An English physician and anatomist, he published "Anatomia cerebri complectens" (1725).

Johann Wilhelm Ritter (1776–1810). Contemporary and friend of Schelling, he taught at Jena and Munich and was primarily interested in electricity, in particular electrochemistry and electrophysiology. He observed thermoelectrical currents, investigated the artificial electrical excitation of muscles, and built the first dry-cell battery and accumulator. His allegiance to Schelling and speculative forays eventually affected his status in the eyes of his scientific peers.

Andreas Röschlaub (1768–1835). German physician, in "Untersuchungen über Pathogenie" (1798) he developed a theory of excitability drawing on John Brown that was much opposed by Alexander von Humboldt and others.

Johann Ulrich Gottlieb Schäffer (1753–1829). Physician in Regensburg.

Jan Swammerdam (or Schwammerdam) (1637–80). A Dutch naturalist who developed the work of William Harvey by using microscopy and innovative laboratory techniques to study the circulatory system. He was the first to observe red-blood corpuscles, composed the first important work of entomology, studied embryology, and was committed to the doctrine of preformation.

Samuel Thomas Sömmering (1755–1830). German anatomist and physician, published *Vom Baue des menschlischen Körpers* (1791), in various divisions, each dedicated to one aspect of human anatomy and physiology, and "Über das Organ der Seele" (1796).

Robert Symmer (1707–63). A Scottish tutor and civil servant, he held that electrical phenomena resulted from an imbalance of two electrical fluids. Although opposed to Benjamin Franklin's views, Symmer's views were supported by Charles-Augustine de Coulomb and others in France.

Felix Vicq' d'Azyr (1748–94). He was a French physician who helped establish the foundation of neuroanatomy. He created one of the principal anatomic folios of the brain.

Alessandro Volta (1745–1827). From 1778 he was professor of experimental physics at Pavia, and some of his best-known contributions to science include the refutation of galvanism as a special form of electricity and the invention of the voltaic pile and the first apparatus to generate an electric current.

William Charles Wells (1757–1817). A Scottish physician, philosopher, and printer, he wrote on meteorology (featuring an important essay on dew) and biology, and in a late text suggested a form of the theory of natural selection.

NOTES

Translator's Introduction

1. See, for example, Sandra G. Harding, *Is Science Multicultural?: Postcolonialisms, Feminisms, and Epistemologies* (Bloomington: Indiana UP, 1998).

2. See, to name but a few, Jesper Hoffmeyer, *Signs of Meaning in the Universe*, trans. Barbara J. Haveland (Bloomington: Indiana UP, 1996); Erich Jantsch, *The Self-Organizing Universe: Scientific and Human Implications of the Emerging Paradigm of Evolution* (Oxford: Pergamon, 1980); Stuart A. Kauffman, *Investigations* (Oxford: Oxford UP, 2000); Richard C. Lewontin, *The Triple Helix: Gene, Organism, and Environment* (Cambridge: Harvard UP, 2000); J. E. Lovelock, *Gaia: A New Look at Life on Earth* (Oxford: Oxford UP, 1987); Lynn Margulis, *Symbiotic Planet: A New Look at Evolution* (New York: Basic, 1998); I. Prigogine and Isabelle Stengers, *The End of Certainty: Time, Chaos, and the New Laws of Nature* (New York: Free P, 1997); Stanley N. Salthe, *Development and Evolution: Complexity and Change in Biology* (Cambridge: MIT P, 1993); and Lee Smolin, *The Life of the Cosmos* (New York: Oxford UP, 1997).

3. Friedrich Heinrich Loschge's review of the *Outline* in 1800 remarks that it is filled with "many sorts of laughable comparisons and combinations" and amounts to nothing more than "a spirited play with concepts"; and in this century Erik Nordenskiold says of Lorenz Oken that "his speculations were as grotesque as they were irrational." Loschge, cited in AA 1,5 52; Nordenskiold cited in Stephen Jay Gould, *Ontogeny and Phylogeny* (Cambridge: Harvard UP, 1977) 416.

4. Cohen, in 1947, describes this classical prejudice against *Naturphilosophie* (cited in ibid., 38). Gould argues that without this speculative element later evolutionary theories of "ontogeny recapitulating phylogeny" would never have developed, and Barry Gower shows how Hans Christian Ørsted's discovery of electromagnetism was facilitated by Schelling (as Ørsted himself admits). See Barry Gower, "Speculation in Physics: The History and Practice of Naturphilosophie," *Studies in the History and Philosophy of Science* 3.4 (1973): 301–56; and H.-J. Treder, "Zum Einfluß von Schellings Naturphilosophie auf die Entwicklung der Physik," *Natur und geschichtlicher Prozeß: Studien zur Naturphilosophie F. W. J. Schellings*, ed. H. J. Sandkühler (Frankfurt a. M.: Suhrkamp, 1984) 326–34.

While among many scientists and philosophers the above prejudice no longer rings true, it is by no means unheard-of today, even among those who one might think would find its holism hospitable. Among philosophers influenced by Schelling, C. S.

Peirce and Henri Bergson leap readily to mind. On the reception of *Naturphilosophie* in nineteenth-century America, see Joseph J. Esposito, *Schelling's Idealism and Philosophy of Nature* (Lewisburg: Bucknell UP, 1977) 186–207.

5. In the literature this is a relatively well-accepted way of considering Schelling's development. Some writers think the changes to be more or less drastic and numerous. Esposito thinks that a major difficulty in Schelling interpretation arises because "between 1797 and 1806 Schelling produces at least six major reformulations of his system," but he is not specific about how many changes take place before 1801 (see ibid., 125.) The editors of the new Schelling critical edition suggest a major change after 1800, when "the relation to the empirical sciences clearly takes a backseat," and his philosophy takes on a "predominantly speculative form" (see Friedrich Wilhelm Joseph Schelling, *Werke: Historisch-kritische Ausgabe*, ed. Hans Michael Baumgartner, Wilhelm G. Jacobs, and Hermann Krings [Stuttgart: Fromann-Holzboog, 1976ff.], I,5–9Suppl. XIV. (Hereafter this edition will be cited as "AA," with series, volume, and page number, for example, AA I,7 45.) Harald Holz, however, disputes that any truly major change takes place between 1796 and 1806, and sees the transcendental philosophy, nature philosophy, and philosophy of identity as "correlative aspects" of one systematic whole (see his "Perspektive Natur," *Schelling: Einführung in seine Philosophie*, ed. Hans Michael Baumgartner [Freiburg/München: Verlag Karl Alber, 1975] 63). Finally, Schelling himself in the 1830's lectures *On the History of Modern Philosophy* only notes that the special method established in the philosophy of nature and *System* of 1800 became the "soul of the system independent of Fichte," and he characterizes his earlier work in terms of the identity philosophy in order, clearly, to contrast it with his then-current move away from such "negative" rationalistic philosophy and toward a "positive" empirical philosophy (see Friedrich Wilhelm Joseph Schelling, *Zur Geschichte der neueren Philosophie*, ed. M. Buhr [Leipzig: Reclam, 1975] 115; *On the History of Modern Philosophy*, trans. A. Bowie, ed. R. Geuss, *Texts in German Philosophy* [Cambridge: Cambridge UP, 1994] 111). We cannot anticipate coming to any conclusions about this issue here, and adopt the given distinction heuristically. (Note: All translations are my own unless otherwise noted, and references to existing English translations have been provided for the reader's convenience.)

6. A more extensive account of these inclinations and the development of the theory of the postulate can be found in Michael Rudolphi, *Produktion und Konstruktion: Zur Genese der Naturphilosophie in Schellings Frühwerk*, ed. Walter E. Eherhardt, vol. 7, *Schellingiana* (Stuttgart/Bad Cannstatt: Fromann-Holzboog, 2001) 51–81.

7. AA 1,1 265–300; and Friedrich Wilhelm Joseph Schelling, *The Unconditional in Human Knowledge: Four Early Essays (1794–1796)*, trans. Fritz Marti (Lewisburg: Bucknell UP, 1979) 38–55.

8. Cf. "How are *a priori* synthetic judgments possible? . . . this question in its highest abstraction is none other than: How is it possible for the absolute I to step out of itself and oppose to itself a not-I?" AA I,2 99; and ibid., 81. Or again: "How could the absolute come out of itself and oppose to itself a world?" AA I,3 78; and ibid., 164.

9. Schelling overcomes the philosophy of reflection, or separation, by returning to the original unity out of which the two terms have emerged as a result of reflection:

"[A]fter we had separated object and representation through freedom, we wanted to unite them again through freedom, we wanted to know that, and why, there is *originally no* separation between them" (AA I,5 73; and Friedrich Wilhelm Joseph Schelling, *Ideas for a Philosophy of Nature*, trans. E. Harris and P. Heath [Cambridge: Cambridge UP, 1988] 13). One of the consequences of the philosophy of reflection is the institution of the Kantian *Ding-an-sich*: "It makes the separation between human beings and the world *permanent*, because it treats the latter as thing-in-itself, which neither intuition nor imagination, neither understanding nor reason can reach" (AA I,5 71–72; and Schelling, *Ideas for a Philosophy of Nature*, 11).

10. AA I,3 103; and Schelling, *The Unconditional in Human Knowledge*, 167.

11. Cf. Hegel's arguments against mere "propositions" as first principles of philosophy in his *Differenz des Fichte'schen und Schelling'schen Systems der Philosophie (1801)* (Leipzig: Philipp Reclam Verlag, 1981) 37–42, and *The Difference between Fichte's and Schelling's System of Philosophy*, trans. H. S. Harris and Walter Cerf (Albany: State U of New York P, 1977) 103–09.

12. AA I,3 193; and Schelling, *The Unconditional in Human Knowledge*, 128. The SW edition has "The first *Postulat* of philosophy" where the AA reads "The first *Resultat* of philosophy." The latter presents the definitive first edition text, but I have tried to preserve the parallel somewhat. The idea of freedom is the "postulate" that demands action as a "result," that which issues from the idea.

13. AA I,3 192; and ibid., 127.

14. AA I,3 193; and ibid., 128.

15. AA I,5 74–75; and Schelling, *Ideas for a Philosophy of Nature*, 14. Or as Ralph Waldo Emerson asks, "Who and what is this criticism that pries into the matter?" (see R. W. Emerson, *Ralph Waldo Emerson: Essays and Lectures*, ed. J. Porte [New York: Library of America, 1983] 953).

16. AA I,3 73; and Schelling, *The Unconditional in Human Knowledge*, 171.

17. AA I,2 166; and ibid., 122.

18. AA I,2 166–67; and ibid., 122–23.

19. AA I,2 169; and ibid., 123.

20. On this phraseology as typical of "deductions" in Kant, see Dieter Henrich, "Kant's Notion of a Deduction and the Methodological Background of the First *Critique*," *Kant's Transcendental Deductions*, ed. E. Foerster (Stanford: Stanford UP, 1989) 44.

21. Despite its seeming clarity, the notion of "development," whether used in transcendental or natural philosophy, is profoundly ambiguous. It is wise to assume that Schelling was aware of this, and sometimes purposely uses the word in contexts where it may be interpreted to mean both a historical or an empirical series of events and a logical or constructive series of categories. Used in the former sense, however, its occurrence is quite rare, and Schelling, like Oken after him, understands "development" as a conceptual, rather than a real empirical unfolding. See the section "Logogenesis, Construction, and Potency in the Philosophy of Nature."

22. Friedrich Wilhelm Joseph Schelling, *Sämmtliche Werke*, ed. K. F. A. Schelling, 14 vols. (Stuttgart: J. G. Cotta'scher Verlag, 1856ff.), I,3 320n.; above p. 195. (Hereafter this edition will be cited as "SW," with volume and page number, for example, SW III 273. Because only the first seven volumes of the critical edition have been published, the SW text will be cited when the work has not yet appeared in the critical edition.)

23. Friedrich Wilhelm Joseph Schelling, *System des transzendentalen Idealismus* (Hamburg: Felix Meiner Verlag, 1992) 45; and Friedrich Wilhelm Joseph von Schelling, *System of Transcendental Idealism (1800)*, trans. Peter Heath (Charlottesville: UP of Virginia, 1978) 32. (Translation modified, bracketed portions added by translator.)

24. AA I,5 93; and Schelling, *Ideas for a Philosophy of Nature*, 30 (translation modified). The terms *natural history*, *genesis*, and *becoming*, like *development* above, always imply conceptual, rather than empirical, evolution. Again, see the section on "Logogenesis."

25. AA I,5 96; and ibid., 32 (translation modified).

26. AA I,5 100; and ibid., 36 (translation modified).

27. AA I,5 106; and ibid., 41 (translation modified). The theory of compulsion or of a "feeling of constraint" that seems both to imply and also to substitute for a relation of logical necessity is never adequately developed by Schelling, and it is partially by virtue of this ambiguity that he is often able to speak of theoretical necessity and practical constraint, and hence theoretical and practical philosophy, as if they were one. The "necessity" in the judgment of purposiveness will be further explored in the section "Transcendental Deductions and the Idea of Nature."

28. AA I,6 69.

29. AA I,7 310; and see above p. 79. Cf. "There has long been a theory that the magnetic, electric, chemical, and, finally, even the organic phenomena, are interwoven into one great interdependent whole. This must be established" (SW III 319; and see above p. 227).

30. AA I,6 192–93.

31. "We can say—at least in a certain sense—that if the universal activity of Nature has the same conditions as the organic, sensibility does not belong exclusively to organic nature, but is a property of the whole of Nature, and that the sensibility of plants and animals is only a modification of the universal sensibility of Nature" AA I,7 183n iii; and see above p. 117).

32. Henrich, "Kant's Notion of a Deduction and the Methodological Background of the First *Critique*," 35. Most of the literature on "deduction" in Kant focuses attention on the deduction of the categories of the understanding, but below I refer to the dependence of Schelling's method on the Kantian deduction of the ideas in the first *Critique* (see, for example, Immanuel Kant, *Kritik der reinen Vernunft*, ed. W. Weischedel, vol. 3–4, *Werkausgabe* [Frankfurt am Main: Suhrkamp Taschenbuch Wissenschaft, 1974] B697–98).

33. Kant, *Kritik der reinen Vernunft*, B117. As an example of an "empirical deduction" Kant cites Locke's empirical psychology (B119).

34. "It is thus conceivable that speculative physics—the soul of real experiment—has, in all time, been the mother of all great discoveries in nature" (SW III 280; and see above p. 199).

35. SW III 279; and see above p. 199. On the relationship between experiment, experience, and speculation in Schelling's philosophy of nature, see Hans Poser, "Spekulative Physik und Erfahrung. Zum Verhältnis von Experiment und Theorie in Schellings Naturphilosophie," *Schelling: Seine Bedeutung für eine Philosophie der Natur und der Geschichte*, ed. L. Hasler (Stuttgart/Bad Canstatt: Fromann-Holzboog, 1981) 129–38.

36. SW III 276; and see above p. 197. This experimental prophecy is meant to confirm that, in Fichte's words, "we do not learn these laws of nature by observation, but rather that the laws provide the basis for all observation" (see J. G. Fichte, "Concerning the Concept of the Wissenschaftslehre," *Fichte: Early Philosophical Writings*, ed. Daniel Breazeale [Ithaca: Cornell UP, 1988] 121n).

37. SW III 276. One should note Kant's implicit endorsement here: "He who would know the world must first manufacture it" (see Immanuel Kant, *Opus Postumum*, trans. Eckhart Förster and Michael Rosen, ed. Paul Guyer and Allan W. Wood, *The Cambridge Edition of the Works of Immanuel Kant* [Cambridge: Cambridge UP, 1993] 240).

38. Cf. AA I,6 91.

39. SW III 277; and see above p. 197.

40. AA I,5 96; and Schelling, *Ideas for a Philosophy of Nature*, 32.

41. There is, admittedly, a logical problem with this argument. If we accept the premise of the reciprocal presupposition of ideal and real, and if human beings are compelled to think or must by necessity think a certain concept when they encounter organisms, this does not of itself account for the specificity of the concept thought (e.g., "purposiveness"), which may be relative to the thinker or culture. Schelling believes that he has shown that teleology is the only possible alternative to mechanism, once mechanism has been proven to be inadequate, and so imagines an exclusive disjunction where to our eyes other (nonmechanistic) concepts may be possible and more plausible in the explanation of living beings.

42. SW III 278; and see above p. 198.

43. SW III 279; and see above pp. 198–199.

44. Hermann Krings, "Die Konstruktion in der Philosophie. Ein Beitrag zu Schellings Logik der Natur," *Aspekte der Kultursoziologie*, ed. J. Stagl (Berlin: D. Reimer, 1982) 350:

> Construction in philosophy is to be interpreted as a logo-genesis (of matter, body, organism, for instance). To conceive nature as constructing activity does not explain nature, rather it is a logic of nature, i.e., a doctrine of those absolute rules according to which a nature, and not only a nature, but also consciousness and self-consciousness, art and history can be thought, but not explained, as a whole constructing itself in ever higher potencies.

See also Hermann Krings, "Natur als Subjekt: Ein Grundzug der Spekulativen Physik Schellings," *Natur und Subjektivität: Zur Auseinandersetzung mit der Naturphilosophie des jungen Schelling*, ed. R. Heckmann, Hermann Krings, and R. W. Meyer (Stuttgart: Fromann-Holzboog, 1983) 111–28.

45. SW IV 4.

46. Very different from the Hegelian dialectic; readers should refer to Schelling's critique of Hegelian method in his *On the History of Modern Philosophy*, 142–43.

47. SW IV 25.

48. AA I,7 356; and see above p. 181.

49. AA I,7 83n; and see above p. 18.

50. Krings outlines the terminology Schelling uses in such constructions in successively more concrete stages: productivity, production, force, product, matter, body, object. On this series of categories, see his "Natur als Subjekt," 118–23.

51. AA I,7 112; and see above p. 49.

52. AA I,7 105n, 106n; and see above p. 40, 41.

53. "Organic nature maintains the whole wealth and variety of its products only by continually changing the relation of those three functions.—In like manner inorganic Nature brings forth the whole wealth of its products only by changing the relation of those three functions of matter to infinity; for magnetism, electricity, and chemical process are the functions of matter generally, and on that ground alone are they categories for the construction of all matter" (see SW III 322n; and above p. 229).

54. Schelling, *Zur Geschichte der neueren Philosophie*, 125; and Schelling, *On the History of Modern Philosophy*, 119.

55. Immanuel Kant, *Prolegomena to any Future Metaphysics*, trans. James W. Ellington (Indianapolis: Hackett, 1977) 98n.

Foreword

1. On George-Louis Lesage, and on mechanistic versus dynamic philosophy, see AA I,5 196–207; and Friedrich Wilhelm Joseph Schelling, *Ideas for a Philosophy of Nature*, trans. E. Harris and P. Heath (Cambridge: Cambridge UP, 1988) 161–69.

Outline of the Whole

1. Schelling uses the latinate term *Aktion*. In an attempt to mirror the foreignness of the term in German, in most cases I have chosen to translate "actant" rather than "actor" or "action" in English, since the latter is too broad here (although it is used more frequently later in a broader sense) and the former too full of intentionality. The "dynamic atom" that Schelling designates by *Aktion* is best understood as an individual "actant," a "natural monad" or "simple productivity."

2. He uses the terms *indecomponible* and *componible*, borrowing directly from the French of Lesage or Pierre Prévost, so I have retained the parallel in the translation.

3. Schelling will use the terms *organisch*, *anorganisch*, and *unorganisch*, translated here as organic, anorganic, and inorganic. Anorganic is normally directly opposed to organic, and refers to the resources in the external world that nourish or inhibit the organic activity of a particular being. Inorganic seems to include this, as well as a reference to "universal" nature, or the largest scale of nonorganic nature, but Schelling is not always clear about this.

4. Intussusception is a piece of medical jargon that means the telescoping of one portion of the intestine into another. Generally meaning "infolding," Schelling also uses the word *involution* as a synonym.

5. *Diremption* (from Latin *dirimo*, to separate, break off, interrupt) throughout translates *Entzweiung* ("bifurcation" or "becoming two"), an important term that designates the ontological dualization or duplicity already inherent in any unity (including the Absolute), the ultimate cause of which remains undetermined in this text.

First Division

1. Very often Schelling uses the word *Prinzip* (principle) as the equivalent of *Ursache* (cause) or *Grund* (ground, reason). Thus it sometimes appears without a definite or indefinite article, as on pp. 19–20.

2. "*Ich tilge sie, und du liegst ganz vor mir.*" From Albrecht von Haller's "Incomplete Poem on Eternity" (1762). See AA I,7 364–65.

3. The full quotation from Lucretius runs: "Something must stand immovable, it must, / Lest all things be reduced to absolute nothing" (see R. Humphries, trans., *Lucretius: The Way Things Are* [Bloomington: Indiana UP, 1969] 42; and Cyril Bailey, ed., *De rerum natura* [Oxford: Oxford UP, 1947], Libr. I, V. 790f).

4. There is no number (5) in the text.

5. The German sentence seems to be grammatically incorrect here. It reads "Die Ausbreitung der Flügel . . . geschieht vermittelst einer schnellen und kräftigen Entwicklung des Gefäßsystems im Centrum, durch ein Zuströmen der Flüssigkeit von innen—nicht etwa durch ein bloßes Auseinanderbreiten des übereinander geschlagenen Schmetterlings, oder durch den Druck der von außen eindringenden Luft" (AA I,7 286). I have compensated in the translation—Trans.

6. "Bestimmung des Begrifs einer Menschenrace" (1785), in Immanuel Kant, *Kants gesammelte Schriften*, 22 vols. (Berlin: Preussische Akademie der Wissenschaften, 1900–42) 8:106 (hereafter AkA).

7. "Bestimmung des Begrifs einer Menschenrace," in Kant AkA VIII 89–106; "Über den Gebrauch teleologischer Prinzipien in der Philosophie" (1788), also in AkA VIII.

8. Most likely, Schaftesbury is referred to on the basis of Kant's own citation in "Über den Gebrauch teleologischer Prinzipien in der Philosophie" (AkA VIII 166).

9. Schelling refers to the text "Exercitationes de generatione animalium" (1651).

10. Schelling presents a viewpoint that he believes would be held by a chemical materialist, e.g., Reil (as just mentioned), but includes some aspects of his own position.

11. Although the quotation marks are discontinued at this point, this staged speech extends to page 60 where the quote is closed.

12. Schelling here speaks in the voice of Brandis (see appendix).

13. Schelling refers most likely to Haller's "De partibus" (1753), translated from the Latin in *Anfangsgründe der Phisiologie des menschlichen Körpers*, 8 vols. (Berlin: Christian Friedrich Boss, 1759–1776).

14. The second parenthesis is missing in the original.

Second Division

1. Schelling goes on to express almost word for word the corpuscular position of Pierre Prévost in his "Magnetische Kräfte" (1794).

2. The present idea is expressed in Franklin's "Letter to Abbé Soulaire," *European Magazine and London Review* 24 (1793): 84–86.

3. The following interpretation is likely derived from Kant's "Universal Natural History" (1755), in AkA I.

4. AA I,5 185; and *Ideas for a Philosophy of Nature*, 144.

5. AkA I 275.

6. AA I,5 176f.; and *Ideas for a Philosophy of Nature*, 132f.

7. pp. 28–31.

8. The specific experiment referred to here is recounted in AA I,5 152 (*Ideas*, 102) and AA I,6 137f., where a white ribbon becomes positively electrified when rubbed with a black, which in its turn becomes negatively electrified.

Third Division

1. Schelling is summarizing his own text here, see pp. 78–104.

2. No closed parenthesis is provided in the AA here, but it seems necessary to supply it.

3. See AA I,6 214 and the note to that passage.

4. See Johann Wolfgang von Goethe and Georg Christoph Tobler, *Goethe's Botany: The Metamorphosis of Plants (1790) and Tobler's Ode to Nature (1782)*, trans. Agnes Arber (Waltham, MA: Chronica Botanica Co., 1946).

5. See AA I,6 200.

6. Schelling is likely quoting from memory. The most similar passage reads "Bodies driven by a compelling force move slowly; but those which move of their own accord possess alertness." Lucius Annaeus Seneca, *Moral Epistles*, trans. Richard M.

Gummere, *The Loeb Classical Library* (Cambridge: Harvard UP, 1917–25, 3 vols.) Volume III, Letter CXXI, p. 401.

7. "Some say that unto bees a share is given / Of the Divine Intelligence." P. Vergilius Maro, *Georgicon*, ed. and trans. J. B. Greenough (*The Perseus Project*, http://www.perseus.tufts.edu, June 2003), Book IV, Vv. 220–21.

8. A second occurrence of "higher" in this sentence has been replaced with "lower" by the translator.

9. The citation could not be found.

10. Schelling's attention to the problems of medical science in this text earned him great respect among many contemporary physicians, many of whom refer to Schelling on the dedication page of their works. See AA I,7 57–59. For Schelling's influence on medical science, see Nelly Tsouyopoulos, "Schellings Krankheitsbegriff und die Begriffsbildung der Modernen Medizin," *Natur und Subjektivität*, ed. R. Heckmann, H. Krings, and R. W. Meyer (Stuttgart: Fromann-Holzboog, 1985), 265–90.

ENGLISH-GERMAN GLOSSARY

Actant, action *Aktion*

Activity *Thätigkeit*

Affectability *Afficirbarkeit*

Anorganic *anorganisch, anorgisch*

Antithesis *Gegensatz*

Appearance *Erscheinung*

Becoming *Werden*

Cohesion *Cohäsion*

Combustion *Verbrennung*

Communication *Mittheilung*

Composable *componible*

Condition *Bedingung*

Configuration *Gestaltung*

Decomposable *decomponible*

Deoxidize *desoxydiren*

Development *Entwicklung*

Diremption *Entzweiung*

Disease *Krankheit*

Division *Vertheilung*

Duplicity *Duplicität*

Excitability *Erregbarkeit*

Excitation *Erregung*

English–German Glossary

Factor *Faktor*

Force of production *Produktionskraft*

Formative drive *Bildungstrieb*

Graduated/graded series (of stages) *Stufenfolge*

Gravitation, universal *Schwere, allgemeine*

Heat-matter *Wärmestoff*

Heterogeneity *Heterogeneität*

Incomposable *incomponible*

Indecomposable *indecomponible*

Indifference *Indifferenz*

Inhibited *Gehemmt*

Inorganic *unorganisch*

Intensity *Intensität*

Interpenetration *Ineinander*

Intussusception *Intussusception*

Irritability *Irritabilität*

Juxtaposition *Außereinander*

Limitation *Einschränkung*

Magnetism *Magnetismus*

Matter *Materie*

Occupation of space *Raumerfüllung*

Opposition *Gegensatz*

Organic *organisch*

Organism *Organismus*

Organism, organization *Organisation*

Oxygen *Sauerstoff*

Phenomenon *Erscheinung*

Point of inhibition *Hemmungspunkt*

Positively/negatively charged *Positiv-/negativ-elektrisch*

Potency *Potenz*

Power *Potenz*

Prehension, prehend *eingreifen in*

Product *Produkt*

Productivity *Produktivität*

Proportion *Verhältnis*

Quality *Qualität*

Ratio *Verhältnis*

Receptivity *Receptivität*

Reciprocal determination *Wechselbestimmung*

Relationship *Verhältnis*

Reproductive force *Reproduktionskraft, Zeugungskraft*

Retarded *Gehemmt*

Sensibility *Sensibilität*

Shape *Gestalt*

Sphere of affinity *Affinitätssphäre*

State of indifference *Indifferenzzustand*

Stimulant, stimulus *Reiz*

Susceptibility *Reizbarkeit*

Technical drive *Kunsttrieb*

Tendency *Tendenz*

Triplicity *Triplicität*

Unconditioned, the *Unbedingte, das*

Vital force *Lebenskraft*

GERMAN-ENGLISH GLOSSARY

Afficirbarkeit affectability

Affinitätssphäre sphere of affinity

Aktion actant, action

Anorganisch, anorgisch anorganic

Außereinander juxtaposition

Bedingung condition

Bildungstrieb formative drive

Cohäsion cohesion

Componible composable

Decomponible decomposable

Desoxydiren deoxidize

Duplicität duplicity

Eingreifen in prehension, prehend

Einschränkung limitation

Entwicklung development

Entzweiung diremption

Erregbarkeit excitability

Erregung excitation

Erscheinung appearance, phenomenon

Faktor factor

Gegensatz opposition, antithesis

Gehemmt inhibited, retarded

Gestalt shape

Gestaltung configuration

Hemmungspunkt point of inhibition

Heterogeneität heterogeneity

Incomponible incomposable

Indecomponible indecomposable

Indifferenz indifference

Indifferenzzustand state of indifference

Ineinander interpenetration

Intensität intensity

Intussusception intussusception

Irritabilität irritability

Krankheit disease

Kunsttrieb technical drive

Lebenskraft vital force

Magnetismus magnetism

Materie matter

Mittheilung communication

Organisch organic

Organisation organism, organization

Organismus organism

Positiv-/negativ-elektrisch positively/negatively charged

Potenz potency, power

Produkt product

Produktionskraft force of production

Produktivität productivity

Qualität quality

Raumerfüllung occupation/filling up of space

Receptivität receptivity

Reiz stimulant, stimulus

Reizbarkeit susceptibility

Reproduktionskraft reproductive force

Sauerstoff oxygen

Schwere, allgemeine gravitation, universal

Sensibilität sensibility

Stufenfolge graduated/graded series (of stages)

Tendenz tendency

Thätigkeit activity

Triplicität triplicity

Unbedingte, das Unconditioned, the

Unorganisch inorganic

Verbrennung combustion

Verhältnis relationship, proportion, ratio

Vertheilung division

Wärmestoff heat-matter

Wechselbestimmung reciprocal determination

Werden becoming

Zeugungskraft reproductive force

PAGE CONCORDANCE

AA	SW	This ed.	AA	SW	This ed.
65	3	3	100	41–42	33–34
67	5	5	101	42–43	34–35
68	5–6	5–6	102	43–44	35–36
69	6–7	6–7	103	44–45	36–37
70	7	7–8	104	45–48	37–39
71	8	8	105	49–50	39–40
72	8–9	8–9	106	50–53	40–42
73	9	9–10	107	53–54	42–43
74	9–10	10–11	108	54–56	43–44
77	11–12	13	109	56–57	44–45
78	12–13	13–14	110	57–59	45–46
79	13–14	14–15	111	59–61	46–47
80	14–15	15–16	112	61–64	48–50
81	16–17	16–17	113	64–65	50
82	17–18	17–18	114	65–67	51–52
83	18–20	18–19	115	67–68	52
84	20–21	19–20	116	68–69	52–53
85	21–22	20	117	69–70	53–54
86	22–24	20–22	118	70–71	54–55
87	24–27	22–24	119	71–72	55–56
88	27–28	24	120	73–74	56–57
89	28–29	24–25	121	74–75	57–58
90	29–30	25–26	122	75–76	58
91	30–32	26–28	123	76–77	58–59
92	32–33	28	124	77–78	59–60
93	33–35	28–29	125	78–79	60
94	35–36	29–30	126	79–81	60–62
95	36–37	30–31	127	82–83	62–63
96	38–39	31–32	128	83–85	63–64
98	39–40	32–33	129	85–86	64–65
99	40–41	33	130	86–87	65–66

131	87–89	66–67	172	144–46	106–07
132	89–92	67–69	173	146–47	107–08
133	92–93	69–70	174	147–49	108–09
134	93–94	70–71	176	149–50	109–10
135	94–95	71–72	177	150–52	110
136	95–96	72–73	178	152–53	110–12
138	96–98	73–74	179	153–54	112–13
139	98–99	74	180	154–56	113–14
140	99–100	74–75	181	156–57	114
141	100–01	75–76	182	157–59	114–16
142	101–02	76–77	183	159–62	116–18
143	102–04	77–78	184	162–63	118–19
144	104–05	78	185	164–65	119–20
145	105–07	79–80	186	165–67	120–21
146	107–09	80–81	187	167–68	121–22
147	109–10	82	188	168–69	122–23
148	110–12	82–83	189	169–70	123–24
149	112–14	83–85	190	170–72	124–25
150	114–15	85–86			
151	115–16	86	191	172–73	125
			192	173–74	125–26
152	117–19	86–88	193	174–76	126–28
153	119–20	88–89	194	176–78	128–29
154	120–21	89	195	178–79	129–30
155	121–22	89–90	196	179–80	130–31
156	123–24	91–92	197	180–81	131
157	124–26	92–93	198	181–82	131–32
158	126–28	93–94	199	182–83	132–33
159	128–29	94–95	200	183–84	133
160	129–30	95–96			
161	130–32	96–97	201	184–85	133–34
162	132–33	97–98	202	185–86	134–35
163	133–34	98	203	186–88	135–36
164	135–36	98–99	204	188–89	136–37
165	136–37	99–100	205	189–90	137
166	137–38	100–01	206	190–91	137–38
167	138–39	101–02	207	191–92	138–39
168	140–41	102–03	208	192–93	139–40
169	141–42	103–04	209	193–94	140
170	142–43	104–05	210	195–96	141
171	143–44	105–06	211	196–97	142

212	197–98	142–43	242	236–37	169–70
213	198–200	143–44	243	237–38	170–71
214	200–01	144–45	244	238–39	171
215	201–02	145–46	245	239–40	170–72
216	202–04	146–47	246	240–41	172–73
217	204–05	147–48	247	241–42	173
218	205–06	148–49	248	242	173–74
219	206–07	149–50	249	243	174–75
220	207–08	150	250	243–44	175
221	208–09	150–51	251	244–45	175–76
222	209–10	151–52	252	245–46	176
223	210–12	152–53	253	246–47	176–77
224	212–14	153–54	254	247–48	177–78
225	214–15	154–55	255	248–49	178
226	215–16	155–56	256	249–50	178–79
227	216–17	156–57	257	250–52	179–81
228	218–19	157	258	252–54	181
229	219	157–58	259	254–55	182
230	220	158–59	260	255–56	182–83
231	220–22	159–60	261	256–57	183–84
232	222–23	160–61	262	257–58	184–85
233	223–25	161–62	263	258–59	185
234	225–27	162–63	264	259–60	185–86
235	227–28	163–64	265	260–61	186–87
236	228–30	164–66	266	261–63	187–88
237	230–31	166	267	263–64	188–89
238	232	166–67	268	264–65	189–90
239	233–34	167–68	269	265–66	190
240	234–35	168–69	270	266–67	190–91
241	235–36	169	271	267–68	191–92

INDEX

Absolute, The, xv, 41n.*
Actant, 5–6, 19–36, 43, 49, 51, 77, 175, 208, 214–15
   simple, 5, 77, 175, 208
Action, xvi, 7–8, 15, 21–25, 27, 31, 35, 42, 48n.†, 51–52, 54, 57, 59–60, 64, 74–75, 81, 88, 91–92, 94–96, 98–100, 103–4, 106, 108–9, 112, 118, 120, 122, 129, 132–33, 140, 151–53, 157–58, 181, 183–86, 203, 226, 230n.||
   chemical, 8, 96, 98, 99n.*, 100, 104, 106, 109
   of gravity, 7–8, 74, 94, 96, 100, 104, 106, 158, 203, 226
   of light, 74, 99nn.*, †, 100, 226
Activity, xviii–xix, xxvi, xxviii, xxxii, 5, 7–8, 14–17, 19–20, 24, 32–35, 39–42, 47, 48n.†, 51n.*, 52–57, 59–60, 62–70, 77, 94n.†, 105, 107, 108–14, 115n.*, 117–24, 126–27, 136–37, 140, 142–44, 151–52, 156–58, 160, 161n.†, 163, 166–67, 169–70, 179n.*, 180, 184, 193, 195, 197, 202–3, 205, 209, 216, 219, 220, 231n.*
   infinite, 33–34, 42, 205
   organic, 8, 57, 60, 63–65, 66n.†, 67n.#, 105, 107, 112–14, 115nn.*, ‡, 116n.‡, 118, 120n.†, 137, 140, 144, 158, 160, 166, 167, 170, 179n.*
   real, 5, 193
   sphere of, 16, 54
Anatomy, xxx, 50, 52n.§, 53
Animal, 8, 36, 39, 40n.*, 43n.*, 45, 47, 56, 58–60, 64, 69, 96, 110, 113, 117n.*, 119n.†, 121, 125, 127n.*, 130–38, 140, 146–47, 149, 154–56, 167, 182–83, 194–95, 213, 231n.*

animal kingdom, 39, 45, 137, 146, 194
Appearance, 9–10, 22n., 25, 32–33, 35, 51, 61n., 74, 87n.*, 100n.†, 103, 108, 115–16, 121n.§, 123, 134–35, 141–42, 143nn.*, †, ||, 144n.†, 149, 152, 153n.*, 157–61, 163, 167, 170, 182–83, 185
Atomism (-ist), xxix, 5–6, 20–22, 24n.*, 26, 27n.†, 173, 195, 203, 208, 211, 213
Attraction, force of, 27n.†, 83, 87n.*, 90, 190. *See also* Gravity
Attractive force, xxix, 17n.‡, 21n.†, 22n.*, 26, 57, 64, 75, 77–78, 85n.*, 90n., 190, 200, 219n.‡. *See also* Gravity

Baader, F. X., 174n., 190n., 191n., 223n.‡
Bacon, Francis, 199n., 227n.†
Becoming, xix, xxix, 5, 15, 28, 33–34, 91, 96n.§, 120, 150, 173, 201, 203–04, 206, 214, 224n.#, 225
Being, xix, xxix, xxxi, 13–14, 19, 107, 160, 187, 202–3, 207
Blumenbach, J. F., 47n.‡, 141n.*, 147
Body, xxix, xxx, 2, 45, 55–56, 58–60, 63–64, 66, 73–74, 79n., 83, 89, 91, 95n.†, 97n., 98–104, 106, 122, 126, 128n.#, 137, 151–54, 156n., 171, 173–74, 176–78, 180–83, 185–86, 189n.†, 190–91, 201, 207n., 208–10, 222–23, 224n.†, 226n.‡, 227–28
   animal, 56, 58, 64, 128n.#, 154n.‡, 183
   chemical, 182–83
   electrical, 156n., 181–82
   organic, 55n., 60, 63, 152
Brandis, Joachim, 125n.‡

261

Index

Brown, John, 48n.†, 66n.#, 106n.*, 111–12, 113n.*, 127n.#, 161, 162n.‡, 164n.||, 166–67, 169

Cause, xiv, xix, xxi, xxv–xxvi, xxxi, 5, 7–10, 16, 30, 48n.†, 51–52, 56, 61, 62n.†, 67, 69, 72, 73n, 74, 79–81, 82n.†, 83–87, 89, 90n., 91–92, 94, 98, 100, 103, 108–18, 122, 126, 128, 130–31, 135–36, 138–40, 145–46, 149–53, 156–58, 161–63, 169, 171–74, 176–77, 179–86, 195, 197, 202–03, 207–8, 211, 215–17, 222, 227, 230n.||, 232
   of duplicity, 113
   of excitability, 8, 108–10, 112, 156, 161–63, 230n.||
   final, xxv–xxvi, 74, 81, 85n.*, 111n.†, 113n.*, 150, 157–58, 171, 177, 180, 183, 195, 197
   of gravity, 73n., 74, 79, 82n.†, 83–85, 94, 103
   of life, xxxi, 110, 112, 122, 146
   of magnetism, 9, 83, 180n.*
Chemistry, xiii, 22, 33, 58, 60, 101n.§, 129, 175, 177
Cohesion, 6, 26, 72n.‡, 176, 211
Coleridge, S. T., xii
Combustion, xxi, 8, 31, 59, 66, 95–96, 97n., 99, 101–4, 107n.||, 151, 177–78, 226, 227n.†
Communication, 10, 185–86
Compulsion, xxv, 6–7, 28, 33–34, 40n.*, 41n.*, 133, 231n.*
Configuration, 6, 34–35
Construction, xxii, xxiv, xxvi–xxviii, xxxii–xxxiii, 5, 10, 13, 17, 19, 22n.*, 23–24, 26, 41n.*, 45, 50, 54n.‡, 60n., 64, 71n.†, 75–77, 78n.*, 79n., 84, 87n.*, 93n.‡, 94n.*, 103n.§, 111–12, 113n.*, 115n.*, 117n.*, 120n.*, 140n.§, 143n.*, 160–61, 164n.||, 167–69, 179n.*, 186, 189–90, 195–98, 206, 209–10, 213, 216–18, 223, 224nn.*, #, 226, 228–29, 230n.#, 231
   of matter, xxvii–xxviii, 5, 10, 22n.*, 23–24, 76, 87n.*, 103n.§, 226n.‡

Contagion, 56n.†, 82, 128, 139
Contraction, 51n.†, 86, 87n.*, 91–93, 117n.*, 120n.§, 121–23, 125, 128n.*, 142n.*, 143, 145, 152–54, 158, 173, 182–83, 185, 213, 215, 218, 228–29
   expansion and, 51n.†, 87n.*, 91, 121, 123, 125, 173, 183, 185, 215, 218, 228–29

Darwin, Erasmus, 121
Deduction, xiv, xv, xviii, xxi–xxviii, xxxii, 7–8, 11, 61n., 77n.†, 106, 113, 119n.†, 123, 141n.*, 198–99, 217, 219n.*, 230n.||
   transcendental, xiv, xv, xxii–xxiii, xxvi, 199
Determination, reciprocal, 8–9, 54n.‡, 62, 64–65, 67n.#, 142, 147, 216
Determinism, xx
Development, stages of, 6, 28, 35–37, 39–48, 49n.*, 53, 138, 214–15, 217
Difference, sexual, 36–37, 39n.*, 41n.*, 42, 43n.‡, 48n.†, 169, 231n.*
   between animal and plant, 59–60, 156
Diremption, 11, 35, 39, 42n.*, 79n., 87n.*, 117n.†, 179, 180n.*, 185, 187, 205, 212, 214–15, 217
Disease, 10, 56n.†, 61n.*, 118, 128, 144, 148n.†, 153n.†, 159–62, 164n.||, 165n.||, 168–71, 213n.
Duplicity, xxvi, 8, 11, 48n.†, 87n.*, 89, 103, 106n.*,†, 107, 108n.†, 109–10, 111n.||, 112n.‡, 113–19, 121–23, 130, 132, 134n., 139, 140n.‡, 152, 157–58, 160, 172, 174, 179, 180n.*, 181, 185, 186nn.*, †, 187, 197, 202–5, 216, 222, 224, 226, 230
   organic, 113–14, 115n.‡, 116, 122–23, 158, 179n.*
   universal, 89, 197

Electricity, xxi, xxv, xxvii–xxviii, xxx–xxxi, 9, 30, 100–3, 106, 113n.*, 121–22, 124n.*, 149, 151–53, 156–57, 161n.||, 163, 173, 177, 181–83, 186, 197, 210, 212, 219n.†, 223, 225–28, 231

Empiricism, 6n., 21n.†, 22, 24, 25n.†, 61n., 66n.#, 163n.||, 200–3
Epigenesis, 37n., 48
Eschenmayer, Karl August, 22n.*, 164n.||
Euler, Leonhard, 96n.§
Evolution, xxvii, xxxii, 6, 11, 16, 18, 35n.*, 37n., 47n.‡, 48n.*, 77, 88, 149, 174–75, 185, 187–89, 190–91, 203–4, 206–9, 211, 213
Excitability, 7–8, 48n.†, 63nn.§, ||, 64, 66n.#, 67n.§, 105–13, 125n.*, 127–28, 149, 153n.†, 156, 160–69, 171
Expansion, 17, 51n.†, 66n.*, 72, 75, 87n.*, 92–93, 173, 183, 188. *See also* Contraction
Experiment, xix, xxiv, xxvi, 30, 83, 98, 119, 124, 129n., 163, 171, 197, 199, 201, 212n., 227n.†
Exteriority, 7, 71, 79n., 82, 112

Fichte, J. G., xiii–xvi, xviii, xxiv, xxvi
Figure, 6, 22, 26–27, 32, 59, 71, 76, 134, 210
Fluid, 6, 27–32, 41n.||, 58, 63, 71, 101, 102n.‡, 118, 121–22, 128n.§, 139, 144, 154, 183, 191, 194, 200, 213
    absolute, 28, 30–31, 58, 71
Fontana, Felice, 152n.‡
Formation, 6–7, 28, 31, 33, 35–49, 55, 57, 64, 76, 84n.§, 85–86, 87n.*, 88–92, 98, 103, 106, 113, 122n.*, 124, 125n.*, 130, 147, 150, 174, 176, 183, 191, 194, 208, 213–15, 219n.‡, 221, 229–30
    organic, 46–47, 48n.†, 49n.§, 86, 150, 194
    process of, 28, 31, 35, 37, 86
Formative drive, 6, 9, 36–48, 111n.‡, 124, 125n.*, 128n.#, 131, 138–40, 150–52, 170n., 172, 228, 231. *See also* Production, force of
Franklin, Benjamin, 84n.‡, 90n.
Freedom, xiv–xviii, xx, xxiv, xxxiii, 32–35, 48n.†, 85, 135, 196, 218
Function, xxx, 9, 16, 36, 50–53, 57, 59–60, 64, 67–69, 101–5, 113, 123, 126, 128n.#, 130, 135, 140n.§, 141–43, 146n.‡, 152, 156, 159–60, 170, 172, 176–78, 182–84, 213, 227n.†, 228n.†, 232
    organic, xxx, 9, 50–53, 113, 128n.#, 135, 142, 146n.‡, 170

Gall, F. J., 45n.#, 108n.†
Galvanism, 8, 30, 63n.||, 82, 88, 102n.†, ‡, 111n.*, 119–21, 124, 128, 129n., 151–52, 154n.‡, 155, 210, 227n.†, 231
Gehler, Johann, 182n.†
Girtanner, Christof, 156
Goethe, J. W. von, xiii, xxi, 30, 125
Gravitation, 7, 73–75, 77n.†, 78–80, 81n.*, 82–83, 84n.§, 87, 88n.||, 89, 92, 93n.‡, 94n.†, 106, 185, 203, 221n.§, 222, 224–25, 226n.‡
Gravity, xxvii–xxviii, 7–8, 11, 23–24, 32, 57, 61n, 72n.*, 73–80, 81nn.*–‡, 83, 84nn.*, §, 88n.†, 93n.‡, 94, 97, 100, 103, 106, 158, 174, 179n.†, 186, 189–91, 203, 222–26, 227n.†
    specific, 23–24, 32, 76, 223
Growth, 9, 37, 47, 129–30

Habermas, Jürgen, xii
Haller, Albrecht von, 62, 153n.*
Harvey, William, 47, 139n.§
Hegel, G. W. F., xii
Heidegger, Martin, xii
Henrich, Dieter, xxii
Herder, J. G., 141n.*
Herschel, F. W., 84, 93n.*
Heterogeneity, 9n., 10–11, 79n., 103, 119n.†, 125, 140, 148, 152, 154, 157–58, 172–74, 176, 178–80, 183–86
Hölderlin, Friedrich, xii
Homogeneity, 9n., 20, 25, 79n., 118, 123–24, 132, 157–58, 160, 178–79, 185–86, 223n.#
Humboldt, Alexander von, 119n.†, 129n.
Hunter, John, 98, 124

Idealism, 136
Ideas for a Philosophy of Nature, xiii, 88, 95, 100n.#, 227n.†
Identity, xxvi, 18n., 40n.*, 48n.†, 49n.§, 79n., 87n.*, 90n., 106nn.*–†, 116–17,

identity (*continued*), 127, 132, 152n.†, 157–58, 179, 180n.*, 181, 190n., 197, 202, 204–6, 216, 218–19, 224–26
Illness. *See* Disease
Indifference, state of, xxx–xxxi, 10, 118, 123, 185–86
    striving toward Indifference, 40n.*, 67n.#, 87n.*, 117nn.*–†, 189n.*, 219–21, 222n.†
Individuality, 6, 21n.*, 24–26, 28–29, 35, 39, 54, 173
Individualization, 39, 40n.*
Induction, xxi–xxii, 147, 166, 184
Ingenhousz, Jan, 155–56
Inhibition, 6, 16–17, 18n., 19–20, 34–35, 36n., 39, 43, 49nn.*, §, 53, 174–75, 206–7, 214, 216
    absolute, 6, 214
Instinct, 9, 132–33, 136
Intensity, xxxi–xxxii, 21n.†, 32, 34, 39, 51, 66–67, 72, 73n., 74, 97, 121n.†, 127, 147, 153n.†, 161, 164–65, 208–9
Intussusception, xxxi, 7, 10, 94, 121–23, 172–74, 186, 230n.||
Involution, 37n., 77, 112, 187–88, 191, 211
Irritability, xxx–xxxi, 8–9, 51n.†, 120n.§, 121–26, 127n.‡, 128, 131, 133, 137, 139, 141–49, 151–57, 166–72, 181–83, 228, 231

Juxtaposition, 79n., 157–58, 174, 179n.†, 187–88

Kant, Immanuel, xiii–xvii, xix, xxiii, xxvi–xxvii, xxxii, 17n.‡, 22n.*, 27n.†, 44, 46, 53, 56n.†, 75n., 76, 77n.†, 78, 85, 86n., 87n.*, 90–91, 97n., 115n.‡, 189, 191, 200, 213, 230n.#, 232
Kielmeyer, K. F., 141n.*
Krings, Hermann, xxvii, xxxii

Lamarck, Jean-Baptiste, xxx
Lambert, J. H., 98
Lesage, George-Louis, 3, 73n., 74, 195
Lichtenberg, G. C., 81n.†, 94, 184
Life, xx–xxii, xxviii, 32, 52, 57–58, 60–63, 65–67, 85, 96, 97n., 110–12, 113n.*, 114, 115n.‡, 117n.†, 123, 125–26, 129, 132, 137–38, 146, 154, 156, 158, 161, 164, 166, 168–72, 186, 216, 226, 229, 230n.§§, 231n.*
Light, xxi, xxvii, xxx, 8–9, 11, 13, 23, 30, 32, 58–60, 73–74, 81n.†, 92, 95n.†, 96–101, 103, 106, 107n.||, 113n.*, 136, 145, 149–52, 156, 157n.†, 163, 176–78, 186, 197, 212, 226–27
Logogenesis, xxvii, xxxii

Magnetism, xxi, xxv, xxvii–xxix, xxxi, 9–10, 74, 82–83, 84n.‡, 90n., 94, 117, 157, 180–85, 197, 224n.#, 226, 228, 231–32
Mechanism, xi, xv, xvii, xx–xxi, xxxii–xxxiii, 14, 22n., 89n.*, 116, 131, 135, 137, 180, 189
Mendelssohn, Moses, 133
Metamorphosis, 9, 36n., 37, 41n.‡, 48n.*, 91–92, 93n.†, 125, 135, 147, 213–14
Metaphysical Foundations of Natural Science, 75n., 76, 189

Natura naturans, xix, xxvi, 202
Natura naturata, xxv, 202
Nature, anorganic, 7–9, 58, 59n.*, 60, 69–70, 71n.†, 92, 117n.*, 139, 172, 179n.*, 180n.†
    inorganic, xxi, xxxi, 104, 105n.†, 115n.*, 181n.*, 207n., 217, 229, 232
    organic, xxxi, 9–10, 35, 37n., 44n.†, 47n.‡, 48n.†, 49nn.*, §, 51, 57–60, 66, 67n.#, 69, 71, 79n., 87n.*, 105n.†, 106n.*, 113–14, 115nn.*, ‡, 116n.‡, 117, 121n.*, 124n.*, 125, 127n.*, 128, 131, 133–34, 138–39, 141, 143, 145–48, 150, 154–55, 157, 159, 166–67, 169, 171n., 172, 181–84, 204, 207n., 216, 218, 226n.‡, 228–29, 232n.†
    as object, xxv, 5, 14n.‡, 17, 32, 202–3, 205, 211, 232
    as subject, xxv, xxxii, 17, 202–3, 205, 211
    universal, 9, 53, 65, 117, 123, 139, 150, 152, 157–58, 171–72, 180–84
Newton, Isaac, xxvii, 23, 30, 75, 91
Nutrition, 9, 125–29, 147, 154, 156

Oken, Lorenz, xii
On the History of Modern Philosophy, xxxi
On the World-Soul, xiii, xxi–xxii, 64n.*, 83, 121n.‡, 129n., 149n., 184n.‡, 223n.§
Organism, xix, xxi, xxiii, xxv, xxxii, 6, 8–9, 18n., 28, 36, 39, 41n.‡, 42nn.*–†, 45–54, 56n.†, 57, 60, 62–64, 67n.*, 69, 86n., 105–9, 112–30, 133–35, 137–39, 141–50, 155–61, 162n.‡, 169–72, 180–82, 184, 195, 213
   absolute, 28, 49, 54
   universal, 6, 54, 67n.#, 86n., 117n.*, 138–39, 143, 150, 158, 184
Organization, xiv, xx–xxi, 3, 8, 11, 27, 31, 45, 52–53, 57–58, 86, 89, 93n.‡, 98, 106, 108, 112, 113n.*, 125, 131–32, 143–44, 146, 148n.†, 149, 157, 163, 169, 174, 184, 187, 191, 198, 199n., 201, 204, 209, 218, 222, 229
Ørsted, Christian, xii
Oxygen, 8, 31n., 69, 95–104, 120n.§, 150–51, 153–56, 177–78, 227n.†

Pallas, P. S., 36
Pfaff, C. H., 156n.
Philosophy of Nature, xi, xiii–xv, xvii–xxi, xxiii–xxvii, xxix, xxxii–xxxiii, 3, 5, 10, 13–15, 16–18, 21n.†, 22, 28, 53, 75n., 76–77, 117n.*, 158, 192–95, 199, 206, 209
Philosophy, practical, xiv–xvi, xx, xxvii
   transcendental, xiii–xiv, xviii–xix, xxii, xxvii, xxx, 13–15, 18, 192–94, 200
Physics, speculative, xxiii–xxiv, xxx, 3, 6n., 93, 195–96, 198–99, 201–2, 217, 232
Physiology, xiii, xxx, 7, 50, 52n.§, 53, 111, 141n.*, 153n.†
Plant, 36–37, 40n.*, 45, 47, 58–60, 96, 114n.‡, 117n.*, 121, 125, 130, 132, 146, 148–49, 153n.†, 155–56, 159, 166–67, 171, 182
Poison, 56, 128
Polarity, xxi, xxv, 30, 82, 83n.*, 124, 139, 157, 181nn.*,†, 184–86
Postulate (noun and verb), xiii–xvi, xviii, xx–xxi, xxiii, xxvi, xxxiii, 5, 7–8, 42n.*, 74, 120, 138, 140, 158, 185, 213, 216–17, 219n.*
Potency, xxvii, xxxi, 33, 48n.†, 60n., 67n.*, 120n.†, 126, 128n.#, 140n.§, 150, 152n.*, 162, 164n.||, 175, 210. *See also* Power, first, second, etc.
   stimulating, 126, 128, 162, 164n.||
Power, first, second, etc., xxxi–xxxii, 112n.‡, 139n.**, 152n.†, 153nn.*, †, 215–17, 225, 228–31
   productive, 76, 193, 206
Preformation, 37n., 47n.‡
Prévost, Pierre, 150n.
Problem, highest for philosophy of nature, 10, 15, 77, 117n.*, 158, 217
Process, chemical, xxviii, xxx–xxxi, 7–10, 30, 48n.†, 57, 60–61, 64, 94–101, 103n.*, 104, 106, 107n.||, 109–10, 111n.†, 126, 128, 140, 150, 153nn.*, †, 171–80, 182–83, 185–86, 212, 225n.#, 226–31
   of combustion. *See* Combustion
   electrical, xxxi, 9, 101, 103, 153–54, 171, 177–78, 180
   of formation. *See* Formation, process of
Product, apparent, 5, 16–19, 34n.
   Nature as, 197, 202, 211
   organic, 48n.†, 56n.†, 60, 67n.#, 107n.||, 117n.†, 131, 134n., 138, 140n.§, 141n.*, 170n., 179n.*, 187n.*, 217, 229, 231
Production, force of, 9, 44n.*, 124–25, 127n.*, 135, 137–41, 147, 149. *See also* Reproductive force
Productivity, xviii–xix, xxiii–xxvii, xxix, 16n.§, 18n., 21n.*, 22n.*, 27n.†, 29n.†, 34n., 43n.*, 48n.†, 49n.§, 50nn.*, ‡, 52nn.*, ||, 62n.§, 63n.‡, 64n.†, 69n.*, 77n.†, 87n.*, 96n.§, 118n.*, 122n.*, 140n.§, 141nn*–‡, 143n.†, 194, 197, 202–8, 211–19, 226, 228–31
   Nature as, 15n.*, 17n.§, 202

Quality, xxvii, 5, 10, 19–24, 28, 59, 96n.†, 101, 106, 126–28, 154–55, 174, 176–77, 185, 189, 208–10, 223n.#, 224n.*

Rationalists, xii
Receptivity, xxii, 7, 24, 34, 54–57, 60–62, 64–67, 112, 114, 115n.*, 127n.**, 136, 153n.†, 156, 160, 166–69, 216
Reciprocal determination. *See* Determination, reciprocal
Reil, J. C., 57n.§
Reimarus, H. S., 133
Reproductive force, xxx–xxxi, 9, 43–45, 125, 127n.*, 128, 131, 135, 139n.††, 141, 147–49, 155, 170, 172–73, 181–83, 231n.‡
Repulsion, 30, 75, 85, 87n.*, 94n.†, 121, 183, 212, 225
Repulsive force, xxix, 17n.‡, 21n.†, 22n.*, 57, 66, 75–76, 78, 85, 90n., 219n.‡, 227n.†
Ridley, Henry, 146n.‡
Ritter, J. W., xii, 61n., 102n.‡, 230
Romantics, xii
Röschlaub, Andreas, 67n.#, 153n.†, 156n.

Shaftesbury, Earl of (Anthony Ashley Cooper), 46n.
Schäffer, J. U. G., 171n.
Secretion, 9, 45, 127–28, 129n., 147, 170
Sensibility, xxii, xxvii, xxx–xxxi, 9, 113–17, 122n.‡, 123–24, 126n.*, 128n.#, 131, 133, 135–37, 139–49, 150, 157–58, 166–71, 179n.*, 181–84, 210, 218, 228, 231
Sexes, differentiation of. *See* Difference, sexual
Sömmering, S. T., 141n.*, 147, 183
Space, xxvii–xxviii, 11, 16–17, 20–26, 72–73, 75–78, 86–87, 89, 91–92, 97, 99, 121, 150, 173, 183, 188–91, 200, 207–9
Sphere of affinity, 10, 81, 86, 95, 177–78
Speculative Physics. *See* Physics, speculative
Spinozism, xvi, 117n.*, 194

Stages, graduated series of, xxvii, xxxii–xxxiii, 6, 9, 37n., 43n.*, 53–54, 69n.*, 141–42, 144n.*, 148n.†, 159, 160n.*, 166, 179n.*, 182–83, 218, 228
State of Indifference. *See* Indifference, state of
Steffens, Henrik, xii
Subject, xv–xvi, xviii, xxi, xxvii–xxviii, 8, 17, 67n.#, 106–7, 110, 112, 114–16, 118, 122–23, 136, 143, 172, 202–3, 205, 211
Schwammerdam, Jan, 37n.
Synthesis, xiv–xvi, xix, xxix, xxxi, 6n., 35n.*, 62, 77, 82, 84n.§, 86n., 87–88, 93n.‡, 187n.†, 192, 207, 211, 214, 220–21, 226, 232
System of Transcendental Idealism, 75n.

Technical drive, 9, 36, 130–38, 140, 146n.‡, 147, 194
Tendency, xxx, 5, 7–8, 15n.§, 16n.†, 18, 25–26, 34, 58, 72, 80–81, 87, 92–94, 100, 109–11, 118, 121–23, 126, 134, 140, 172, 187, 190–91, 195, 205, 212, 214, 222–23
Tillich, Paul, xii
Time, xxvii, 11, 40, 91, 187–88, 190–91, 203, 220

Unconditioned, The, xv–xvi, xxiii, 13–17, 19–20, 22, 201, 207
Universe, The, 8, 11, 61n., 85, 89, 92, 93nn.†–‡, 94n.†, 98, 112, 113n.*, 117n.*, 130, 157, 163, 174, 176, 178–79, 184–87, 190–91, 207n., 222

Vicq' d'Azyr, Felix, 155
Virgil, 138n.*
Vital force, 61, 63, 111, 160
Volta, Alessandro, 102n.†, 119, 210n.†

Wells, W. C., 119n.†

www.ingramcontent.com/pod-product-compliance
Lightning Source LLC
Chambersburg PA
CBHW021214240426
43672CB00026B/80